IARC MONOGRAPHS
ON THE
EVALUATION OF THE CARCINOGENIC RISK
OF CHEMICALS TO MAN:

Some aromatic amines and related nitro compounds - hair dyes, colouring agents and miscellaneous industrial chemicals

Volume 16

This volume represents the views and expert opinions of an IARC Working Group on the Evaluation of the Carcinogenic Risk of Chemicals to Man which met in Lyon, 7-14 June 1977

January 1978

IARC MONOGRAPHS

In 1971, the International Agency for Research on Cancer (IARC) initiated a programme on the evaluation of the carcinogenic risk of chemicals to humans, involving the production of critically evaluated monographs on individual chemicals.

The role of the monograph programme is to collect all available relevant experimental and epidemiological data about groups of chemicals for which there is known or possible human exposure, to evaluate these data in terms of human risk with the help of international working groups of acknowledged experts in chemical carcinogenesis and related fields, and to publish the conclusions of these working groups as a series of monographs.

ISBN 92 832 1216 9

PRINTED IN SWITZERLAND

IARC WORKING GROUP ON THE EVALUATION OF THE CARCINOGENIC RISK OF CHEMICALS TO MAN: SOME AROMATIC AMINES AND RELATED NITRO COMPOUNDS - HAIR DYES, COLOURING AGENTS AND MISCELLANEOUS INDUSTRIAL CHEMICALS

Lyon, 7-14 June 1977

Members

Dr E. Arrhenius, Division of Cellular Toxicology, Environmental Toxicology Unit, University of Stockholm, Wallenberg Laboratory, S-106 91 Stockholm, Sweden (*Chairman*)

Professor E. Boyland, TUC Centenary Institute of Occupational Health, London School of Hygiene and Tropical Medicine, Keppel Street, London WC1E 7HT, UK

Dr S.M. Brown, Associate Research Epidemiologist, University of California, School of Public Health, Earl Warren Hall, Berkeley, California 94720, USA (*Rapporteur section 3.3*)

Dr A.W. Craig, Paterson Laboratories, Christie Hospital and Holt Radium Institute, Manchester M20 9BX, UK

Dr J.M. Davies, Research Worker, Division of Epidemiology, Institute of Cancer Research, Fulham Road, London SW3 6JB, UK

Dr F.K. Dzhioev, N. Petrov Research Institute of Oncology, 68 Leningradskaya Street, Pesochny-2, Leningrad 188646, USSR

Dr W.G. Flamm, Assistant Director, Building 31, Room 11A05, National Cancer Institute, Bethesda, Maryland 20014, USA

Dr R.C. Garner, Cancer Research Unit, University of York, Heslington, York YO1 5DD, UK

Dr E. Kriek, Chemical Carcinogenesis Division, The Netherlands Cancer Institute, Antoni van Leeuwenhoekhuis, Sarphatistraat 108, Amsterdam-C, The Netherlands

Dr R. Krowke, Institut für Toxikologie und Embryonal-Pharmakologie der Freien Universität Berlin, Garystrasse 9, 1000 Berlin 33, FRG

Professor N. Loprieno, Laboratorio di Genetica, Istituto di Antropologia e Paleontologia Umana dell'Università degli Studi, Via S. Maria 53, 56100 Pisa, Italy

Dr H.-G. Neumann, Institut für Pharmakologie und Toxikologie der Universität Würzburg, Versbacher Landstrasse 9, 87 Würzburg, FRG

Dr M.D. Reuber, Pathologist, Chemical Carcinogenesis Group, Frederick Cancer Research Center, PO Box B, Frederick, Maryland 21701, USA

Dr B. Teichmann, Zentralinstitut für Krebsforschung, Akademie der Wissenschaften der DDR, Lindenberger Weg 80, 1115 Berlin-Buch, GDR (*Vice-Chairman*)

Representative from Stanford Research Institute International

Mrs S. Urso, Research Analyst, Chemical-Environmental Program, Stanford Research Institute International, Menlo Park, California 94025, USA (*Rapporteur sections 2.1 and 2.2*)

Observers[1]

Secretariat

Dr H. Bartsch, Unit of Chemical Carcinogenesis (*Rapporteur section 3.2*)

Dr J. Cooper, Unit of Epidemiology and Biostatistics

Dr L. Griciute, Chief, Unit of Environmental Carcinogens

Dr J.E. Huff, Unit of Chemical Carcinogenesis

Dr T. Kuroki, Unit of Chemical Carcinogenesis

Mrs D. Mietton, Unit of Chemical Carcinogenesis (*Library assistant*)

Dr R. Montesano, Unit of Chemical Carcinogenesis (*Rapporteur section 3.1*)

Mrs C. Partensky, Unit of Chemical Carcinogenesis (*Technical editor*)

Mrs I. Peterschmitt, Unit of Chemical Carcinogenesis, WHO, Geneva (*Bibliographic researcher*)

Dr V. Ponomarkov, Unit of Chemical Carcinogenesis

Dr R. Saracci, Unit of Epidemiology and Biostatistics

Dr L. Tomatis, Chief, Unit of Chemical Carcinogenesis (*Head of the Programme and Secretary*)

Dr G. Vettorazzi, Food Safety Unit, WHO, Geneva

Mr E.A. Walker, Unit of Environmental Carcinogens (*Rapporteur sections 1 and 2.3*)

Mrs E. Ward, Montignac, France (*Editor*)

Mr J.D. Wilbourn, Unit of Chemical Carcinogenesis (*Co-secretary*)

---

[1]Unable to attend: Mrs M.-T. van der Venne, Commission of the European Communities, Health and Safety Directorate, Bâtiment Jean Monnet, Avenue Alcide-de-Gasperi, Kirchberg/Luxembourg, Great Duchy of Luxembourg

Note to the reader

Every effort is made to present the monographs as accurately as possible without unduly delaying their publication. Nevertheless, mistakes have occurred and are still likely to occur. In the interest of all users of these monographs, readers are requested to communicate any errors observed to the Unit of Chemical Carcinogenesis of the International Agency for Research on Cancer, Lyon, France, in order that these can be included in corrigenda which will appear in subsequent volumes.

Since the monographs are not intended to be a review of the literature and contain only data considered relevant by the Working Group, it is not possible for the reader to determine whether a certain study was considered or not. However, research workers who are aware of important published data that may change the evaluation are requested to make them available to the above-mentioned address, in order that they can be considered for a possible re-evaluation by a future Working Group.

CONTENTS

Miscellaneous industrial chemicals:

BACKGROUND AND PURPOSE OF THE IARC PROGRAMME ON THE EVALUATION OF THE CARCINOGENIC RISK OF CHEMICALS TO MAN

The International Agency for Research on Cancer (IARC) initiated in 1971 a programme to evaluate the carcinogenic risk of chemicals to man. This programme was supported by a Resolution of the Governing Council at its Ninth Session concerning the role of IARC in providing government authorities with expert, independent scientific opinion on environmental carcinogenesis.

In view of the importance of the project and in order to expedite production of monographs, the National Cancer Institute of the United States has provided IARC with additional funds for this purpose.

The objective of the programme is to elaborate and publish in the form of monographs critical reviews of carcinogenicity and related data in the light of the present state of knowledge, with the final aim of evaluating the data in terms of possible human risk, and at the same time to indicate where additional research efforts are needed.

SCOPE OF THE MONOGRAPHS

The monographs summarize the evidence for the carcinogenicity of individual chemicals and other relevant information on the basis of data compiled, reviewed and evaluated by a Working Group of experts. No recommendations are given concerning preventive measures or legislation, since these matters depend on risk-benefit evaluations, which seem best made by individual governments and/or international agencies such as WHO and ILO.

Since 1971, when the programme was started, fifteen volumes have been published[1-15]. As new data on chemicals for which monographs have already been prepared and new principles for evaluation become available, re-evaluations will be made at future meetings, and revised monographs will be published as necessary.

The monographs are distributed to international and governmental agencies, are available to industries and scientists dealing with these chemicals and are offered to any interested reader through their world-wide distribution as WHO publications. They also form the basis of advice from IARC on carcinogenesis from these substances.

MECHANISM FOR PRODUCING THE MONOGRAPHS

As a first step, a list of chemicals for possible consideration by the Working Group is established. IARC then collects pertinent references regarding physico-chemical characteristics*, production and use*, occurrence and analysis* and biological data** on these compounds. The material on biological aspects is summarized by an expert consultant or an IARC staff member, who prepares the first draft, which in some cases is sent to another expert for comments. The drafts are circulated to all members of the Working Group about one month before the meeting. During the meeting, further additions to and deletions from the data are agreed upon, and a final version of comments and evaluation on each compound is adopted.

Priority for the preparation of monographs

Priority is given mainly to chemicals belonging to particular chemical groups and for which there is at least some suggestion of carcinogenicity from observations in animals and/or man and evidence of human exposure. However, *the inclusion of a particular compound in a volume does not necessarily mean that it is considered to be carcinogenic. Equally, the fact that a substance has not yet been considered does not imply that it is without carcinogenic hazard.*

---

*Data provided by Chemical Industries Center, Stanford Research Institute International, Menlo Park, California, USA

**In the collection of original data reference was made to the series of publications 'Survey of Compounds which have been Tested for Carcinogenic Activity'[16-22]. Most of the information on mutagenicity and teratogenicity was provided by the Environmental Mutagen Information Center and the Environmental Teratology Information Center, Oak Ridge, Tennessee, USA.

Data on which the evaluation is based

With regard to the biological data, only published articles and papers already accepted for publication are reviewed. The monographs are not intended to be a full review of the literature, and they contain only data considered relevant by the Working Group. Research workers who are aware of important data (published or accepted for publication) that may influence the evaluation are invited to make them available to the Unit of Chemical Carcinogenesis of the International Agency for Research on Cancer, Lyon, France.

The Working Group

The tasks of the Working Group are five-fold: (1) to ascertain that all data have been collected; (2) to select the data relevant for the evaluation; (3) to determine whether the data, as summarized, will enable the reader to follow the reasoning of the committee; (4) to judge the significance of results of experimental and epidemiological studies; and (5) to make an evaluation of carcinogenicity.

The members of the Working Group who participated in the consideration of particular substances are listed at the beginning of each publication. The members serve in their individual capacities as scientists and not as representatives of their governments or of any organization with which they are affiliated.

GENERAL PRINCIPLES FOR THE EVALUATION

The general principles for evaluation of carcinogenicity were elaborated by previous Working Groups and also applied to the substances covered in this volume.

Terminology

The term 'chemical carcinogenesis' in its widely accepted sense is used to indicate the induction or enhancement of neoplasia by chemicals. It is recognized that, in the strict etymological sense, this term means the induction of cancer; however, common usage has led to its employment to denote the induction of various types of neoplasms. The terms 'tumourigen', 'oncogen' and 'blastomogen' have all been used synonymously with 'carcinogen', although occasionally 'tumourigen' has been used specifically to denote the induction of benign tumours.

Response to carcinogens

In general, no distinction is made between the induction of tumours and the enhancement of tumour incidence, although it is noted that there may be fundamental differences in mechanisms that will eventually be elucidated. The response of experimental animals to a carcinogen may take several forms: a significant increase in the incidence of one or more of the same types of neoplasms as found in control animals; the occurrence of types of neoplasms not observed in control animals; and/or a decreased latent period for the production of neoplasms as compared with that in control animals.

Purity of the compounds tested

In any evaluation of biological data, particular attention must be paid to the purity of the chemicals tested and to their stability under conditions of storage or administration. Information on purity and stability is given, when available, in the monographs.

Qualitative aspects

In many instances, both benign and malignant tumours are induced by chemical carcinogens. There are so far few recorded instances in which only benign tumours are induced by chemicals that have been studied extensively. Their occurrence in experimental systems has been taken to indicate the possibility of an increased risk of malignant tumours also.

In experimental carcinogenesis, the type of cancer seen may be the same as that recorded in human studies, e.g., bladder cancer in man, monkeys, dogs and hamsters after administration of 2-naphthylamine. In other instances, however, a chemical may induce other types of neoplasms at different sites in various species, e.g., benzidine induces hepatic carcinomas in rats but bladder carcinomas in man.

Quantitative aspects

Dose-response studies are important in the evaluation of human and animal carcinogenesis: the confidence with which a carcinogenic effect can be established is strengthened by the observation of an increasing incidence of neoplasms with increasing exposure. In addition, such studies form the only basis on which a minimal effective dose can be established, allowing some comparison with data for human exposure.

Comparison of compounds with regard to potency can only be made when the substances have been tested simultaneously.

Animal data in relation to the evaluation of risk to man

At the present time no attempt can be made to interpret the animal data directly in terms of human risk, since no objective criteria are available to do so. The critical assessments of the validity of the animal data given in these monographs are intended to assist national and/or international authorities in making decisions concerning preventive measures or legislation. In this connection, attention is drawn to WHO recommendations in relation to food additives[23], drugs[24] and occupational carcinogens[25].

Evidence of human carcinogenicity

Evaluation of the carcinogenic risk to man of suspected environmental agents rests on purely observational studies. Such studies must cover a sufficient variation in levels of human exposure to allow a meaningful relationship between cancer incidence and exposure to a given chemical to be established. Difficulties arise in isolating the effects of individual agents, however, since people are usually exposed to multiple carcinogens.

The initial suggestion of a relationship between an agent and disease often comes from case reports of patients with similar exposures. Variations and time trends in regional or national cancer incidences, or their correlation with regional or national 'exposure' levels, may also provide valuable insights. Such observations by themselves cannot, however, in most circumstances be regarded as conclusive evidence of carcinogenicity.

The most satisfactory epidemiological method is to compare the cancer risk (adjusted for age, sex and other confounding variables) among groups of cohorts, or among individuals exposed to various levels of the agent in question, and among control groups not so exposed. Ideally, this is accomplished directly, by following such groups forward in time (prospectively) to determine time relationships, dose-response relationships and other aspects of cancer induction. Large cohorts and long observation periods are required to provide sufficient cases for a statistically valid comparison.

An alternative to prospective investigation is to assemble cohorts from past records and to evaluate their subsequent morbidity or mortality by means of medical histories and death certificates. Such occupational carcinogens as nickel, 2-naphthylamine, asbestos and benzidine have been confirmed by this method. Another method is to compare the past exposures of a defined group of cancer cases with those of control samples from the hospital or general population. This does not provide an absolute measure of carcinogenic risk but can indicate the relative risks associated with different levels of exposure. Indirect means (e.g., interviews or tissue residues) of measuring exposures which may have commenced many years before can constitute a major source of error. Nevertheless, such 'case-control' studies can often isolate one factor from several suspected agents and can thus indicate which substance should be followed up by cohort studies.

EXPLANATORY NOTES ON THE MONOGRAPHS

In sections 1, 2 and 3 of each monograph, except for minor remarks, the data are recorded as given by the author, whereas the comments by the Working Group are given in section 4, headed 'Comments on Data Reported and Evaluation'.

Chemical and Physical Data (section 1)

The Chemical Abstracts Services Registry Number and the latest Chemical Abstracts Name are recorded in this section. Other synonyms and trade names are given, but this list is not intended to be comprehensive.

It should also be noted that some of the trade names are those of mixtures in which the compound being evaluated is only one of the active ingredients.

Chemical and physical properties include, in particular, data that might be relevant to carcinogenicity (for example, lipid solubility) and those that concern identification. All chemical data in this section refer to the pure substance, unless otherwise specified.

Production, Use, Occurrence and Analysis (section 2)

The purpose of this section is to indicate the extent of possible human exposure. With regard to data on production, use and occurrence, IARC has collaborated with the Stanford Research Institute International, USA, with the support of the National Cancer Institute of the USA. Since cancer is a delayed toxic effect, past use and production data are also provided.

The United States, Europe and Japan are reasonably representative industrialized areas of the world, and if data on production or use are available from these countries they are reported. It should *not*, however, be inferred that these nations are the sole or even the major sources of any individual chemical.

Production data are obtained from both governmental and trade publications in the three geographic areas. In some cases, separate production data on chemicals manufactured in the US were not available, for proprietary reasons. However, the fact that a manufacturer acknowledges production of a chemical to the US International Trade Commission implies that annual production of that chemical is greater than 450 kg or that its annual sales exceed $1000. Information on use and occurrence is obtained by a review of published data, complemented by direct contact with manufacturers of the chemical in question; however, information on only some of the uses is available, and this section cannot be considered to be comprehensive. In an effort to provide estimates of production in some European countries, Stanford Research Institute International in Zurich sent general questionnaires to some of those European companies thought to produce the compounds being evaluated. Information from the replies to these questionnaires has been compiled by country and included in the individual monographs.

Statements concerning regulations in some countries are mentioned as examples only. They may not reflect the most recent situation, since such legislation is in a constant state of change; nor should it be taken to imply that other countries do not have similar regulations. In the case of drugs, mention of the therapeutic uses of such chemicals does not necessarily represent presently accepted therapeutic indications, nor does it imply judgement as to their clinical efficacy.

The purpose of the section on analysis is to give the reader a general indication, rather than a complete review, of methods cited in the literature. No attempt is made to evaluate the methods quoted.

Biological Data Relevant to the Evaluation of Carcinogenic Risk to Man (section 3)

The monographs are not intended to consider all reported studies. Some studies were purposely omitted (a) because they were inadequate, as judged from previously described criteria[26-29] (e.g., too short a duration, too few animals, poor survival or too small a dose); (b) because they only confirmed findings which have already been fully described; or (c) because they were judged irrelevant for the purpose of the evaluation. However, in certain cases, reference is made to studies which did not meet established criteria of adequacy, particularly when this information was considered a useful supplement to other reports or when it was the only data available. Their inclusion does not, however, imply acceptance of the adequacy of their experimental design.

In general, the data recorded in this section are summarized as given by the author; however, certain shortcomings of reporting or of experimental design that were commented upon by the Working Group are given in square brackets.

Carcinogenicity and related studies in animals: Mention is made of all routes of administration by which the compound has been adequately tested and of all species in which relevant tests have been carried out. In most cases, animal strains are given (general characteristics of mouse strains have been reviewed[30]). Quantitative data are given to indicate the order of magnitude of the effective doses. In general, the doses are indicated as they appear in the original paper; sometimes conversions have been made for better comparison. When the carcinogenicity of known metabolites has been tested this also is reported.

Other relevant biological data: LD_{50} data are given when available, and other data on toxicity are included when considered relevant. The metabolic data included is restricted to studies showing the metabolic fate of the chemical in animals and man, and comparisons of animal and human data are made when possible. Other metabolic information (e.g., absorption, storage and excretion) is given when the Working Group considered that it would be useful for the reader to have a better understanding of the fate of the compound in the body.

Teratogenicity data from studies in experimental animals and in humans are included for some of the substances considered; however, they are not meant to represent a thorough review of the literature.

Mutagenicity data are also included; the reasons for including them and the principles adopted by the Working Group for their selection are outlined below.

Many, but not all, mutagens are carcinogens and *vice versa*; the exact level of correlation is still under investigation. Nevertheless, practical use may be made of the available mutagenicity test procedures that combine microbial, mammalian or other animal cell systems as genetic targets with an *in vitro* or *in vivo* metabolic activation system. The results of relatively rapid and inexpensive mutagenicity tests on non-human organisms may help to pre-screen chemicals and may also aid in the selection of the most relevant animal species in which to carry out long-term carcinogenicity tests on these chemicals.

The role of genetic alterations in chemical carcinogenesis is not yet fully understood, and therefore consideration must be given to a variety of changes. Although nuclear DNA has been defined as the main cellular target for the induction of genetic changes, other relevant targets have been recognized, e.g., mitochondrial DNA, enzymes involved in DNA synthesis, repair and recombination, and the spindle apparatus. Tests to detect the genetic activity of chemicals, including gene mutation, structural and numerical chromosomal changes and mitotic recombination, are available for non-human models; but not all such tests can be applied at present to human cells.

Ideally, an appropriate mutagenicity test system would include the full metabolic competency of the intact human. Since the development or application of such a system appears to be impossible, a battery of test systems is necessary in order to establish the mutagenic potential of chemicals. There are many genetic indicators and metabolic activation systems available for detecting mutagenic activity; they all, however, have individual advantages and limitations.

Since many chemicals require metabolism to an active form, test systems which do not take this into account may fail to reveal the full range of genetic damage. Furthermore, since some reactive metabolites with a limited lifespan may fail to reach or to react with the genetic indicator, either because they are further metabolized to inactive compounds or because they react with other cellular constituents, mutagenicity tests in intact animals may give false negative results.

It is difficult in the present state of knowledge to select specific mutagenicity tests as being the most appropriate for the pre-screening of substances for possible carcinogenic activity. However, greater reliance may be placed on data obtained from those test systems which (a) permit identification of the nature of induced genetic changes, and (b) demonstrate that the changes are transmitted to subsequent generations. Mutagenicity tests using organisms that are well understood genetically, e.g., *Escherichia coli*, *Salmonella typhimurium*, *Saccharomyces cerevisiae* and *Drosophila melanogaster*, meet these requirements.

Although a correlation has often been observed between the ability of a chemical to cause chromosome breakage and its ability to induce gene mutation, data on chromosomal breakage alone do not provide adequate evidence for mutagenicity, and therefore less weight should be given to pre-screening that is based on the use of peripheral leucocyte cultures.

Because of the complexity of factors that can contribute to reproductive failure, as well as the insensitivity of the method, the dominant lethal test in the mammal does not provide reliable data on mutagenicity.

A large-scale systematic screening of compounds to assess a correlation between mutagenicity and carcinogenicity has so far been carried out only with the bacterial/mammalian liver microsome system. Notwithstanding the demonstration of the mutagenicity of many known carcinogens to *Salmonella typhimurium* in the presence of liver microsomal systems, the possibility of false-negative and false-positive results must not be overlooked. False negatives might arise as a consequence of mutagen specificity or from failure to achieve optimal conditions for activation *in vitro*. Alternative test systems must be used if there appear to be substantial reasons for suspecting that a chemical which is apparently non-mutagenic in a bacterial test system may nevertheless be potentially carcinogenic. Conversely, some chemicals found to be mutagenic in this test may not in fact have mutagenic activity in other systems.

For more detailed information, see references 31-38.

Observations in man: Case reports of cancer and epidemiological studies are summarized in this section.

Comments on Data Reported and Evaluation (section 4)

This section gives the critical view of the Working Group on the data reported.

Animal data: The animal species mentioned are those in which the carcinogenicity of the substances was clearly demonstrated. The route of administration used in experimental animals that is similar to the possible human exposure (ingestion, inhalation and skin exposure) is given particular mention. Tumour sites are also indicated.

Experiments involving a possible action of the vehicle or a physical effect of the agent, such as in studies by subcutaneous injection or bladder implantation, are included; however, the results of such tests require careful consideration, particularly if they are the only ones raising a suspicion of carcinogenicity. If the substance has produced tumours after prenatal exposure or in single-dose experiments, this also is indicated. This sub-section should be read in the light of comments made in the section, 'Animal Data in Relation to the Evaluation of Risk to Man' of this introduction.

Human data: In some cases, a brief statement is made on possible human exposure. The significance of epidemiological studies and case reports is discussed, and the data are interpreted in terms of possible human risk.

References

1. IARC (1972) IARC Monographs on the Evaluation of Carcinogenic Risk of Chemicals to Man, 1, Lyon

2. IARC (1973) IARC Monographs on the Evaluation of Carcinogenic Risk of Chemicals to Man, 2, Some Inorganic and Organometallic Compounds, Lyon

3. IARC (1973) IARC Monographs on the Evaluation of Carcinogenic Risk of Chemicals to Man, 3, Certain Polycyclic Aromatic Hydrocarbons and Heterocyclic Compounds, Lyon

4. IARC (1974) IARC Monographs on the Evaluation of Carcinogenic Risk of Chemicals to Man, 4, Some Aromatic Amines, Hydrazine and Related Substances, *N*-Nitroso Compounds and Miscellaneous Alkylating Agents, Lyon

5. IARC (1974) IARC Monographs on the Evaluation of Carcinogenic Risk of Chemicals to Man, 5, Some Organochlorine Pesticides, Lyon

6. IARC (1974) IARC Monographs on the Evaluation of Carcinogenic Risk of Chemicals to Man, 6, Sex Hormones, Lyon

7. IARC (1974) IARC Monographs on the Evaluation of Carcinogenic Risk of Chemicals to Man, 7, Some Anti-thyroid and Related Substances, Nitrofurans and Industrial Chemicals, Lyon

8. IARC (1975) IARC Monographs on the Evaluation of Carcinogenic Risk of Chemicals to Man, 8, Some Aromatic Azo Compounds, Lyon

9. IARC (1975) IARC Monographs on the Evaluation of Carcinogenic Risk of Chemicals to Man, 9, Some Aziridines, *N*-, *S*- and *O*-Mustards and Selenium, Lyon

10. IARC (1976) IARC Monographs on the Evaluation of Carcinogenic Risk of Chemicals to Man, 10, Some Naturally Occurring Substances, Lyon

11. IARC (1976) IARC Monographs on the Evaluation of Carcinogenic Risk of Chemicals to Man, 11, Cadmium, Nickel, Some Epoxides, Miscellaneous Industrial Chemicals and General Considerations on Volatile Anaesthetics, Lyon

12. IARC (1976) IARC Monographs on the Evaluation of Carcinogenic Risk of Chemicals to Man, 12, Some Carbamates, Thiocarbamates and Carbazides, Lyon

13. IARC (1977) IARC Monographs on the Evaluation of Carcinogenic Risk of Chemicals to Man, 13, Some Miscellaneous Pharmaceutical Substances, Lyon

14. IARC (1977) IARC Monographs on the Evaluation of Carcinogenic Risk of Chemicals to Man, 14, Asbestos, Lyon

15. IARC (1977) IARC Monographs on the Evaluation of the Carcinogenic Risk of Chemicals to Man, 15, Some Fumigants, the Herbicides 2,4-D and 2,4,5-T, Chlorinated Dibenzodioxins and Miscellaneous Industrial Chemicals, Lyon

16. Hartwell, J.L. (1951) Survey of Compounds which have been Tested for Carcinogenic Activity, Washington DC, US Government Printing Office (Public Health Service Publication No. 149)

17. Shubik, P. & Hartwell, J.L. (1957) Survey of Compounds which have been Tested for Carcinogenic Activity, Washington DC, US Government Printing Office (Public Health Service Publication No. 149: Supplement 1)

18. Shubik, P. & Hartwell, J.L. (1969) Survey of Compounds which have been Tested for Carcinogenic Activity, Washington DC, US Government Printing Office (Public Health Service Publication No. 149: Supplement 2)

19. Carcinogenesis Program National Cancer Institute (1971) Survey of Compounds which have been Tested for Carcinogenic Activity, Washington DC, US Government Printing Office (Public Health Service Publication No. 149: 1968-1969)

20. Carcinogenesis Program National Cancer Institute (1973) Survey of Compounds which have been Tested for Carcinogenic Activity, Washington DC, US Government Printing Office (Public Health Service Publication No. 149: 1961-1967)

21. Carcinogenesis Program National Cancer Institute (1974) Survey of Compounds which have been Tested for Carcinogenic Activity, Washington DC, US Government Printing Office (Public Health Service Publication No. 149: 1970-1971)

22. Carcinogenesis Program National Cancer Institute (1976) Survey of Compounds which have been Tested for Carcinogenic Activity, Washington DC, US Government Printing Office (Public Health Service Publication No. 149: 1972-1973)

23. WHO (1961) Fifth Report of the Joint FAO/WHO Expert Committee on Food Additives. Evaluation of carcinogenic hazard of food additives. Wld Hlth Org. techn. Rep. Ser., No. 220, pp. 5, 18, 19

24. WHO (1969) Report of a WHO Scientific Group. Principles for the testing and evaluation of drugs for carcinogenicity. Wld Hlth Org. techn. Rep. Ser., No. 426, pp. 19, 21, 22

25. WHO (1964) Report of a WHO Expert Committee. Prevention of cancer. Wld Hlth Org. techn. Rep. Ser., No. 276, pp. 29, 30

26. WHO (1958) Second Report of the Joint FAO/WHO Expert Committee on Food Additives. Procedures for the testing of international food additives to establish their safety and use. Wld Hlth Org. techn. Rep. Ser., No. 144

27. WHO (1961) Fifth Report of the Joint FAO/WHO Expert Committee on Food Additives. Evaluation of carcinogenic hazard of food additives. Wld Hlth Org. techn. Rep. Ser., No. 220

28. WHO (1967) Scientific Group. Procedures for investigating intentional and unintentional food additives. Wld Hlth Org. techn. Rep. Ser., No. 348

29. Berenblum, I., ed. (1969) Carcinogenicity testing. UICC techn. Rep. Ser., 2

30. Committee on Standardized Genetic Nomenclature for Mice (1972) Standardized nomenclature for inbred strains of mice. Fifth listing. Cancer Res., 32, 1609-1646

31. Bartsch, H. & Grover, P.L. (1976) Chemical carcinogenesis and mutagenesis. In: Symington, T. & Carter, R.L., eds, Scientific Foundations of Oncology, Vol. IX, Chemical Carcinogenesis, London, Heinemann Medical Books Ltd, pp. 334-342

32. Hollaender, A., ed. (1971) Chemical Mutagens: Principles and Methods for Their Detection, Vols 1-3, New York, Plenum Press

33. Montesano, R. & Tomatis, L., eds (1974) Chemical Carcinogenesis Essays, Lyon (IARC Scientific Publications No. 10)

34. Ramel, C., ed. (1973) Evaluation of genetic risks of environmental chemicals: report of a symposium held at Skokloster, Sweden, 1972. Ambio Special Report, No. 3

35. Stoltz, D.R., Poirier, L.A., Irving, C.C., Stich, H.F., Weisburger, J.H. & Grice, H.C. (1974) Evaluation of short-term tests for carcinogenicity. Toxicol. appl. Pharmacol., 29, 157-180

36. WHO (1974) Report of WHO Scientific Group. Assessment of the carcinogenicity and mutagenicity of chemicals. Wld Hlth Org. techn. Rep. Ser., No. 546

37. Montesano, R., Bartsch, H. & Tomatis, L., eds (1976) Screening Tests in Chemical Carcinogenesis, Lyon (IARC Scientific Publications No. 12)

38. Committee 17 (1975) Environmental mutagenic hazards. Science, 187, 503-514

GENERAL REMARKS ON THE SUBSTANCES CONSIDERED

This volumes includes monographs on a number of aromatic amines and some related nitro compounds that have not been considered in previous IARC monographs. Some compounds that belong to this chemical group are known carcinogens (IARC, 1972, 1974, 1975), and for those that are considered here, carcinogenic effects or a suspicion of carcinogenicity have been reported.

Of the compounds considered in this volume, a large number are colourants - several of them are used as hair dyes, as industrial dyes (e.g., in printing inks, stains, textiles, paper, leather, foods and cosmetics), as dye intermediates or as intermediates in other chemical industries, or as additives in the manufacture of plastics and polymers.

Table 1 summarizes information on the types of products or processes in which these compounds may be used. This table is not intended to be a primary source, and the reader should also refer to the sections on production and use in the individual monographs. The table indicates only those reported commercial uses (past or present) that are mentioned in the monographs. There may well be other significant uses of these substances about which no information was available.

While data on chemical and physical properties, technical products and impurities, production and use, and analytical methods refer to the pure chemical, biological data in many instances are from experiments in which the chemical was administered as a constituent of a mixture, the exact composition of which was not known. In these cases, an evaluation of the carcinogenicity of the pure compound was not possible. Evaluations of the carcinogenicity of other chemicals could also not be made, either because the available data were inadequate or because they were given only in abstracts, which did not allow a full evaluation of the results. The Working Group was aware that for some compounds final results were available but unpublished, and they could only express their regret at the delay in publication.

Table 1

Summary of reported commercial uses
(past and present) mentioned in the monographs

Compound	Dye and pigment intermediate	Other intermediates	Textile dyeing	Leather dyeing	Printing inks and paper dyes	Miscellaneous dyeing applications[1]	Hair dyes and hair colourants	Soaps, cosmetics and perfumes	Food processing	Rubber manufacture	Plastics (polymers) manufacture	Other applications
Hair dyes:												
4-Amino-2-nitrophenol						+	+					
2,4-Diaminoanisole (sulphate)	+	+				+	+					
1,2-Diamino-4-nitrobenzene							+					+
1,4-Diamino-2-nitrobenzene						+	+					
2,4-Diaminotoluene	+	+	+	+		+	+				+	
2,5-Diaminotoluene (sulphate)	+					+	+					
meta-Phenylenediamine (hydrochloride)	+	+	+				+			+	+	+
para-Phenylenediamine (hydrochloride)	+	+	+			+	+			+	+	+
Colouring agents:												
Acridine orange	+		+	+		+						
Benzyl violet 4B			+	+	+	+	+	+	+			
Blue VRS	+		+	+	+	+		+	+		+	
Brilliant blue FCF diammonium and disodium salts	+		+	+	+	+	+	+	+	+	+	+
Fast green FCF						+	+	+	+			
Guinea green B	+		+	+	+	+		+	+			+
Light green SF			+	+	+	+		+	+			
Rhodamine B	+		+	+	+	+		+				+
Rhodamine 6G	+		+	+	+			+				+
Miscellaneous industrial chemicals:												
5-Aminoacenaphthene	No known commercial significance											
para-Aminobenzoic acid	+	+						+				+
Anthranilic acid	+	+						+	+			+
para-Chloro-*ortho*-toluidine (hydrochloride)	+	+	+									
Cinnamyl anthranilate								+	+			
N,N'-Diacetylbenzidine	+	+										
4,4'-Diaminodiphenyl ether		+									+	
3,3'-Dichloro-4,4'-diaminodiphenyl ether	No known commercial significance											
2,4'-Diphenyldiamine	No known commercial significance[2]											
5-Nitroacenaphthene	+				+							
N-Phenyl-2-naphthylamine	+	+								+	+	+
4,4'-Thiodianiline	+											
ortho-Toluidine (hydrochloride)	+		+							+		+
2,4-Xylidine (hydrochloride)	+											+
2,5-Xylidine (hydrochloride)	+											

[1]E.g., the dyeing of animal furs, wood, pharmaceutical products and miscellaneous household products, and use as a biological stain

[2]May occur as an impurity in benzidine, which may be used in dye and pigment production (see also IARC, 1972, p. 80)

The following compounds were considered but not included in this volume because of lack of data on carcinogenicity and/or of epidemiological studies: 2-amino-4-nitrophenol, 2-amino-5-nitrophenol, 5-chloro-*ortho*-toluidine, 2,5-diaminoanisole, *ortho*-phenylenediamine, *N*-phenyl-*para*-phenylenediamine and 2,6-xylidine. The Working Group was aware that carcinogenicity studies are in progress on some of these compounds, and these may be reconsidered when the results of such studies become available (IARC, 1976). It was noted that 2-amino-4-nitrophenol, 2-amino-5-nitrophenol, 2,5-diaminoanisole and *ortho*-phenylenediamine have positive mutagenic activity in the *Salmonella*/microsome test (Ames *et al.*, 1975).

In order to be toxic or carcinogenic, aromatic amines usually require metabolic activation. The potential risk from structurally related compounds depends largely on their metabolic fate in the test system used and on the reactivity of the ultimate carcinogenic reactants. The Working Group noted the absence of pertinent data on the metabolism of most of the compounds considered. Their metabolic activation, in particular *N*-oxygenation and the subsequent formation of reactive metabolites capable of reacting with cellular macromolecules (e.g., nucleic acids and proteins), has not yet been adequately investigated.

Many of the compounds considered change on exposure to light, due to photooxidation; in general, their salts are more stable. When such chemicals are being used for experiments in biological systems, rigid proof of their purity is mandatory.

Hair dye formulations and mode of use (Wall, 1972)

Many of the aromatic amines considered in these monographs are used in commercially available products for dyeing hair. Indeed, much of the information regarding the toxicity and carcinogenicity of these aromatic amines is derived from studies in which the aromatic amine was only one constituent among many in a hair-dye formulation and in which the doses that were administered topically constituted only an unknown small fraction of the LD_{50}. Results from such studies were considered to be neither appropriate nor adequate for evaluating the carcinogenicity of the individual aromatic amines described in the monographs, although such data are of use in evaluating the carcinogenicity of hair-dye formulations.

Hair dyes can be divided into three types, depending on whether they colour the hair temporarily, semi-permanently or permanently. The dyes, dye intermediates and other ingredients involved vary according to the type of hair product.

(a) Temporary dyes

These consist of a miscellaneous group of chemicals designed to effect a change in the shade of the hair and are generally applied in the form of shampoos; if they are not re-applied, they are removed by washing at the next shampoo. Compounds included in this volume that are used in temporary hair-dye formulations are benzyl violet 4B, brilliant blue FCF disodium salt and fast green FCF. A commercial formulation may contain between 0.5 to 2.0% of the permitted colour, to give the appropriate shade, together with surface-active agents. These colour shampoos may also contain nitro compounds as colourants, together with anionic detergents, and compounds such as urea that increase solubility.

(b) Semi-permanent dyes

Such dyes remain on the hair for several washings and are only gradually washed out by successive shampooing. Many of the aromatic, nitro and amino compounds used in the formulation of permanent dyes (see following sub-section) may be used in semi-permanent dyes, without the addition of an oxidizing agent.

An example of such a formulation is:

Monoethanolamine lauryl sulphate	20.0%
Ethylene glycol monostearate	5.0%
Diethanolamine fatty-acid salts	3.0%
Diamino-nitrobenzenes (dyes)	1.5%
Perfume) -to- Water)	100.0%

(c) Permanent dyes (oxidation dyes)

With this group of dyes, development of the colour shade depends on the addition of an oxidizing agent (usually hydrogen peroxide). These dyes can penetrate the hair shaft and, when oxidized, are irreversibly bound within it, allowing the colour to remain. There are a vast number of formulations used as permanent dyes; 2-amino-4-nitrophenol, 2-amino-5-nitrophenol,

4-amino-2-nitrophenol, 2,4-diaminoanisole, 2,5-diaminoanisole, 1,2-diamino-4-nitrobenzene, 1,4-diamino-2-nitrobenzene, 2,4-diaminotoluene, 2,5-diamino-toluene, *meta*-phenylendiamine, *ortho*-phenylenediamine, *para*-phenylenediamine and *N*-phenyl-*para*-phenylenediamine are found in such dye formulations, often as mixtures in order to give the desired shade. Typical formulations contain between 1% and 4% of different dye intermediates.

An example of such a formulation is:

Oleic acid	20.0%
Oleyl alcohol	15.0%
Solubilized lanolin	3.0%
Propylene glycol	12.0%
Isopropanol	10.0%
EDTA	0.5%
Sodium sulphite	0.5%
Ammonium hydroxide (28% w/v)	10.0%
Deionized water	26.0%
Dye intermediates	3.0%

The oxidizing agent rapidly oxidizes the aromatic group of the dye to form products which bind to the hair shaft, at the same time lowering absorption by the skin of the specific aromatic amines. However, it should be recognized that some of the products formed by the oxidation process, such as the three-ringed 'Bandrowski-base', may also be absorbed through the skin and may thus pose an additional health hazard.

Further information on the ingredients used in hair rinses and hair dyes is given by Gosselin *et al.* (1976).

Epidemiological evidence relating to the possible carcinogenic effect of exposure to hair dyes

Workers in several occupations and industries which have been the object of epidemiological study are likely to have been exposed to some of the substances considered in this volume of monographs. Such occupations include dye-workers, printers and leather- and textile-workers. However, except for studies of hair-service occupations, such studies have not been reviewed, since exposure cannot be ascribed specifically to the substances in question; workers in such occupations might have been exposed to other substances, including demonstrated carcinogens (such as benzidine and

2-naphthylamine). The problem of mixed exposure also applies to beauticians, hairdressers and barbers, who are also exposed, for example, to aerosol propellants (e.g., vinyl chloride and fluorocarbons). Studies of these occupations have been reviewed here since the specificity of exposure is somewhat greater. The epidemiological evidence presently available on the substances considered deals mainly with possible risks of bladder cancer; there are as yet relatively few data for cancers at other sites. These studies relate to exposures occurring at different periods over the last 30 years or more, during which time there have been changes in the usage of hair dyes in terms of both type and quantity (Wall, 1972).

The term 'hairdresser' may have different meanings in different countries. In the US, it applies unequivocally to beauticians, i.e., those involved in the care of hair, primarily women's hair, in beauty salons; it excludes barbers. In the UK, the term is more inclusive; it is used to denote ladies' hairdressers, and it may also mean barbers.

Occupational exposure

(a) Bladder tumours

Wynder *et al.* (1963) studied the distribution of smoking habits and occupations in patients with carcinomas of the bladder. They combined interview data from two hospital-based series covering a total of 300 male patients in New York City during the period 1957-1961. A case-control design was used, and controls were age- and sex-matched; 93% of cases and 82% of controls were smokers. Among patients, there were four who had worked as hairdressers (all smokers); there were none among the controls [No statistical analysis of these data was provided by the authors; the relative risk is approximately 9].

Anthony & Thomas (1970) studied a series of bladder tumour cases from Leeds, UK. Of 1030 patients (812 males and 218 females) interviewed in an 8-year series (1959-1967), 43% had known smoking histories. Occupations were defined according to the Registrar General's classification of occupations (General Register Office, 1966), looking separately at predominant occupation, occupation ever undertaken and 20 or more years in occupation. Of several alternative analyses presented by the authors, the most reliable appears to be based upon a comparison of the male cases with a control group who had benign surgical disease and were matched to the cases for sex, age, place of residence and amount smoked. Among the cases, there were four hairdressers, of whom three had more than 20 years in the occupation. The relative risks of bladder tumour for hairdressers were, therefore, 4 (predominant occupation), 4 (occupation ever undertaken) and 3 (20 or more years in the occupation). Among the 218 female cases, there were no hairdressers, although 0.6 were expected. None of these differences were significant at the P=0.05 level.

In a letter to the Editor of Science, Menkart (1975) argued that the results of Anthony & Thomas (1970) and Wynder *et al.* (1963) could not reflect occupational exposure to hair dyes. His semantic argument is, however, incorrect in the light of the discussion of the term 'hairdresser' above.

Dunham *et al.* (1968) compared most recent occupations of 265 white, male patients with bladder cancer and those of comparable controls in New Orleans, Louisiana during 1958-1964. They found four barbers in this group, while 1.45 were expected (relative risk, 2.76). During the last 3¼ years of this study, 132 cases were compared with 136 controls for history of 'use of tonics, lotions and other preparations for the hair and scalp'. The percentage of controls who used such preparations (36%) was slightly higher than that of cases (32%).

Cole *et al.* (1972) examined occupational risk among 461 cases (356 males, 105 females) aged 20-89 (a systematic sample of all adult cases) with transitional- and squamous-cell carcinomas of the lower urinary tract in a defined population in Boston, Massachusetts, during an 18-month period in 1968-1969. Controls were comparable with respect to age and sex. Occupational risk was assessed for each of 42 selected occupational titles derived from a modification of the standard US Bureau of the Census classification. Among men, there was no increased risk for barbers (4 observed cases, 7.2 expected). Among women, risk was apparently not increased for hairdressers (1 observed case, 0.9 expected).

In an abstract, Howe *et al.* (1977) reported that a case-control study of 532 Canadians with bladder cancer had shown that barbers were one of six specific occupational groups who were at significantly increased risk.

(b) Other tumours

The Registrar General (1971) analysed numbers of cancer deaths during 1959-1963 among men and single women described on their death certificates as hairdressers, manicurists or beauticians. These were related to the numbers of expected deaths derived from 1961 census data and national death rates. Cancer at all sites combined, cancers of the lung and stomach and leukaemia were examined for both men and women; the results did not indicate an elevated mortality. For single women, there were 21 observed deaths from breast cancer *versus* 12 expected. For cancers of the cervix uteri and other parts of the uterus, the ratios of observed:expected deaths were 3:2 and 4:2, respectively. Mortality from neoplasms at the same sites was similarly examined for the period 1949-1953 (Registrar General, 1958). For male barbers and hairdressers, the only indication of excess mortality was from cancer of the lung and bronchus (114 observed, 99 expected). For single women hairdressers and manicurists, the numbers of observed deaths exceeded those expected for cancers at all sites combined (43:37) and for cancers of lung and bronchus (4:2), breast (13:9) and cervix uteri (4:1); there were no excesses of cancers of the stomach, cancers of other parts of the uterus or leukaemia.

Milham (1976) presented the results of a mortality study by occupation carried out in Washington state. These were based on a proportional mortality analysis of statements of occupation in death certificates for all white male residents over the age of 20 who died in the state during the period 1950-1971. For barbers, all causes of death were examined; mortality was stated to be significantly increased ($P<0.05$) for oesophageal cancer and multiple myeloma.

Viadana *et al.* (1976) reported an increase in laryngeal cancer in barbers in a hospital-based case-control study in Buffalo, New York, between 1956 and 1965; they studied men in 13 groups occupationally exposed to 'inhalation of chemicals or to combustion products'. There were 11,591 white males (cases plus controls), of whom 76% had cancer. These men were scored for each occupation ever held and for occupations held for five or more years for the 17 sites of cancer for which there were five or more cases. Results were standardized separately for age and smoking. Barbers had an age-adjusted relative risk of 2.83 (P<0.05) for laryngeal cancer (10 cases); this excess persisted in the 5 or more years of exposure category (2.49, P>0.05) (8 cases). When standardized for smoking, but not age, the relative risk for laryngeal cancer in barbers was higher and statistically significant (3.39 for ever exposed, 3.19 for 5 or more years exposed; P<0.05 for both).

Garfinkel *et al.* (1977) assessed the risk of any cancer for beauticians in Alameda County, California during the period 1958-1962 by analysis of death certificates. Of 25 beauticians in a series of all cancer deaths, 24 were women. These died of cancers at the following sites: lung (6), breast (5), cervix (4), ovary (3), brain (1), bladder (1), stomach (1), synovium (1) and unspecified (2). The one male died of liver cancer. Using a case-control analysis, the 3460 adult female cancer patients who died were compared with 1000 controls who had no mention of cancer on their death certificates. The relative risk for death from cancer for female beauticians was 1.73 (P=0.43). The hypothesis that female beauticians were at increased risk for lung cancer was then tested, using a matched pairs design. Adult female non-cancer controls were matched to lung cancer cases for age, race, date of death and county of residence. Of 176 cases, six were beauticians, compared with one among 176 controls. The relative risk was 6.0, with one-tailed P=0.06, suggesting an increased risk of death from lung cancer among female beauticians.

Clemmesen (1977) has studied the incidence of malignant neoplasms among hairdressers throughout Denmark during 5-year periods from 1943-1972, on the basis of Danish Cancer Registry data. Using as denominators the numbers of men and women enumerated as hairdressers in successive censuses, Clemmesen calculated expected numbers of cases using age- and sex-specific registration rates. For male hairdressers, the expected total of malignant neoplasms during 1943-1972 was 517.4, and the observed number was 447. There was no excess for any of the 11 groupings of specified cancer sites.

For females, there appeared to be a large overall excess, with 872 malignant neoplasms observed and only 475.4 expected. An excess appeared for each 5-year period and for each of 14 site groupings [The relatively consistent excess for all periods and all sites in women in this study suggests the possibility of a bias in the reporting of occupation. Additional information is needed before these findings can be interpreted (see also Danish Ministry of the Environment, 1977)].

Non-occupational exposure

Shafer & Shafer (1976) asked 100 consecutive breast cancer patients in a clinical practice for a history of hair-dye use, obtaining such history from next-of-kin for deceased patients. Only women who had used permanent hair colouring formulations for five or more years were considered to have such a history. Of 100 patients, 87 had been regular users of permanent dyes, compared with 26% of age-comparable controls. No other information was provided concerning the number, nature or selection of the controls or their relationship to cases with respect to potential confounding variables [The lack of information concerning both the cases and the controls leaves the results in doubt].

Kinlen *et al.* (1977) have reported a case-control study on 191 breast cancer patients interviewed in hospital in 1975-1976 in Oxford, UK, and on 506 non-cancer controls, matched by age, marital status and social class. The analysis used by the authors showed no statistically significant differences in the use of hair-care preparations between cases and controls. Similarly, no statistically significant differences between cases and controls were found when considering in detail the use of permanent and semi-permanent hair dyes, taking into account time prior to diagnosis and duration of use. Because of the small number of hairdressers among those studied (2 among cases and 10 among controls), no analysis of occupational risk was attempted.

Conclusions

The epidemiological evidence suggests an elevated risk for both users of hair dyes and those with occupational exposure (barbers and hairdressers) to hair preparations.

For users of hair dyes, the results are equivocal, since only one site (breast) has been studied; the Working Group considered that the study with negative results for this site (Kinlen *et al.*, 1977) was more convincing than the positive one (Shafer & Shafer, 1976).

For persons with occupational exposure to hair-care products, including dyes, there is more evidence for an increased risk of cancer at certain sites. The results of four of five case-control studies of bladder cancer are consistent with an increased risk for hairdressers (Anthony & Thomas, 1970; Wynder *et al.*, 1963) and barbers (Dunham *et al.*, 1968; Howe *et al.*, 1977). One case-control study reported an elevated risk of lung cancer in female beauticians (Garfinkel *et al.*, 1977); another reported an increased risk of laryngeal cancer in male barbers (Viadana *et al.*, 1976). Additional data from population-based death and cancer registries are also consistent with the hypothesis that there is an increased occupational risk among barbers and beauticians for cancers at different sites (Clemmesen, 1977; Milham, 1976; Registrar General, 1958, 1971).

Further epidemiological studies, which should include workers employed in the production of hair dyes, are necessary before any firm conclusions can be drawn.

References

Ames, B.N., Kammen, H.O. & Yamasaki, E. (1975) Hair dyes are mutagenic: identification of a variety of mutagenic ingredients. Proc. nat. Acad. Sci. (Wash.), 72, 2423-2427

Anthony, H.M. & Thomas, G.M. (1970) Tumors of the urinary bladder: an analysis of the occupations of 1,030 patients in Leeds, England. J. nat. Cancer Inst., 45, 879-895

Clemmesen, J. (1977) Statistical studies in the aetiology of malignant neoplasms. V. Acta path. microbiol. scand., Suppl. (in press)

Cole, P., Hoover, R. & Friedell, G.H. (1972) Occupation and cancer of the lower urinary tract. Cancer, 29, 1250-1260

Danish Ministry of the Environment (1977) Hair Dyes, Copenhagen, p. 38

Dunham, L.J., Rabson, A.S., Stewart, H.L., Frank, A.S. & Young, J.L., Jr (1968) Rates, interview, and pathology study of cancer of the urinary bladder in New Orleans, Louisiana. J. nat. Cancer Inst., 41, 683-709

Garfinkel, J., Selvin, S. & Brown, S.M. (1977) Brief communication: possible increased risk of lung cancer among beauticians. J. nat. Cancer Inst., 58, 141-143

General Register Office (1966) Classification of Occupations, 1966, HMSO, London

Gosselin, R.E., Hodge, H.C., Smith, R.P. & Gleason, M.N. (1976) Clinical Toxicology of Commercial Products, 4th ed., Baltimore, Williams & Wilkins, Section VI, pp. 116-119

Howe, G.R., Chambers, L., Gordon, P., Morrison, B. & Miller, A.B. (1977) An epidemiological study of bladder cancer (Abstract). Amer. J. Epidemiol., 106, 239

IARC (1972) IARC Monographs on the Evaluation of Carcinogenic Risk of Chemicals to Man, 1, Lyon

IARC (1974) IARC Monographs on the Evaluation of Carcinogenic Risk of Chemicals to Man, 4, Some Aromatic Amines, Hydrazine and Related Substances, *N*-Nitroso Compounds and Miscellaneous Alkylating Agents, Lyon

IARC (1975) IARC Monographs on the Evaluation of Carcinogenic Risk of Chemicals to Man, 8, Some Aromatic Azo Compounds, Lyon

IARC (1976) Information Bulletin on the Survey of Chemicals Being Tested for Carcinogenicity, No. 6, Lyon

Kinlen, L.J., Harris, R., Garrod, A. & Rodriguez, K. (1977) Use of hair dyes by patients with breast cancer: a case-control study. Brit. med. J., ii, 366-368

Menkart, J. (1975) Excess bladder cancer in beauticians? Science, 190, 96-98

Milham, S., Jr (1976) Occupational Mortality in Washington State, 1950-1971, Vol. 1, Cincinnati, Ohio, US Department of Health, Education, and Welfare, National Institute for Occupational Safety and Health, p. 36

Registrar General (1958) The Registrar General's Decennial Supplement, England and Wales, 1951, Occupational Mortality, Part 2, Vol. 2, Tables, London, HMSO

Registrar General (1971) The Registrar General's Decennial Supplement, England and Wales, 1961, Occupational Mortality Tables, London, HMSO

Shafer, N. & Shafer, R.W. (1976) Potential of carcinogenic effects of hair dyes. N.Y. State J. Med., 76, 394-396

Viadana, E., Bross, I.D.J. & Houten, L. (1976) Cancer experience of men exposed to inhalation of chemicals or to combustion products. J. occup. Med., 18, 787-792

Wall, F.E. (1972) Bleaches, hair colorings and dye removers. In: Balsam, M.S. & Sagarin, E., eds, Cosmetics Science and Technology, 2nd ed., Vol. 2, New York, Wiley-Interscience, pp. 279-343

Wynder, E.L., Onderdonk, J. & Mantel, N. (1963) An epidemiological investigation of cancer of the bladder. Cancer, 16, 1388-1407

THE MONOGRAPHS

HAIR DYES

4-AMINO-2-NITROPHENOL

1. Chemical and Physical Data

1.1 Synonyms and trade names

Colour Index No.: 76555

Colour Index Name: Oxidation Base 25

Chem. Abstr. Services Reg. No.: 119-34-6

Chem. Abstr. Name: 4-Amino-2-nitrophenol

4-Hydroxy-3-nitroaniline; 2-nitro-4-aminophenol; *ortho*-nitro-*para*-aminophenol

Fourrine 57; Fourrine Brown PR; Fourrine Brown propyl

1.2 Chemical formula and molecular weight

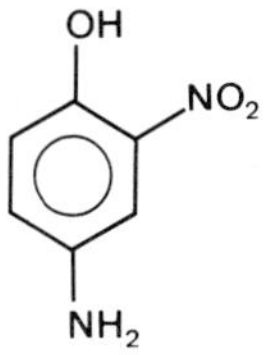

OH

NO_2

NH_2

$C_6H_6N_2O_3$ Mol. wt: 154.1

1.3 Chemical and physical properties of the pure substance

From Weast (1976), unless otherwise specified

(a) Description: Dark-red plates or needles

(b) Melting-point: 131°C

(c) Spectroscopy data: λ_{max} 234 nm in methanol; infra-red and nuclear magnetic resonance spectra have also been tabulated (Grasselli, 1973).

(d) Solubility: Soluble in hot water, ethanol and ether

(e) Stability: See 'General Remarks on the Substances Considered', p. 27.

1.4 Technical products and impurities

No data were available to the Working Group.

2. Production, Use, Occurrence and Analysis

For background information on this section, see preamble, p. 15.

2.1 Production and use

(a) Production

4-Amino-2-nitrophenol was prepared by Friedländer & Zeitlin in 1894 by reacting 3-nitro-1-azidobenzene with dilute sulphuric acid (Prager *et al.*, 1930). It may also be prepared by (1) heating 3-nitrophenylhydroxylamine with sulphuric acid; (2) nitrating *N*-acetyl-4-aminophenol followed by hydrolysis with dilute sulphuric acid; or (3) reducing 4'-hydroxy-3'-nitroazobenzene-4-sulphonic acid (Morse, 1963).

Commercial production in the US of an unspecified isomer of aminonitrophenol was first reported in 1921 (US Tariff Commission, 1922). Production of 4-amino-2-nitrophenol, specifically, was first reported in 1941 (US Tariff Commission, 1945). Only one US company reported an undisclosed amount (see preamble, p. 15) in 1972, the last year in which US production was reported (US Tariff Commission, 1974).

One or two Japanese companies occasionally produce negligible quantities (less than 100 kg per year) of 4-amino-2-nitrophenol.

No data on its production in Europe were available to the Working Group.

(b) Use

4-Amino-2-nitrophenol is used in dyeing furs a beige or deep reddish brown (The Society of Dyers and Colourists, 1971).

It is also used in hair dyes and produces red and gold shades on the hair. It has been used in dark-brown (at levels of 0.1%), medium-red (1.5%) and auburn (0.4%) dye formulations and in bleach toner formulations for silver (at levels of 0.002%), platinum-blond (0.02%), ash-blond (0.05%) and golden-blond (0.2%) shades (Wall, 1972).

4-Amino-2-nitrophenol has been used in hair dyes in Japan since 1955.

2.2 Occurrence

4-Amino-2-nitrophenol is not known to occur as a natural product. No data on its occurrence in the environment were available to the Working Group.

2.2 Analysis

Two-dimensional thin-layer chromatography on silica gel G has been used to detect dye intermediates, including 4-amino-2-nitrophenol, in hair dyes at concentrations as low as 0.02% (Kottemann, 1966).

Thin-layer chromatography on polystyrene-based anion exchangers and on microcrystalline cellulose and Cellex D anion exchangers has been used to separate primary aromatic amines, including 4-amino-2-nitrophenol (Lepri *et al.*, 1974).

3. Biological Data Relevant to the Evaluation of Carcinogenic Risk to Man

3.1 Carcinogenicity and related studies in animals[1]

Skin application

Mouse: Groups of 32 A or 32 DBAf young adult mice of both sexes were given twice weekly skin applications on the clipped dorsal skin of 0.4 ml (A mice), or 0.4 ml reduced to 0.2 ml at 24 weeks (DBAf mice), of a 10% solution of a commercially available hair dye, colourant 'RB', containing, among other constituents, CI Acid Black 107 (an azo-dye metal complex) and 4-amino-2-nitrophenol in 50% aqueous acetone. The same numbers of control animals received acetone only. When the experiment was terminated at 80 weeks, the incidence of tumours in treated DBAf mice was 5 (3 lymphomas after 41, 47 and 71 weeks, 1 ovarian granulosa-cell tumour at 79 weeks,

---

[1]The Working Group was aware of carcinogenicity tests in progress on 4-amino-2-nitrophenol involving oral administration in the diet in mice and rats and skin application in mice and rabbits (IARC, 1976).

and 1 ovarian cystadenocarcinoma at 80 weeks) compared with 1 in controls (1 lymphoid tumour after 72 weeks). In A mice, 8 treated animals developed lymphomas after 38-80 weeks compared with 7 control animals after 61-80 weeks; no differences in the incidence of liver or lung tumours occurred in treated or control mice. In DBAf mice, 75% of the treated animals were still alive at 64 weeks, compared with 78% at 80 weeks among the controls (22 treated animals died between 60-80 weeks without tumours compared with 25 controls). In A mice, 19 treated mice died without tumours between 60-80 weeks, compared with 16 controls (Searle, 1977; Searle & Jones, 1977; Venitt & Searle, 1976).

3.2 Other relevant biological data

(a) Experimental systems

The oral LD_{50} of 4-amino-2-nitrophenol in oil-in-water emulsion in rats is 3300 mg/kg bw; the i.p. LD_{50} of the compound in dimethylsulphoxide in rats is 302 mg/kg bw (Burnett *et al.*, 1977).

Six male and six female adult white rabbits received twice weekly skin applications of 1 ml/kg bw of a composite hair-dye preparation containing among other constituents 0.3% 4-amino-2-nitrophenol, mixed 1:1 with 6% hydrogen peroxide for 13 weeks. No changes in body weight gain, blood or urine parameters were observed (Burnett *et al.*, 1976).

The teratogenicity of the same dye formulation was tested in a group of 20 mated Charles River CD female rats. It was applied topically to a shaved site in the dorsoscapular region at a dose of 2 ml/kg bw (equivalent to 6 mg/kg bw 4-amino-2-nitrophenol) on days 1, 4, 7, 10, 13, 16 and 19 of gestation; just prior to use it was mixed with an equal volume of 6% hydrogen peroxide. The mothers were killed on day 20 of gestation, and no embryotoxic or teratogenic effects were observed (Burnett *et al.*, 1976).

This compound induced reverse mutations in *Salmonella typhimurium* TA 1538 in the absence of liver activation. Addition of a liver post-mitochondrial fraction from phenobarbital pretreated rats increased the mutagenic activity (Garner & Nutman, 1977).

No dominant lethals were induced in Charles River rats injected intraperitoneally three times weekly for 8 weeks with 20 mg/kg bw 4-amino-2-nitrophenol before mating (no positive control substance was tested) (Burnett *et al.*, 1977).

(b) Man

No data were available to the Working Group.

3.3 Case reports and epidemiological studies

No data were available to the Working Group.

4. Comments on Data Reported and Evaluation

4.1 Animal data

In the only available study, 4-amino-2-nitrophenol mixed with CI Acid Black 107 (an azo-dye metal complex) was tested in mice by skin application. No data on 4-amino-2-nitrophenol alone were available. No evaluation of its carcinogenicity can be made.

4.2 Human data

No case reports or epidemiological studies were available to the Working Group (see, however, 'General Remarks on the Substances Considered', p. 29).

5. References

Burnett, C., Goldenthal, E.I., Harris, S.B., Wazeter, F.X., Strausburg, J., Kapp, R. & Voelker, R. (1976) Teratology and percutaneous toxicity studies on hair dyes. J. Toxicol. environm. Hlth, 1, 1027-1040

Burnett, C., Loehr, R. & Corbett, J. (1977) Dominant lethal mutagenicity study on hair dyes. J. Toxicol. environm. Hlth, 2, 657-662

Garner, R.C. & Nutman, C.A. (1977) Testing of some azo dyes and their reduction products for mutagenicity using *Salmonella typhimurium* TA 1538. Mutation Res. (in press)

Grasselli, J.G., ed. (1973) CRC Atlas of Spectral Data and Physical Constants for Organic Compounds, Cleveland, Ohio, Chemical Rubber Co., p. B-755

IARC (1976) Information Bulletin on the Survey of Chemicals Being Tested for Carcinogenicity, No. 6, Lyon, pp. 189, 230

Kottemann, C.M. (1966) Two-dimensional thin layer chromatographic procedure for the identification of dye intermediates in arylamine oxidation hair dyes. J. Ass. off. analyt. Chem., 49, 954-959

Lepri, L., Desideri, P.G. & Coas, V. (1974) Chromatographic and electrophoretic behaviour of primary aromatic amines on anion-exchange thin layers. J. Chromat., 90, 331-339

Morse, S.K. (1963) Aminophenols. In: Kirk, R.E. & Othmer, D.F., eds, Encyclopedia of Chemical Technology, 2nd ed., Vol. 2, New York, John Wiley and Sons, p. 218

Prager, B., Jacobson, P., Schmidt, P. & Stern, D., eds (1930) Beilsteins Handbuch der Organischen Chemie, 4th ed., Vol. 13, Syst. No. 1852, Berlin, Springer-Verlag, p. 520

Searle, C.E. (1977) Evidence regarding the possible carcinogenicity of mutagenic hair dyes and constituents. In: Mécanismes d'Altération et de Réparation du DNA, Relations avec la Mutagénèse et la Cancérogénèse Chimique, Menton, 1976, Colloques Internationaux du CNRS, No. 256, Paris, Centre National de la Recherche Scientifique, pp. 407-415

Searle, C.E. & Jones, E.L. (1977) Effects of repeated applications of two semi-permanent hair dyes to the skin of A and DBAf mice. Brit. J. Cancer , 36, 467-478

The Society of Dyers and Colourists (1971) Colour Index, 3rd ed., Vol. 3, Yorkshire, UK, p. 3264

US Tariff Commission (1922) Census of Dyes and other Synthetic Organic Chemicals, 1921, Tariff Information Series No. 26, Washington DC, US Government Printing Office, p. 24

US Tariff Commission (1945) Synthetic Organic Chemicals, US Production and Sales, 1941-43, Report No. 153, Second Series, Washington DC, US Government Printing Office, p. 70

US Tariff Commission (1974) Synthetic Organic Chemicals, US Production and Sales, 1972, TC Publication 681, Washington DC, US Government Printing Office, p. 85

Venitt, S. & Searle, C.E. (1976) Mutagenicity and possible carcinogenicity of hair colourants and constituents. In: Rosenfeld, C. & Davis, W., eds, Environmental Pollution and Carcinogenic Risks, Lyon (IARC Scientific Publications No. 13), pp. 263-271

Wall, F.E. (1972) Bleaches, hair colorings and dye removers. In: Balsam, M.S. & Sagarin, E., eds, Cosmetics Science and Technology, 2nd ed., Vol. 2, New York, Wiley-Interscience, pp. 308, 313, 316, 317

Weast, R.C., ed. (1976) CRC Handbook of Chemistry and Physics, 57th ed., Cleveland, Ohio, Chemical Rubber Co., p. C-428

2,4-DIAMINOANISOLE (SULPHATE)

1. Chemical and Physical Data

2,4-Diaminoanisole

1.1 Synonyms and trade names

Colour Index No.: 76050

Colour Index Name: Oxidation Base 12

Chem. Abstr. Services Reg. No.: 615-05-4

Chem. Abstr. Name: 4-Methoxy-1,3-benzenediamine

2,4-Diamino-anisol; *meta*-diaminoanisole; 1,3-diamino-4-methoxybenzene; 2,4-diaminosole base; 4-methoxy-*meta*-phenylenediamine

Furro L; Pelagol DA; Pelagol Grey L; Pelagol L

1.2 Chemical formula and molecular weight

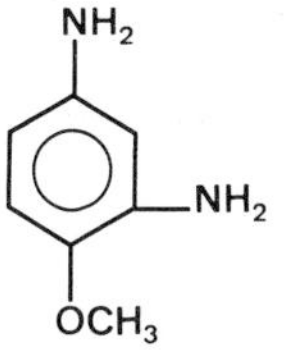

NH_2

NH_2

OCH_3

$C_7H_{10}N_2O$ Mol. wt: 138.2

1.3 Chemical and physical properties of the pure substance

From Weast (1976), unless otherwise specified

(a) Description: Colourless needles

(b) Melting-point: 67-68°C

(c) Spectroscopy data: λ_{max} 635, 587, 545 and 511 nm (Allan & Mužík, 1953); infra-red and nuclear magnetic resonance spectra have been tabulated (Grasselli, 1973).

(d) Solubility: Soluble in water, ethanol and ether

(e) Stability: See 'General Remarks on the Substances Considered', p. 27.

1.4 Technical products and impurities

In the US, 2,4-diaminoanisole is available with a purity of approximately 99% (Mobay Chemical Corporation, 1975).

In Japan, it is available with a minimum purity of 98.5% and contains 2,4-diaminophenol and nitroaminoanisole as impurities.

2,4-Diaminoanisole sulphate

1.1 Synonyms and trade names

Colour Index No.: 76051

Colour Index Name: Oxidation Base 12A

Chem. Abstr. Services Reg. No.: 39156-41-7

Chem. Abstr. Name: 4-Methoxy-1,3-benzenediamine sulphate

2,4-Diamino-anisol sulphate; 2,4-diaminosole sulphate; 4-methoxy-1,3-benzenediamine sulphate; 4-methoxy-*meta*-phenylenediamine sulphate

BASF Ursol SLA; Fouramine BA; Fourrine 76; Fourrine SLA; Furro SLA; Nako TSA; Pelagol BA; Pelagol Grey SLA; Pelagol SLA; Renal SLA; Ursol SLA; Zoba SLE

1.2 Empirical formula and molecular weight

$$C_7H_{10}N_2O.H_2SO_4$$

Mol. wt: 236.3

1.3 Chemical and physical properties of the pure substance

(a) Description: Off-white to violet powder (Ashland Chemical Company, 1975)

(b) Solubility: Soluble in water and ethanol (The Society of Dyers and Colourists, 1971a)

1.4 Technical products and impurities

Commercial grade 2,4-diaminoanisole sulphate is available in the US as the dihydrate with a minimum purity of 80% and a maximum of 1.0% ash and 75 mg/kg iron (Ashland Chemical Company, 1975).

2. Production, Use, Occurrence and Analysis

For background information on this section, see preamble, p. 15.

2.1 Production and use

(a) Production

2,4-Diaminoanisole was prepared by the reduction of 2,4-dinitroanisole with iron and acetic acid by the Badische Anilin- & Soda-Fabrik AG in 1913 (Richter, 1933). 2,4-Diaminoanisole is produced commercially in Japan by methylation of 2,4-dinitro-1-chlorobenzene followed by reduction with iron.

Commercial production of 2,4-diaminoanisole in the US was first reported in 1933 (US Tariff Commission, 1934) and that of 2,4-diaminoanisole sulphate in 1967 (US Tariff Commission, 1969). However, commercial production of 2,4-diaminoanisole has not been reported in the US since 1940 (US Tariff Commission, 1941), and production of 2,4-diaminoanisole sulphate has not been reported since 1971 (US Tariff Commission, 1973).

US imports of 2,4-diaminoanisole through principal US customs districts were reported as 16 thousand kg in 1974 (US International Trade Commission, 1976) and 5 thousand kg in 1975 (US International Trade Commission, 1977a). Separate data for imports of the sulphate are not available.

It is believed that 10-100 thousand kg 2,4-diaminoanisole and its salts are imported into Switzerland annually.

2,4-Diaminoanisole has been produced commercially in Japan since 1974; until 1973, Japan imported 2,4-diaminoanisole from the Federal

Republic of Germany. In 1975, one Japanese company produced about 7 thousand kg. There are no Japanese imports or exports of this chemical.

(b) Use

2,4-Diaminoanisole is used as a colour modifier in certain permanent hair-dye compositions (Anon., 1976), i.e., drab-brown, red, blond and blue-grey dye formulations (Wall, 1972). It has been estimated that about 600 mg 2,4-diaminoanisole are used in each dye application (Ames *et al.*, 1975). It is also used as a colour modifier in blue and grey hair-dye formulations, at levels of 0.25% in blue-violet shades and 0.55% in steel-grey shades. It is also used as a component of bleach-toner formulations in platinum-blond (0.02%) and ash-blond (0.01%) shades (Wall, 1972).

2,4-Diaminoanisole is used as an intermediate for the production of a dye, C.I. Basic Brown 2, which is produced commercially by one company in the US (The Society of Dyers and Colourists, 1971b; US International Trade Commission, 1977b). C.I. Basic Brown 2 is mainly used to dye acrylic fibres, but it has also been used to colour cotton, viscose, wool, nylon, polyester and leather and as an ingredient of shoe polishes and suede dressings (The Society of Dyers and Colourists, 1971c).

2,4-Diaminoanisole and 2,4-diaminoanisole sulphate are also used in dyeing furs (The Society of Dyers and Colourists, 1971d).

2,4-Diaminoanisole is believed to be used in Japan principally as an intermediate in the production of pharmaceuticals, although about 100 kg are used in hair-dye compositions.

2,4-Diaminoanisole and its salts are believed to be used as dyestuff intermediates in Switzerland. The use of 2,4-diaminoanisole in hair-dye formulations will be forbidden in Italy after 1977 (Dal Falco, 1976).

2.2 Occurrence

2,4-Diaminoanisole and 2,4-diaminoanisole sulphate are not known to occur as natural products. No data on their occurrence in the environment were available to the Working Group.

2.3 Analysis

Primary aromatic amines, including 2,4-diaminoanisole, have been identified by measurement of the absorption spectra of the azo dye resulting from diazotization and coupling of the amine (Allan & Muzik, 1953).

Commercial hair dyes containing phenols and amines, including 2,4-diaminoanisole, have been analysed by isolation of the basic fraction and separation by paper chromatography using eleven reagents for identification of the components (Turi, 1957).

Gas chromatography has been used to separate and identify oxidation dyes, including 2,4-diaminoanisole, in hair colourants (Goldstein *et al.*, 1968).

Primary aromatic amines, including 2,4-diaminoanisole, have been separated and identified using thin-layer chromatography on polystyrene-based anion exchangers and on microcrystalline cellulose and Cellex D anion exchangers by means of aqueous eluants (Lepri *et al.*, 1974), on silica gel buffered with sodium acetate (Bassl *et al.*, 1967) and on Kieselgel DG (Kurteva & Takeva, 1971). Emulsified hair-dye components, including 2,4-diaminoanisole, have been separated by thin-layer chromatography on silica gel G plates with a butyl alcohol-ethyl alcohol-water-acetic acid mixture as eluant and aqueous ferric chloride as indicator (Zelazna & Legatowa, 1971). Thin-layer chromatography in two dimensions has been used to separate and identify dye intermediates, including 2,4-diaminoanisole, in aromatic amine oxidation hair dyes (Kottemann, 1966).

3. Biological Data Relevant to the Evaluation of Carcinogenic Risk to Man

3.1 Carcinogenicity and related studies in animals[1]

(a) Skin application

Mouse: Two groups of 50 male and 50 female 6-8-week old Swiss-Webster mice received weekly or fortnightly applications of 0.05 ml of a

---

[1]The Working Group was aware of carcinogenicity tests in progress in mice and rats involving oral administration of 2,4-diaminoanisole sulphate in the diet (IARC, 1976).

freshly prepared 1:1 mixture of a hair-dye formulation, containing among its constituents 0.38% 2,4-diaminoanisole sulphate, 3% 2,5-diaminotoluene sulphate and 1.5% *para*-phenylenediamine, and 6% hydrogen peroxide. The mixture was painted on shaved skin in the midscapular region for 18 months, at which time the surviving animals were killed. The incidence of lung tumours did not differ significantly from that in untreated control animals (Burnett *et al.*, 1975) [The Working Group noted the high incidence of tumours in the mice treated with the dye base alone, which made up the greatest part of the hair-dye formulation].

Rat: Two experimental preparations in a carboxymethyl cellulose gel were tested: formulation 1 contained 0.75% 2,4-diaminoanisole and 3% 2,5-diaminotoluene (both as sulphates), and formulation 2 was a control without added dye intermediate. Each formulation was mixed with an equal volume of 6% hydrogen peroxide immediately before use, and 0.5 ml of the mixture was applied to the dorsal skin. Two groups, each of 50 male and 50 female Sprague-Dawley rats, 12 weeks old, were treated twice weekly for 2 years with formulation 1 or left untreated; a third group of 25 male and 25 female rats were treated with formulation 2. No significant differences were detected in tumour types or incidences between the experimental and control groups (Kinkel & Holzmann, 1973) [The Working Group noted the absence of information about survival times and the high incidence of tumours in controls].

3.2 Other relevant biological data

(a) Experimental systems

The oral LD_{50} of 2,4-diaminoanisole sulphate in oil-in-water emulsion in rats is >4000 mg/kg bw; the i.p. LD_{50} in dimethylsulphoxide is 372 mg/kg bw (Burnett *et al.*, 1977).

A hair-dye preparation, containing among its constituents 0.38% 2,4-diaminoanisole sulphate, 3% 2,5-diaminotoluene sulphate and 1.5% *para*-phenylenediamine, was mixed with an equal volume of 6% hydrogen peroxide and applied to the skin of Swiss-Webster mice. Weekly doses sufficient to cause moderate reversible alopecia in about 50% of the animals of each sex within 5 months resulted in no changes in liver weight:body weight ratios, and the results of haematological examinations were within normal limits (Burnett *et al.*, 1975). No toxic effects were observed in additional experiments with complex mixtures containing 2,4-diaminoanisole sulphate (Burnett *et al.*, 1976).

A dose of 0.5 ml of a preparation containing 0.75% 2,4-diaminoanisole and 3% 2,5-diaminotoluene sulphates (percentages calculated as free base), in carboxymethyl cellulose gel, mixed 1:1 with 6% hydrogen peroxide and applied to the shaved dorsal skin of Sprague-Dawley rats twice weekly for 2 years, had no effect on growth rates, mean lifespan, haematological parameters or liver function (bromosulphthalein excretion or serum transaminase levels) (Kinkel & Holzmann, 1973).

Three commercially available hair-dye formulations, containing 0.02, 2 or 4% 2,4-diaminoanisole sulphate and several aromatic amine derivatives among their constituents, were tested for teratogenicity in groups of 20 mated Charles River CD female rats. Each formulation was applied topically to a shaved site in the dorsoscapular region at a dose of 2 ml/kg bw on days 1, 4, 7, 10, 13, 16 and 19 of gestation; just prior to its use, each formulation was mixed with an equal volume of 6% hydrogen peroxide. The mothers were killed on day 20 of gestation. There were no soft tissue anomalies in the live foetuses, but in 3 of 20 litters in the group given the formulation containing the highest concentration of 2,4-diaminoanisole sulphate (4%), skeletal changes were seen in 9 of the 169 live foetuses. On comparison with 3 control groups, this finding was statistically significant ($P<0.05 \rightarrow P<0.01$) (Burnett *et al.*, 1976).

It was reported in an abstract that in rats given single i.p. injections of 50 mg/kg bw ^{14}C-labelled 2,4-diaminoanisole, 94% of the dose was excreted within 48 hours (85% in the urine and 9% in the faeces). In 24-hour urine samples, radioactivity was found in unconjugated metabolites, glucuronides, sulphates and water-soluble, non-extractable material. The metabolites were 4-acetylamino-2-aminoanisole, 2,4-diacetylaminoanisole and a small amount of diacetylaminophenol; 2,4-diacetylaminophenol, hydroxylated derivatives of 2,4-diacetylaminoanisole and a small amount of 4-acetylamino-2-aminoanisole were present as glucuronides (Grantham *et al.*, 1977).

2,4-Diaminoanisole induces reverse mutations in *Salmonella typhimurium* TA 1538 after incubation with liver post-mitochondrial supernatant fraction from rats pretreated with polychlorinated biphenyls (Aroclor 1254) (Ames *et al.*, 1975). 2,4-Diaminoanisole sulphate was weakly mutagenic in *Drosophila melanogaster*, inducing sex-linked recessive lethals when fed at 15.1 mM, the highest concentration tested (Blijleven, 1977). In studies on thymidine kinase mutation in mouse lymphoma cells (L51178Y), reported in an abstract, the compound was found to be inactive (no liver fraction was added) (Palmer *et al.*, 1976).

No dominant lethals were induced in Charles River CD rats injected intraperitoneally three times weekly for 8 weeks with 20 mg/kg bw 2,4-diaminoanisole before mating (no positive control substance was tested) (Burnett *et al.*, 1977).

This compound did not induce micronucleated cells in bone marrow when 1000 mg/kg bw were administered orally to two groups of 5 male and 5 female rats in two doses separated by an interval of 24 hours (Hossack & Richardson, 1977).

(b) Man

No data were available to the Working Group.

3.3 Case reports and epidemiological studies

No data were available to the Working Group.

4. Comments on Data Reported and Evaluation

4.1 Animal data

2,4-Diaminoanisole sulphate present in a commercial hair-dye formulation or in an experimental formulation similar to those used as hair dyes was inadequately tested by skin application in mice and rats. No data on 2,4-diaminoanisole or its sulphate alone were available. No evaluation of the carcinogenicity of these compounds could be made.

4.2 Human data

No case reports or epidemiological studies were available to the Working Group (see, however, 'General Remarks on the Substances Considered', p. 29).

5. References

Allan, Z.J. & Mužík, F. (1953) Aromatic diazo compounds. XII. Identification of primary aromatic amines by means of the absorption spectra of the corresponding azo dyes. Chem. Listy, 47, 380-391

Ames, B.N., Kammen, H.O. & Yamasaki, E. (1975) Hair dyes are mutagenic: identification of a variety of mutagenic ingredients. Proc. nat. Acad. Sci. (Wash.), 72, 2423-2427

Anon. (1976) Modern trends in hair colourants. Soap, Perfumery and Cosmetics, May, pp. 189-194

Ashland Chemical Company (1975) Technical Data, 2,4-Diaminoanisole Sulfate, Columbus, Ohio, Fine Chemicals Department

Bassl, A., Heckemann, H.J. & Baumann, E. (1967) Thin-layer chromatography of primary aromatic amines. I. J. Prakt. Chem., 36, 265-273

Blijleven, W.G.H. (1977) Mutagenicity of four hair dyes in *Drosophila melanogaster*. Mutation Res., 48, 181-186

Burnett, C., Lanman, B., Giovacchini, R., Wolcott, G., Scala, R. & Keplinger, M. (1975) Long-term toxicity studies on oxidation hair dyes. Fd Cosmet. Toxicol., 13, 353-357

Burnett, C., Goldenthal, E.I., Harris, S.B., Wazeter, F.X., Strausburg, J., Kapp, R. & Voelker, R. (1976) Teratology and percutaneous toxicity studies on hair dyes. J. Toxicol. environm. Hlth, 1, 1027-1040

Burnett, C., Loehr, R. & Corbett, J. (1977) Dominant lethal mutagenicity study on hair dyes. J. Toxicol. environm. Hlth, 2, 657-662

Dal Falco (Il Ministro per la Sanita) (1976) Decreto ministeriale, 18 giugno 1976, Art. 1. Gazzetta Ufficiale Della Repubblica Italiana, N. 166, 5025

Goldstein, S., Kopf, A.A. & Feinland, R. (1968) Analysis of oxidation dyes in hair colorants by thin-layer and gas chromatography. In: Proceedings of the Joint Conference on Cosmetic Sciences, Washington DC, Toilet Goods Ass. Inc., pp. 19-38

Grantham, P.H., Mohan, L.C., Benjamin, T.T., Roller, P.P. & Weisburger, E.K. (1977) Metabolism of the dyestuff intermediate, 2,4-diaminoanisole, in the rat (Abstract No. 69). Toxicol. appl. Pharmacol. (in press)

Grasselli, J.G., ed. (1973) CRC Atlas of Spectral Data and Physical Constants for Organic Compounds, Cleveland, Ohio, Chemical Rubber Co., p. B-223

Hossack, D.J.N. & Richardson, J.C. (1977) Examination of the potential mutagenicity of hair dye constituents using the micronucleus test. Experientia, 33, 377-378

IARC (1976) Information Bulletin on the Survey of Chemicals Being Tested for Carcinogenicity, No. 6, Lyon, p. 285

Kinkel, H.J. & Holzmann, S. (1973) Study of long-term percutaneous toxicity and carcinogenicity of hair dyes (oxidizing dyes) in rats. Fd Cosmet. Toxicol., 11, 641-648

Kottemann, C.M. (1966) Two-dimensional thin layer chromatographic procedure for the identification of dye intermediates in arylamine oxidation hair dyes. J. Ass. off. analyt. Chem., 49, 954-959

Kurteva, R. & Takeva, S. (1971) Thin-layer chromatography of aromatic diamines and their N-derivatives. God. Nauchnoizsled. Inst. Khim. Prom., 9, 53-57

Lepri, L., Desideri, P.G. & Coas, V. (1974) Chromatographic and electrophoretic behaviour of primary aromatic amines on anion-exchange thin layers. J. Chromat., 90, 331-339

Mobay Chemical Corporation (1975) 2,4-Diaminoanisole. Pittsburgh, Penn., Industrial Chemicals Division

Palmer, K.A., DeNunzio, A. & Green, S. (1976) The mutagenic assay of some hair dye components using the thymidine kinase locus of L5178Y mouse lymphoma cells (Abstract No. 39). Toxicol. appl. Pharmacol., 37, 108

Richter, F., ed. (1933) Beilsteins Handbuch der Organischen Chemie, 4th ed., 1st Suppl., Vol. 13, Syst. No. 1854, Berlin, Springer-Verlag, p. 204

The Society of Dyers and Colourists (1971a) Colour Index, 3rd ed., Vol. 4, Yorkshire, UK, p. 4644

The Society of Dyers and Colourists (1971b) Colour Index, 3rd ed., Vol. 4, Yorkshire, UK, p. 4821

The Society of Dyers and Colourists (1971c) Colour Index, 3rd ed., Vol. 1, Yorkshire, UK, p. 1683

The Society of Dyers and Colourists (1971d) Colour Index, 3rd ed., Vol. 3, Yorkshire, UK, p. 3262

Turi, C.J. (1957) Analysis of hair dyes. II. Rend. ist. super. sanità, 20, 570-589

US International Trade Commission (1976) Imports of Benzenoid Chemicals and Products, 1974, USITC Publication 762, Washington DC, US Government Printing Office, p. 22

US International Trade Commission (1977a) Imports of Benzenoid Chemicals and Products, 1975, USITC Publication 806, Washington DC, US Government Printing Office, p. 21

US International Trade Commission (1977b) Synthetic Organic Chemicals, US Production and Sales, 1975, USITC Publication 804, Washington DC, US Government Printing Office, p. 62

US Tariff Commission (1934) Production and Sales of Dyes and Other Synthetic Organic Chemicals, 1933, Report No. 89, Second Series, Washington DC, US Government Printing Office, p. 9

US Tariff Commission (1941) Synthetic Organic Chemicals, US Production and Sales, 1940, Report No. 148, Second Series, Washington DC, US Government Printing Office, p. 11

US Tariff Commission (1969) Synthetic Organic Chemicals, US Production and Sales, 1967, TC Publication 295, Washington DC, US Government Printing Office, p. 81

US Tariff Commission (1973) Synthetic Organic Chemicals, US Production and Sales, 1971, TC Publication 614, Washington DC, US Government Printing Office, p. 43

Wall, F.E. (1972) Bleaches, hair colorings and dye removers. In: Balsam, M.S. & Sagarin, E., eds, Cosmetics Science and Technology, 2nd ed., Vol. 2, New York, Wiley-Interscience, pp. 309, 310, 313, 317

Weast, R.C., ed. (1976) CRC Handbook of Chemistry and Physics, 57th ed., Cleveland, Ohio, Chemical Rubber Co., p. C-156

Zelazna, K. & Legatowa, B. (1971) Identification of basic dyes in emulsified hair dyes by thin layer chromatography. Rocz. Panstw. Zakl. Hig., 22, 427-430

1,2-DIAMINO-4-NITROBENZENE

1. Chemical and Physical Data

1.1 Synonyms and trade names

Colour Index Number: 76020

Chem. Abstr. Services Reg. No.: 99-56-9

Chem. Abstr. Name: 4-Nitro-1,2-benzenediamine

2-Amino-4-nitroaniline; 4-nitro-1,2-diaminobenzene; 4-nitro-1,2-phenylenediamine; 4-nitro-*ortho*-phenylenediamine; *para*-nitro-*ortho*-phenylenediamine

1.2 Chemical formula and molecular weight

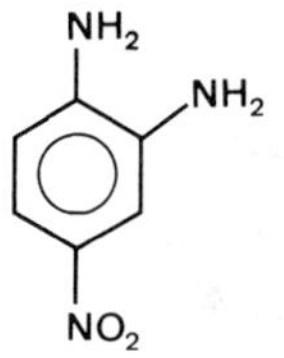

NH_2 NH_2 NO_2

$C_6H_7N_3O_2$ Mol. wt: 153.1

1.3 Chemical and physical properties of the pure substance

From Weast (1976), unless otherwise specified

(a) Description: Dark-red needles

(b) Melting-point: 199-200°C

(c) Spectroscopy data: λ_{max} 380 nm (E_1^1 = 801) in 20% sulphuric acid; the infra-red spectrum has been tabulated by Grasselli (1973).

(d) Solubility: Sparingly soluble in water (Windholz, 1976); soluble in acetone (Searle, 1977) and in aqueous acids

(e) Stability: See 'General Remarks on the Substances Considered', p. 27.

1.4 Technical products and impurities

1,2-Diamino-4-nitrobenzene is available in the US as a commercial grade with the following typical specifications: assay, 98.0% min.; melting range, 200-204°C; residue on ignition, 0.3% max.; iron, 150 mg/kg max.; loss on drying, 0.5% max. (Ashland Chemical Company, 1975).

2. Production, Use, Occurrence and Analysis

For background information on this section, see preamble, p. 15.

2.1 Production and use

(a) Production

1,2-Diamino-4-nitrobenzene was prepared by Heim in 1888 by reducing 2,4-dinitroaniline with alcoholic ammonium sulphide (Prager *et al.*, 1930).

Commercial production in the US of an unspecified isomer of diamino-nitrobenzene was first reported in 1936 (US Tariff Commission, 1938) and that of 1,2-diamino-4-nitrobenzene, specifically, in 1946 (US Tariff Commission, 1948). Although only one US company reported an undisclosed amount (see preamble, p. 15) in 1975 (US International Trade Commission, 1977a), it is believed that two companies currently manufacture this chemical.

US imports of 1,2-diamino-4-nitrobenzene through principal customs districts were reported to be 1900 kg in 1973 (US Tariff Commission, 1974) and 1100 kg in 1975 (US International Trade Commission, 1977b).

No evidence was found that 1,2-diamino-4-nitrobenzene is produced on a commercial scale in Japan, although certain companies produce quantities on request.

No data on its production in Europe were available to the Working Group.

(b) Use

1,2-Diamino-4-nitrobenzene is used in semi-permanent (Anon., 1976) and permanent hair colouring products. It produces brown, red and blond shades on the hair. It has typically been used in medium-brown (at levels of 0.15%) and medium-red (0.5%) dye formulations. It can also been used in bleach-toner formulations for reddish-blond shades (at levels of 0.03%) (Wall, 1972).

1,2-Diamino-4-nitrobenzene is also reportedly used as a reagent for α-keto acids (Windholz, 1976) and as a colorimetric reagent for the determination of ascorbic and dehydroascorbic acids in foods (Bourgeois *et al.*, 1975).

The use of 1,2-diamino-4-nitrobenzene in hair-dye formulations will be forbidden in Italy after 1977 (Dal Falco, 1976).

2.2 Occurrence

1,2-Diamino-4-nitrobenzene is not known to occur as a natural product. No data on its occurrence in the environment were available to the Working Group.

2.3 Analysis

Paper chromatography using six solvent systems and 15 reagents has been used to separate and identify 30 aminobenzene derivatives, including 1,2-diamino-4-nitrobenzene (Reio, 1970).

Thin-layer chromatography using various adsorbents, eluants and indicators has been used for the analysis of 1,2-diamino-4-nitrobenzene in hair-dye compositions: on silica gel plates impregnated with rice starch using four eluants with *para*-dimethylaminobenzaldehyde as an indicator (Urquizo, 1969), on silica gel G plates with ferric chloride as indicator (Zelazna & Legatowa, 1971) and with a sodium metabisulphite-saturated methanol solution as eluant, followed by elution and spectrophotometric determination (Legatowa, 1973).

Primary aromatic amines have been separated and identified in samples containing 0.5-2 μg of each amine by thin-layer chromatography on polystyrene-based anion exchangers, microcrystalline cellulose or Cellex D anion exchangers, and by electrophoresis on polystyrene-based anion exchangers with various solvents (Lepri *et al.*, 1974).

Two-dimensional thin-layer chromatography on silica gel G has been used to separate and identify dye intermediates, including 1,2-diamino-4-nitrobenzene, with a limit of 0.02% (Kottemann, 1966).

3. Biological Data Relevant to the Evaluation of Carcinogenic Risk to Man

3.1 Carcinogenicity and related studies in animals[1]

Skin application

Mouse: Groups of 48 DBAf or 52 A young adult mice of both sexes were given twice weekly skin applications on the clipped dorsal skin of 0.4 ml reduced to 0.2 ml at 24 weeks (DBAf mice), or 0.4 ml (A mice), of a 10% solution of a commercially available hair dye, colourant 'GS', containing, among other constituents, 1,2-diamino-4-nitrobenzene and 1,4-diamino-2-nitrobenzene in 50% aqueous acetone; 32 control mice of each strain received acetone only. When the experiment was terminated at 80 weeks, 5 lymphomas and 6 tumours of the female reproductive tract (4 ovarian cystadenomas and 2 uterine fibrosarcomas) had developed in the treated DBAf mice within 26-80 weeks; in the DBAf control mice, 1 lymphoma, 1 liver tumour and 1 lung tumour were found. No differences occurred in the incidence of lymphomas, liver or lung tumours between treated and control A mice. Of the treated animals, 19 DBAf mice and 27 A mice survived 60 or more weeks and died without tumours (Searle, 1977; Searle & Jones, 1977; Venitt & Searle, 1976).

3.2 Other relevant biological data

(a) Experimental systems

The oral LD_{50} of 1,2-diamino-4-nitrobenzene in oil-in-water emulsion in rats is 3720 mg/kg bw; the i.p. LD_{50} of the compound in dimethylsulphoxide in rats is >1600 mg/kg bw (Burnett *et al.*, 1977).

---

[1]The Working Group was aware of carcinogenicity tests in progress on 1,2-diamino-4-nitrobenzene involving skin application in mice and oral administration in the diet in mice and rats (IARC, 1976).

A composite mixture containing 10 dyes used as semi-permanent hair colourants, which contained 0.16% 1,2-diamino-4-nitrobenzene and a total dye content of 5.82%, was fed in the diet to groups of 6 male and 6 female pure-bred beagle dogs, 6-8 months of age, at concentrations that gave doses of 0.54 and 2.68 mg/kg bw/day, respectively, of the 1,2-diamino-4-nitrobenzene. A control group was maintained on normal diet. Haematological tests and urine analyses showed no significant changes, other than a blue-brown colouration of the urine (Wernick *et al.*, 1975).

Six male and six female adult white rabbits received twice weekly skin applications of 1 ml/kg bw of a composite hair-dye preparation containing among other constituents 0.25% 1,2-diamino-4-nitrobenzene (but which also contained 6% 2,4-diaminotoluene sulphate), mixed 1:1 with 6% hydrogen peroxide solution for 13 weeks. Haematological tests and urine analyses showed no significant changes (Burnett *et al.*, 1976).

The teratogenicity of the same dye formulation was tested in a group of 20 mated Charles River CD female rats. It was applied topically to a shaved site in the dorsoscapular region at a dose of 2 ml/kg bw on days 1, 4, 7, 10, 13, 16 and 19 of gestation; just prior to use it was mixed with an equal volume of 6% hydrogen peroxide. The mothers were killed on day 20 of gestation, and no significant changes were found in soft tissues or skeletal elements of the foetuses when compared with control groups (Burnett *et al.*, 1976).

1,2-Diamino-4-nitrobenzene induced morphological transformation in a mouse cell line (C3H/10T½CL8) (Benedict, 1976) [The Working Group noted that the transformed cells were not tested for tumourigenicity *in vivo*].

1,2-Diamino-4-nitrobenzene has been reported to induce reverse mutations in *Salmonella typhimurium* TA 1538 in the absence of liver activation. Addition of liver post-mitochondrial supernatant fraction from rats pre-treated with polychlorinated biphenyl (Aroclor 1254) or phenobarbital reduced mutagenic activity (Ames *et al.*, 1975; Garner & Nutman, 1977; Searle *et al.*, 1975). The compound induced sex-linked recessive lethals in *Drosophila* when fed at a concentration of 1.2 mM (Blijleven, 1977). Results reported in an abstract indicate that the compound induced mutations at the thymidine kinase locus in a dose-dependent manner in a mouse lymphoma cell line (L51178Y) (no liver fraction was added) (Palmer *et al.*, 1976).

No dominant lethals were induced in Charles River rats injected intraperitoneally three times weekly for 8 weeks with 20 mg/kg bw 1,2-diamino-4-nitrobenzene before mating (no positive control substance was tested) (Burnett *et al.*, 1977).

1,2-Diamino-4-nitrobenzene induced chromatid breaks in a hamster cell line A(T_1) Cl-3 (Benedict, 1976) and chromosome aberrations in CHMP/E cells (Kirland & Venitt, 1976). No chromosomal aberrations were observed when this compound was incubated with cultured human peripheral lymphocytes (Searle *et al.*, 1975).

This compound did not induce micronucleated cells in bone marrow when 5000 mg/kg bw were administered orally to two groups of 5 male and 5 female rats in two doses separated by an interval of 24 hours (Hossack & Richardson, 1977).

(b) Man

No data were available to the Working Group.

3.3 Case reports and epidemiological studies

No data were available to the Working Group.

4. Comments on Data Reported and Evaluation

4.1 Animal data

In the only available study, a hair-dye formulation containing 1,2-diamino-4-nitrobenzene mixed with 1,4-diamino-2-nitrobenzene was tested in mice by skin application. No data on 1,2-diamino-4-nitrobenzene alone were available. No evaluation of the carcinogenicity of this compound can be made.

4.2 Human data

No case reports or epidemiological studies were available to the Working Group (see, however, 'General Remarks on the Substances Considered', p. 29).

5. References

Ames, B.N., Kammen, H.O. & Yamasaki, E. (1975) Hair dyes are mutagenic: identification of a variety of mutagenic ingredients. Proc. nat. Acad. Sci. (Wash.), 72, 2423-2427

Anon. (1976) Modern trends in hair colourants. Soap, Perfumery and Cosmetics, May, 189-194

Ashland Chemical Company (1975) Technical Data, 4-Nitro-*o*-phenylene diamine, Fine Chemicals Department, Columbus, Ohio

Benedict, W.F. (1976) Morphological transformation and chromosome aberrations produced by two hair dye components. Nature (Lond.), 260, 368-369

Blijleven, W.G.H. (1977) Mutagenicity of four hair dyes in *Drosophila melanogaster*. Mutation Res., 48, 181-186

Bourgeois, C.F., Czornomaz, A.M., George, P., Belliot, J.P., Mainguy, P.R. & Watier, B. (1975) Specific determination of vitamin C (ascorbic and dehydroascorbic acids) in foods. Analusis, 3, 540-548

Burnett, C., Goldenthal, E.I., Harris, S.B., Wazeter, F.X., Strausburg, J., Kapp, R. & Voelker, R. (1976) Teratology and percutaneous toxicity studies on hair dyes. J. Toxicol. environm. Hlth, 1, 1027-1040

Burnett, C., Loehr, R. & Corbett, J. (1977) Dominant lethal mutagenicity study on hair dyes. J. Toxicol. environm. Hlth, 2, 657-662

Dal Falco (Il Ministro per la Sanita) (1976) Decreto ministeriale, 18 giugno 1976, Art. 1. Gazzetta Ufficiale Della Repubblica Italiana, N. 166, 5025

Garner, R.C. & Nutman, C.A. (1977) Testing of some azo dyes and their reduction products for mutagenicity using *Salmonella typhimurium* TA 1538. Mutation Res. (in press)

Grasselli, J.G., ed. (1973) CRC Atlas of Spectral Data and Physical Constants for Organic Compounds, Cleveland, Ohio, Chemical Rubber Co., p. B-223

Hossack, D.J.N. & Richardson, J.C. (1977) Examination of the potential mutagenicity of hair dye constituents using the micronucleus test. Experientia, 33, 377-378

IARC (1976) Information Bulletin on the Survey of Chemicals Being Tested for Carcinogenicity, No. 6, Lyon, pp. 105, 193

Kirkland, D.J. & Venitt, S. (1976) Cytotoxicity of hair colourant constituents: chromosome damage induced by two nitrophenylenediamines in cultured Chinese hamster cells. Mutation Res., 40, 47-56

Kottemann, C.M. (1966) Two-dimensional thin layer chromatographic procedure for the identification of dye intermediates in arylamine oxidation hair dyes. J. Ass. off. analyt. Chem., 49, 954-959

Legatowa, B. (1973) Determination of aromatic amines and aminophenols in hair dyes. Rocz. Panstw. Zakl. Hig., 24, 393-402

Lepri, L., Desideri, P.G. & Coas, V. (1974) Chromatographic and electrophoretic behaviour of primary aromatic amines on anion-exchange thin layers. J. Chromat., 90, 331-339

Palmer, K.A., DeNunzio, A. & Green, S. (1976) The mutagenic assay of some hair dye components using the thymidine kinase locus of L5117BY mouse lymphoma cells (Abstract No. 39). Toxicol. appl. Pharmacol., 37, 108

Prager, B., Jacobson, P., Schmidt, P. & Stern, D., eds (1930) Beilsteins Handbuch der Organischen Chemie, 4th ed., Vol. 13, Syst. No. 1755, Berlin, Springer-Verlag, p. 29

Reio, L. (1970) Third supplement for the paper chromatographic separation and identification of phenol derivatives and related compounds of biochemical interest using a 'reference system'. J. Chromat., 47, 60-85

Searle, C.E. (1977) Evidence regarding the possible carcinogenicity of mutagenic hair dyes and constituents. In: Mécanismes d'Altération et de Réparation du DNA, Relations avec la Mutagénèse et la Cancérogénèse Chimique, Menton, 1976, Colloques Internationaux du CNRS, No. 256, Paris, Centre National de la Recherche Scientifique, pp. 407-415

Searle, C.E. & Jones, E.L. (1977) Effects of repeated applications of two semi-permanent hair dyes to the skin of A and DBAf mice. Brit. J. Cancer, 36, 467-478

Searle, C.E., Harnden, D.G., Venitt, S. & Gyde, O.H.B. (1975) Carcinogenicity and mutagenicity tests of some hair colourants and constituents. Nature (Lond.), 255, 506-507

Urquizo, S. (1969) Identification of some aromatic amines and phenolic compounds by thin-layer chromatography. Ann. Fals. Expert. Chim., 62, 27-31

US International Trade Commission (1977a) Synthetic Organic Chemicals, US Production and Sales, 1975, USITC Publication 804, Washington DC, US Government Printing Office, p. 39

US International Trade Commission (1977b) Imports of Benzenoid Chemicals and Products, 1975, USITC Publication 806, Washington DC, US Government Printing Office, p. 23

US Tariff Commission (1938) Dyes and Other Synthetic Organic Chemicals in the US, 1936, Report No. 125, Second Series, Washington DC, US Government Printing Office, p. 20

US Tariff Commission (1948) Synthetic Organic Chemicals, US Production and Sales, 1946, Report No. 159, Second Series, Washington DC, US Government Printing Office, p. 74

US Tariff Commission (1974) Imports of Benzenoid Chemicals and Products, 1973, TC Publication 688, Washington DC, US Government Printing Office, p. 21

Venitt, S. & Searle, C.E. (1976) Mutagenicity and possible carcinogenicity of hair colourants and constituents. In: Rosenfeld, C. & Davis, W., eds, Environmental Pollution and Carcinogenic Risks, Lyon (IARC Scientific Publications No. 13), pp. 263-271

Wall, F.E. (1972) Bleaches, hair colorings and dye removers. In: Balsam, M.S. & Sagarin, E., eds, Cosmetics Science and Technology, 2nd ed., Vol. 2, New York, Wiley-Interscience, pp. 308, 313, 317

Weast, R.C., ed. (1976) CRC Handbook of Chemistry and Physics, 57th ed., Cleveland, Ohio, Chemical Rubber Co., p. C-156

Wernick, T., Lanman, B.M. & Fraux, J.L. (1975) Chronic toxicity, teratologic, and reproduction studies with hair dyes. Toxicol. appl. Pharmacol., 32, 450-460

Windholz, M., ed. (1976) The Merck Index, 9th ed., Rahway, NJ, Merck & Co., p. 860

Zelazna, K. & Legatowa, B. (1971) Identification of basic dyes in emulsified hair dyes by thin layer chromatography. Rocz. Panstw. Zakl. Hig., 22, 427-430

1,4-DIAMINO-2-NITROBENZENE

1. Chemical and Physical Data

1.1 Synonyms and trade names

Colour Index No.: 76070

Colour Index Name: Oxidation Base 22

Chem. Abstr. Services Reg. No.: 5307-14-2

Chem. Abstr. Name: 2-Nitro-1,4-benzenediamine

4-Amino-2-nitroaniline; 2-nitro-1,4-phenylenediamine; nitro-*para*-phenylenediamine; 2-nitro-*para*-phenylenediamine; *ortho*-nitro-*para*-phenylenediamine

Dye GS; Durafur Brown 2R; Fouramine 2R; Fourrine 36; Fourrine Brown 2R; Ursol Brown RR; Zoba Brown RR

1.2 Chemical formula and molecular weight

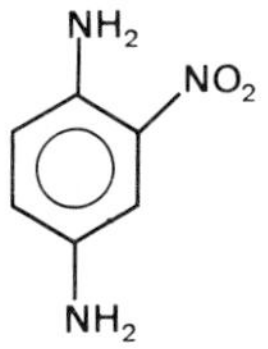

$C_6H_7N_3O_2$ Mol. wt: 153.1

1.3 Chemical and physical properties of the pure substance

(a) Description: Almost black needles with dark-green lustre (Prager *et al.*, 1930)

(b) Melting-point: 137°C (Prager *et al.*, 1930)

(c) Spectroscopy data: Infra-red and nuclear magnetic resonance spectra have been published (Aldrich Chemical Company, 1974).

(d) Solubility: Soluble in water and ethanol (The Society of Dyers and Colourists, 1971a)

(e) Stability: See 'General Remarks on the Substances Considered', p. 27.

1.4 Technical products and impurities

1,4-Diamino-2-nitrobenzene is available in the US as a commercial grade with the following typical specifications: assay, 97.0% min.; melting range, 135-138°C; loss on drying, 1.0% max.; iron content, 100 mg/kg max.; residue on ignition, 0.3% (Ashland Chemical Company, 1975).

1,4-Diamino-2-nitrobenzene is available in Japan with a minimum purity of 99% and contains nitroaminoacetanilide isomers as impurities.

2. Production, Use, Occurrence and Analysis

For background information on this section, see preamble, p. 15.

2.1 Production and use

(a) Production

1,4-Diamino-2-nitrobenzene was first prepared by hydrolysis of 1,4-diamino-5-acetyl-4-nitrobenzene (Prager *et al.*, 1930). A similar method is used commercially in Japan, where 1,4-diamino-2-nitrobenzene is prepared by acetylating *para*-phenylenediamine (see also monograph on *para*-phenylenediamine, p. 125) with acetic anhydride followed by nitration and hydrolysis.

The commercial production of unspecified isomers of diaminonitrobenzene in the US was first reported in 1936 (US Tariff Commission, 1938) and that of 1,4-diamino-2-nitrobenzene, specifically, in 1960 (US Tariff Commission, 1961). In 1973, the latest year in which information was supplied, only one US company reported an undisclosed amount (see preamble, p. 15) (US International Trade Commission, 1975). US imports of 1,4-diamino-2-nitrobenzene through principal US customs districts were reported to be about 1400 kg in 1974 (US International Trade Commission, 1976) and 200 kg in 1975 (US International Trade Commission, 1977).

1,4-Diamino-2-nitrobenzene was first produced commercially in Japan in 1971. The one Japanese manufacturer reported production of 650-750 kg in 1976. Japanese imports of 1,4-diamino-2-nitrobenzene amounted to about 100-200 kg in 1975 (Japan Tariff Association, 1976).

No data on its production in Europe were available to the Working Group.

(b) Use

1,4-Diamino-2-nitrobenzene is used in dyeing furs brown and reddish brown (The Society of Dyers and Colourists, 1971b) and as a dye intermediate in semi-permanent (Anon., 1976) and permanent hair-colouring products. It produces brown and red shades on the hair. It has been used in dark-brown (at levels of 0.1%), medium-brown (0.15%), medium-red (2.5%) and ash-blond (0.065%) dye formulations. It can also be used in bleach-toner formulations for golden-blond (0.008%) and red-blond (0.02%) shades (Wall, 1972).

In Japan, 1,4-diamino-2-nitrobenzene is used as a dye intermediate in hair dyes.

Its use in hair-dye formulations will be forbidden in Italy after 1977 (Dal Falco, 1976).

2.2 Occurrence

1,4-Diamino-2-nitrobenzene is not known to occur as a natural product. No data on its occurrence in the environment were available to the Working Group.

2.3 Analysis

Various eluants have been used on silica gel plates impregnated with rice starch to separate and identify aromatic amines and phenols, including 1,4-diamino-2-nitrobenzene (Urquizo, 1969). Silica gel G plates have been used to separate and identify a mixture of basic dyes, including 1,4-diamino-2-nitrobenzene (Zelazna & Legatowa, 1971); a mixture of aromatic amines and aminophenols, including 1,4-diamino-2-nitrobenzene, was determinated spectrophotometrically after elution from the plates (Legatowa, 1973). A two-dimensional thin-layer chromatographic method has been used to separate and identify dye intermediates, including 1,4-diamino-2-nitrobenzene, with a limit of 0.02% (Kottemann, 1966). Thin-layer chromatography has been used for the analysis of 1,4-diamino-2-nitrobenzene in hair-dye compositions. Mixtures of aromatic diamines, including 1,4-diamino-2-nitrobenzene, in sample sizes of 10-500 μg have been separated and determined on silica gel plates with a standard deviation of ±10% (Pinter & Kramer, 1967).

3. Biological Data Relevant to the Evaluation of Carcinogenic Risk to Man

3.1 Carcinogenicity and related studies in animals[1]

Skin application

Mouse: Groups of 48 DBAf or 52 A young adult mice of both sexes were given twice weekly skin applications on the clipped dorsal skin of 0.4 ml reduced to 0.2 ml at 24 weeks (DBAf mice), or 0.4 ml (A mice), of a 10% solution of a commercially available hair dye, colourant 'GS', containing, among other constituents, 1,4-diamino-2-nitrobenzene and 1,2-diamino-4-nitrobenzene in 50% aqueous acetone; 32 control mice of each strain received acetone only. When the experiment was terminated at 80 weeks, 5 lymphomas and 6 tumours of the female reproductive tract (4 ovarian cystadenomas and 2 uterine fibrosarcomas) had developed in the treated DBAf mice within 26-80 weeks; in the DBAf control mice, 1 lymphoma, 1 liver tumour and 1 lung tumour were found. No differences occurred in the incidence of lymphomas, liver or lung tumours between treated and control A mice. Of the treated animals, 19 DBAf mice and 27 A mice survived 60-80 weeks and died without tumours (Searle, 1977; Searle & Jones, 1977; Venitt & Searle, 1976).

3.2 Other relevant biological data

(a) Experimental systems

The oral LD_{50} of 1,4-diamino-2-nitrobenzene in oil-in-water emulsion in Charles River CD rats is 3080 mg/kg bw; the i.p. LD_{50} of the compound

---

[1]The Working Group was aware of carcinogenicity tests in progress on 1,4-diamino-2-nitrobenzene involving oral administration in the diet in mice and rats (IARC, 1976).

in dimethylsulphoxide in rats is 348 mg/kg bw (Burnett *et al.*, 1977). The oral LD_{50} of the compound in water in male Wistar rats is 2100 mg/kg bw (Gloxhuber *et al.*, 1972).

Ten male and 10 female rats were fed diets containing 500 mg/kg of diet 1,4-diamino-2-nitrobenzene for 13 weeks; no changes in body weight, blood or urine parameters or in the histological appearance of a range of tissues were found in comparison with controls (Gloxhuber *et al.*, 1972).

A preparation containing a mixture of 10 commercially available semi-permanent hair colourants, including 0.24% 1,4-diamino-2-nitrobenzene, was fed in the diet to two groups of 6 male and 6 female pure-bred beagle dogs, 6-8 months of age, at concentrations of 19.5 and 97.5 mg/kg bw/day, respectively. One group, maintained on a normal diet, served as controls. Haematological tests and urine analyses showed no significant changes, other than a blue-brown colouration of the urine from treated animals (Wernick *et al.*, 1975).

Six male and six female adult white rabbits received twice weekly skin applications of 1 ml/kg of bw of a composite hair-dye preparation, containing among other constituents 1.1% 1,4-diamino-2-nitrobenzene, 3% *para*-phenylenediamine and 2% 2,4-diaminoanisole sulphate, mixed 1:1 with 6% hydrogen peroxide solution for 13 weeks. No significant differences were found in haematological or urine parameters or relative organ weights between treated and control groups (Burnett *et al.*, 1976).

The teratogenicity of the same dye formulation was tested in a group of 20 mated Charles River CD female rats. It was applied topically to a shaved site in the dorsoscapular region at a dose of 2 ml/kg bw on days 1, 4, 7, 10, 13, 16 and 19 of gestation; just prior to use it was mixed with an equal volume of 6% hydrogen peroxide. The mothers were killed on day 20 of gestation, and no significant embryotoxic or teratogenic effects were found compared with controls (Burnett *et al.*, 1976).

The compound induced morphological transformation in a mouse cell line (C3H/10T½CL8) (Benedict, 1976) [The Working Group noted that the transformed cells were not tested for tumourigenicity *in vivo*].

1,4-Diamino-2-nitrobenzene has been reported to induce reverse mutations in *Salmonella typhimurium* TA 1538 in the absence of liver activation. Addition of liver post-mitochondrial supernatant fraction from rats pretreated with polychlorinated biphenyl (Aroclor 1254) or phenobarbital reduced mutagenic activity (Garner & Nutman, 1977; Searle *et al.*, 1975). Results reported in an abstract indicate that the compound induced mutations at the thymidine kinase locus in a dose-dependent manner in a mouse lymphoma cell line (L51178Y) (no liver fraction was added) (Palmer *et al.*, 1976).

No dominant lethals were induced in Charles River CD rats injected intraperitoneally three times weekly for 8 weeks with 20 mg/kg bw 1,4-diamino-2-nitrobenzene before mating (no positive control substance was tested) (Burnett *et al.*, 1977).

1,4-Diamino-2-nitrobenzene induced chromatid breaks and chromosome aberrations in a hamster cell line A(T_1) Cl-3 (Benedict, 1976) and chromosome aberrations in CHMP/E cells (Kirkland & Venitt, 1976). Chromatid breaks and gaps have also been found in cultured human peripheral lymphocytes after exposure to 1,4-diamino-2-nitrobenzene (Searle *et al.*, 1975).

This compound did not induce micronucleated cells in bone marrow when 2000 mg/kg bw were administered orally to two groups of 5 male and 5 female rats in two doses separated by an interval of 24 hours (Hossack & Richardson, 1977).

(b) Man

No data were available to the Working Group.

3.3 Case reports and epidemiological studies

No data were available to the Working Group.

4. Comments on Data Reported and Evaluation

4.1 Animal data

In the only available study, a hair-dye formulation containing 1,4-diamino-2-nitrobenzene mixed with 1,2-diamino-4-nitrobenzene was tested in mice by skin application. No data on 1,4-diamino-2-nitrobenzene alone were available. No evaluation of the carcinogenicity of this compound can be made.

4.2 Human data

No case reports or epidemiological studies were available to the Working Group (see, however, 'General Remarks on the Substances Considered', p. 29).

5. References

Aldrich Chemical Company, Inc. (1974) The 1975-1976 Aldrich Catalog/ Handbook of Organic and Biochemicals, Milwaukee, Wisconsin, p. 547

Anon. (1976) Modern trends in hair colourants. Soap, Perfumery and Cosmetics, May, 189-194

Ashland Chemical Company (1975) Technical Data, 2-Nitro-*p*-phenylene diamine, Fine Chemicals Department, Columbus, Ohio

Benedict, W.F. (1976) Morphological transformation and chromosome aberrations produced by two hair dye components. Nature (Lond.), 260, 368-369

Burnett, C., Goldenthal, E.I., Harris, S.B., Wazeter, F.X., Strausburg, J., Kapp, R. & Voelker, R. (1976) Teratology and percutaneous toxicity studies on hair dyes. J. Toxicol. environm. Hlth, 1, 1027-1040

Burnett, C., Loehr, R. & Corbett, J. (1977) Dominant lethal mutagenicity study on hair dyes. J. Toxicol. environm. Hlth, 2, 657-662

Dal Falco (Il Ministro per la Sanita) (1976) Decreto ministeriale, 18 giugno 1976, Art. 1. Gazzetta Ufficiale Della Repubblica Italiana, N. 166, 5025

Garner, R.C. & Nutman, C.A. (1977) Testing of some azo dyes and their reduction products for mutagenicity using *Salmonella typhimurium* TA 1538. Mutation Res. (in press)

Gloxhuber, C., Potokar, M., Reese, G. & Flemming, P. (1972) Toxikologische Prüfung direktziehender Haarfarbstoffe. J. Soc. Cosmetic Chemists, 23, 259-269

Hossack, D.J.N. & Richardson, J.C. (1977) Examination of the potential mutagenicity of hair dye constituents using the micronucleus test. Experientia, 33, 377-378

IARC (1976) Information Bulletin on the Survey of Chemicals Being Tested for Carcinogenicity, No. 6, Lyon, p. 192

Japan Tariff Association (1976) Japan Exports and Imports, Commodity by Country, Tokyo

Kirkland, D.J. & Venitt, S. (1976) Cytotoxicity of hair colourant constituents: chromosome damage induced by two nitrophenylenediamines in cultured Chinese hamster cells. Mutation Res., 40, 47-56

Kottemann, C.M. (1966) Two-dimensional thin layer chromatographic procedure for the identification of dye intermediates in arylamine oxidation hair dyes. J. Ass. off. analyt. Chem., 49, 954-959

Legatowa, B. (1973) Determination of aromatic amines and aminophenols in hair dyes. Rocz. Panstw. Zakl. Hig., 24, 393-402

Palmer, K.A., DeNunzio, A. & Green, S. (1976) The mutagenic assay of some hair dye components using the thymidine kinase locus of L5117BY mouse lymphoma cells (Abstract No. 39). Toxicol. appl. Pharmacol., 37, 108

Pinter, I. & Kramer, M. (1967) The determination of some aromatic diamines used in hair dyes by thin-layer chromatography. Parfum. Cosmet. Savons, 10, 257-260

Prager, B., Jacobson, P., Schmidt, P. & Stern, D., eds (1930) Beilsteins Handbuch der Organischen Chemie, 4th ed., Vol. 13, Syst. No. 1776, Berlin, Springer-Verlag, p. 120

Searle, C.E. (1977) Evidence regarding the possible carcinogenicity of mutagenic hair dyes and constituents. In: Mécanismes d'Altération et de Réparation du DNA, Relations avec la Mutagénèse et la Cancérogénèse Chimique, Menton, 1976, Colloques Internationaux du CNRS, No. 256, Paris, Centre National de la Recherche Scientifique, pp. 407-415

Searle, C.E. & Jones, E.L. (1977) Effects of repeated applications of two semi-permanent hair dyes to the skin of A and DBAf mice. Brit. J. Cancer, 36, 467-478

Searle, C.E., Harnden, D.G., Venitt, S. & Gyde, O.H.B. (1975) Carcinogenicity and mutagenicity tests of some hair colourants and constituents. Nature (Lond.), 255, 506-507

The Society of Dyers and Colourists (1971a) Colour Index, 3rd ed., Vol. 4, Yorkshire, UK, p. 4644

The Society of Dyers and Colourists (1971b) Colour Index, 3rd ed., Vol. 3, Yorkshire, UK, p. 3264

Urquizo, S. (1969) Identification of some aromatic amines and phenolic compounds by thin-layer chromatography. Ann. Fals. Expert. Chim., 62, 27-31

US International Trade Commission (1975) Synthetic Organic Chemicals, US Production and Sales, 1973, USITC Publication 728, Washington DC, US Government Printing Office, p. 43

US International Trade Commission (1976) Imports of Benzenoid Chemicals and Products, 1974, USITC Publication 762, Washington DC, US Government Printing Office, p. 24

US International Trade Commission (1977) Imports of Benzenoid Chemicals and Products, 1975, USITC Publication 806, Washington DC, US Government Printing Office, p. 23

US Tariff Commission (1938) Dyes and Other Synthetic Organic Chemicals in the US, 1936, Report No. 125, Second Series, Washington DC, US Government Printing Office, p. 20

US Tariff Commission (1961) Synthetic Organic Chemicals, US Production and Sales, 1960, TC Publication 34, Washington DC, US Government Printing Office, p. 104

Venitt, S. & Searle, C.E. (1976) Mutagenicity and possible carcinogenicity of hair colourants and constituents. In: Rosenfeld, C. & Davis, W., eds, Environmental Pollution and Carcinogenic Risks, Lyon (IARC Scientific Publications No. 13), pp. 263-271

Wall, F.E. (1972) Bleaches, hair colorings and dye removers. In: Balsam, M.S. & Sagarin, E., eds, Cosmetics Science and Technology, 2nd ed., Vol. 2, New York, Wiley-Interscience, pp. 308, 313, 317

Wernick, T., Lanman, B.M. & Fraux, J.L. (1975) Chronic toxicity, teratologic, and reproduction studies with hair dyes. Toxicol. appl. Pharmacol., 32, 450-460

Zelazna, K. & Legatowa, B. (1971) Identification of basic dyes in emulsified hair dyes by thin layer chromatography. Rocz. Panstw. Zakl. Hig., 22, 427-430

2,4-DIAMINOTOLUENE

1. Chemical and Physical Data

1.1 Synonyms and trade names

Colour Index No.: 76035

Colour Index Names: Oxidation Base 20; Developer 14

Chem. Abstr. Services Reg. No.: 95-80-7

Chem. Abstr. Name: 4-Methyl-1,3-benzenediamine

3-Amino-*para*-toluidine; 5-amino-*ortho*-toluidine; 1,3-diamino-4-methylbenzene; 2,4-diamino-1-methylbenzene; diaminotoluene; 2,4-diaminotoluol; 4-methyl-*meta*-phenylenediamine; MTD; toluene-2,4-diamine; 2,4-toluenediamine; *meta*-toluenediamine; 2,4-toluylenediamine; *meta*-toluylenediamine; tolylene-2,4-diamine; 2,4-tolylenediamine; 4-*meta*-tolylenediamine; *meta*-tolylenediamine

Azogen Developer H; Benzofur MT; Developer B; Developer DB; Developer DBJ; Developer H; Developer MC; Developer MT; Developer MT-CF; Developer MTD; Developer T; Eucanine GB; Fouramine; Fouramine J; Fourrine 94; Fourrine M; Nako TMT; Pelagol Grey J; Pelagol J; Pontamine Developer TN; Renal MD; Tertral G; Zoba GKE

1.2 Chemical formula and molecular weight

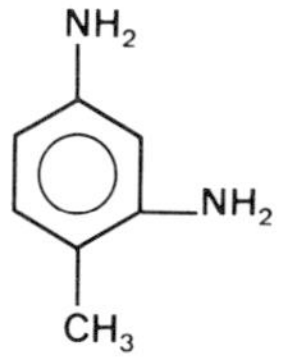

NH_2

NH_2

CH_3

$C_7H_{10}N_2$ Mol. wt: 122.2

1.3 Chemical and physical properties of the pure substance

From Weast (1976), unless otherwise specified

(a) Description: Colourless needles

(b) Boiling-point: 292°C

(c) Melting-point: 99°C

(d) Spectroscopy data: λ_{max} 294 nm in methanol; infra-red, nuclear magnetic resonance and mass spectral data have been tabulated (Grasselli, 1973).

(e) Solubility: Very soluble in hot water, ethanol, ether and hot benzene

(f) Volatility: Vapour pressure is 1 mm at 106.5°C and 10 mm at 151.7°C.

(g) Stability: See 'General Remarks on the Substances Considered', p. 27.

1.4 Technical products and impurities

2,4-Diaminotoluene is available in the US as a commercial grade; however, most of this chemical is used as an intermediate, without isolation, and is usually produced as part of a mixture of about 80% 2,4- and 20% 2,6-diaminotoluene. It is also produced in smaller amounts as part of a mixture containing about 65% 2,4- and 35% 2,6-diaminotoluene.

2,4-Diaminotoluene available in Japan has the following specifications: purity, above 99%; freezing point, above 96.5°C; moisture content, below 0.15%; insoluble matter in hydrochloric acid, below 0.2%. 2,4-Dinitrotoluene and diaminotoluene isomers are present as impurities.

2. Production, Use, Occurrence and Analysis

For background information on this section, see preamble, p. 15.

2.1 Production and use

(a) Production

2,4-Diaminotoluene was prepared by Hofmann in 1861 by reducing 2,4-dinitrotoluene with iron in acetic acid (Prager *et al.*, 1930). It is produced commercially in the US and Japan by nitrating toluene to produce a dinitrotoluene mixture containing approximately 80% 2,4-isomer and 20%

2,6-isomer (with small amounts of 2,3- and 3,4-dinitrotoluene), followed by catalytic reduction to a mixture of the corresponding diamines.

2,4-Diaminotoluene has been produced commercially in the US for over fifty years (US Tariff Commission, 1919). In 1975, six US companies reported production of 86.7 million kg (US International Trade Commission, 1977a); figures for the portion of 2,4-diaminotoluene produced as an unisolated intermediate are not available. US imports of 2,4-diaminotoluene through principal US customs districts amounted to 35 thousand kg in 1974 (US International Trade Commission, 1976a) and 20 thousand kg in 1975 (US International Trade Commission, 1977b). Although some 2,4-diaminotoluene is believed to be exported from the US, data are not available.

An estimated 180-200 million kg 2,4-diaminotoluene are produced annually in western Europe. Less than 100 thousand kg 2,4-diaminotoluene are produced annually in the UK, and less than 25 thousand kg are imported.

In Japan, 2,4-diaminotoluene has been produced commercially for over thirty years. Japanese production of 2,4-diaminotoluene during 1971-1975 amounted to approximately 120 thousand kg annually. One Japanese company currently manufactures this chemical, and there are no imports or exports (Japan Dyestuff Industry Association, 1976).

(*b*) Use

2,4-Diaminotoluene is used as an intermediate to make toluene diisocyanate, which is used in the production of polyurethane; it is also used to make dyes for textiles, leather and furs and in hair-dye formulations.

Most of the 2,4-diaminotoluene produced in the US is used as part of an unisolated mixture of 80% 2,4- and 20% 2,6-diaminotoluene for the manufacture of toluene diisocyanate (TDI). Some 2,4-diaminotoluene is isolated for conversion to pure toluene 2,4-diisocyanate; a mixture of 65% 2,4- and 35% 2,6-diaminotoluene is also used for conversion to toluene diisocyanate.

In 1974, eight US companies reported production of 233.6 million kg of an 80/20 TDI mixture (80% 2,4-isomer and 20% 2,6-isomer) (US International Trade Commission, 1976b), and in 1975 seven companies produced 217.4 million kg of this mixture (US International Trade Commission, 1977a).

In 1974, TDI was used in the US for the production of: flexible and rigid polyurethane foams (142 million kg TDI consumed); polyurethane coatings (5.5-7.3 million kg); cast elastomers, including fabric coatings (3.6-4.1 million kg); and polyurethane and other adhesives (0.9 million kg).

2,4-Diaminotoluene can be used for the production of about 60 dyes (The Society of Dyers and Colourists, 1971a), 28 of which are currently believed to be of commercial world significance (The Society of Dyers and Colourists, 1975). The following eight dyes, believed to be produced commercially from 2,4-diaminotolune, were produced in the US in 1975: C.I. Basic Brown 4 (96.2 thousand kg produced), Basic Orange 1, Direct Brown 154, Direct Black 4, Direct Black 9, Leuco Sulphur Brown 10, Leuco Sulphur Brown 26 and Sulphur Black 2 (US International Trade Commission, 1977a). These dyes are generally used to colour silk, wool, paper and leather. Some are also used to dye cotton, bast fibres and cellulosic fibres, in spirit varnishes and wood stains, as indicators, in the manufacture of pigments, and as biological stains (The Society of Dyers and Colourists, 1971b,c).

2,4-Diaminotoluene is used as a developer for direct dyes, particularly to obtain black, dark-blue and brown shades and to obtain navy-blue and black colours on leather (The Society of Dyers and Colourists, 1971c). It is also used in dyeing furs (The Society of Dyers and Colourists, 1971d). In the US, 2,4-diaminotoluene was used in hair-dye formulations to produce drab-brown, drab-blond, blue and grey shades on the hair (Wall, 1972); this use was forbidden in 1971 (Ames *et al.*, 1975).

In western Europe, almost all of the 2,4-diaminotoluene produced is used as an unisolated intermediate for the production of TDI.

In Japan, 2,4-diaminotoluene is used as a dye intermediate.

2.2 Occurrence

2,4-Diaminotoluene is not known to occur as a natural product. Although the isocyanate was detected in air samples from a polyurethane plant, no 2,4-diaminotoluene was detected (Ciosek *et al.*, 1969). It is a hydrolysis product of toluene diisocyanate (Lowe, 1970).

2.3 Analysis

Non-aqueous titration methods have been used to determine amines, including 2,4-diaminotoluene (Kreshkov *et al.*, 1966; Shvyrkova & Burakova, 1974). It has been determined in air after trapping out in acetone at $-14^{\circ}C$, followed by diazotization and coupling with *N*-(1-naphthyl)ethylenediamine; the limit of detection was 0.02 mg/m^3 (Ciosek *et al.*, 1969). 2,4-Diaminotoluene may also be determined colorimetrically by reaction with thiotrithiazyl chloride (Levin *et al.*, 1967).

Nuclear magnetic resonance spectrometry has been used to determine the relative amounts of diaminotoluene isomers in isomeric mixtures, with an error of 2.5% or less (Mathias, 1966).

Thin-layer chromatography has been used to separate and identify primary aromatic amines, including 2,4-diaminotoluene: on silica gel buffered with sodium acetate, using visualization with a *para*-dimethylaminocinnamaldehyde developer (Bassl *et al.*, 1967), on silica gel followed by elution with methyl nitrite and absorbance measurement at 855 nm (Biernacka *et al.*, 1974) and on polystyrene-based anion exchangers or on microcrystalline cellulose and Cellex D anion exchangers (Lepri *et al.*, 1974). Two-dimensional thin-layer chromatography has been used to separate and identify dye intermediates, including 2,4-diaminotoluene, in arylamine oxidation hair dyes (Kottemann, 1966).

Gas chromatography has been used to determine isomeric diaminotoluenes in their mixtures, using a flame ionization detector (Boufford, 1968) or a thermal conductivity detector, with an accuracy of ± 0.3% for 2,4-diaminotoluene (Willeboordse *et al.*, 1968). Gas chromatography of isomeric diaminotoluenes, including 2,4- and 2,5-diaminotoluenes, as their *N*-trifluoroacetyl derivatives has been used for quality control in their production, with an accuracy of ± 0.45% (Brydia & Willeboordse, 1968).

3. Biological Data Relevant to the Evaluation of Carcinogenic Risk to Man

3.1 Carcinogenicity and related studies in animals

(a) Oral administration[1]

Rat: Two groups, each of 12 male Wistar rats, were fed diets containing 0.06% or 0.1% 2,4-diaminotoluene for 30-36 weeks. For comparison, a third group of 6 animals were fed the basal diet alone (10% casein). In the group fed 0.06%, 11 animals survived for 35 weeks, and of these 7 had multiple hepatocellular carcinomas with metastases in the lymph nodes and omentum. Of the group fed 0.1%, 9 animals survived for 33 weeks, and all had multiple hepatocellular carcinomas; multiple metastases were present in the lymph nodes, omentum, lungs or epididymis in 6 of the 9 survivors. No tumours were found in any of the 6 controls. In a separate experiment in rats fed various combinations of 0.1% 2,4-diaminotoluene and 0.25% ethionine in the diet, there was evidence of an increased incidence of hepatocellular carcinomas over that in rats treated with 2,4-diaminotoluene alone (Ito *et al.*, 1969).

(b) Skin application

Mouse: Groups of 28 male and 28 female 4-7-week old Swiss-Webster mice were given weekly applications on shaved skin in the intrascapular region of 0.05 ml of either a 6% solution of 2,4-diaminotoluene alone, hair-dye formulations containing among their constituents either 0.2% or 0.6% 2,4-diaminotoluene and 3% 2,5-diaminotoluene sulphate and 1.5% *para*-phenylenediamine, or a hair-dye base without the dye components; the treatment was discontinued after various times, depending on toxicity. An untreated control group was also included. The hair dye was mixed with an equal volume of 6% hydrogen peroxide just before use, and the 2,4-diaminotoluene was mixed either with an equal volume of 6% hydrogen

---

[1]The Working Group was aware of carcinogenicity studies in progress in mice and rats involving oral administration of 2,4-diaminotoluene in the diet (IARC, 1976).

peroxide or with an equal volume of water (Giles *et al.*, 1976) [A large number of animals were unaccounted for in the final analysis of tumour incidence, thus making the published data inadequate for evaluation of the carcinogenicity of this chemical; see also Bridges & Green (1976)].

Two groups of 50 male and 50 female 6-8-week old Swiss-Webster mice received weekly or fortnightly applications of 0.05 ml of a freshly prepared 1:1 mixture of a hair-dye formulation, containing among its constituents 0.2% 2,4-diaminotoluene, 3% 2,5-diaminotoluene sulphate and 1.5% *para*-phenylenediamine, and 6% hydrogen peroxide. The mixture was painted on shaved skin in the midscapular region for 18 months, at which time the surviving animals were killed. The incidence of lung tumours did not differ significantly from that in untreated control mice (Burnett *et al.*, 1975) [The Working Group noted the high incidence of tumours in the mice treated with the dye base alone, which made up the greatest part of the hair-dye formulation].

(c) Subcutaneous and/or intramuscular administration

Rat: Twenty rats of mixed strain and sex were injected subcutaneously with 0.5 ml of a 0.4% solution of 2,4-diaminotoluene in propylene glycol at weekly intervals. Eleven rats that died within 8 months showed no marked tissue changes. All of the 9 surviving rats, which received 29-44 weekly injections, developed subcutaneous sarcomas. No contemporary controls were used; however, in another group of 12 rats, s.c. injection of xanthene in propylene glycol (11 injections over 10 months) did not induce local sarcomas (Umeda, 1955).

3.2 Other relevant biological data

(a) Experimental systems

It was reported in an abstract that when ^{14}C-2,4-diaminotoluene was administered intraperitoneally to male Fischer rats, the radioactivity in blood and plasma reached a maximum at 1 hour and decreased rapidly for 7 hours; 98.6% had been excreted after 5 days (76.5% in urine and 22.2% in faeces) (Grantham *et al.*, 1975).

The metabolism of 2,4-diaminotoluene has been investigated in several rodent species: the major urinary metabolite was 2,4-diamino-5-hydroxytoluene; *N*-acetyl and glucuronide conjugates were also found (Waring & Pheasant, 1976). Data reported in an abstract indicate that in rats the major unconjugated metabolites in the urine are 4-acetylamino-2-aminotoluene, 2,4-diacetylaminotoluene and 2,4-diacetylaminobenzoic acid (Grantham *et al.*, 1975). An *N*-acetyltransferase, which converts 2,4-diaminotoluene into its 2- or 4-acetyl derivative, has been found in the cytosol of liver from several species and to a lesser extent in that of kidney, intestinal mucosa and lung (Glinsukon *et al.*, 1975, 1976). When [7-^{14}C]-2,4-diaminotoluene was administered to rats, liver nuclear DNA, RNA and microsomal and soluble proteins were found to be labelled. There were indications that this was due to covalent binding (Hiasa, 1970).

2,4-Diaminotoluene, when administered to rats, guinea-pigs and rabbits, produced methaemoglobinaemia; the level could be correlated with the total amount of free aminophenol excreted in the urine (Waring & Pheasant, 1976).

The carcinogenicity of 2,4-diaminotoluene in male Wistar rats fed diets containing 0.06% or 0.1% was antagonized by concomitant feeding of 3-methylcholanthrene, α-naphthyl isothiocyanate or *para*-hydroxypropiophenone (Ito *et al.*, 1969).

No data on the embryotoxicity or teratogenicity of 2,4-diaminotoluene were available to the Working Group.

It induced morphological transformation in Syrian golden hamster embryo cells (Pienta *et al.*, 1977) [The Working Group noted that transformed cells were not tested for growth *in vivo*].

It also induced reverse mutations in *Salmonella typhimurium* TA 1538 and TA 98 in the presence of a rat liver post-mitochondrial supernatant fraction from animals pretreated with polychlorinated biphenyl (Aroclor 1254) (Ames *et al.*, 1975; Shah *et al.*, 1977).

2,4-Diaminotoluene was shown to be a weak mutagen in *Drosophila melanogaster*, inducing sex-linked recessive lethals when fed a concentration of 15.2 mM (Blijleven, 1977).

(b) Man

No data were available to the Working Group.

3.3 Case reports and epidemiological studies

No data were available to the Working Group.

4. Comments on Data Reported and Evaluation[1]

4.1 Animal data

2,4-Diaminotoluene is carcinogenic in rats after its oral administration, producing hepatocellular carcinomas, and after its subcutaneous injection, inducing local sarcomas.

4.2 Human data

No case reports or epidemiological studies were available to the Working Group (see, however, 'General Remarks on the Substances Considered', p. 29).

---

[1]See also the section, 'Animal Data in Relation to the Evaluation of Risk to Man' in the introduction to this volume, p. 13.

5. References

Ames, B.N., Kammen, H.O. & Yamasaki, E. (1975) Hair dyes are mutagenic: identification of a variety of mutagenic ingredients. Proc. nat. Acad. Sci. (Wash.), 72, 2423-2427

Bassl, A., Heckemann, H.J. & Baumann, E. (1967) Thin-layer chromatography of primary aromatic amines. J. Prakt. Chem., 36, 265-273

Biernacka, T., Sekowska, B. & Michonska, J. (1974) Determination of 2,4- and 2,6-diaminotoluene in technical products by IR spectroscopy. Chem. Anal. (Warsaw), 19, 619-632

Blijleven, W.G.H. (1977) Mutagenicity of four hair dyes in *Drosophila melanogaster*. Mutation Res., 48, 181-186

Boufford, C.E. (1968) Determination of isomeric diaminotoluenes by direct gas liquid chromatography. J. Gas Chromat., 6, 438-440

Bridges, B.A. & Green, M.H.L. (1976) Carcinogenicity of hair dyes by skin painting in mice. J. Toxicol. environm. Hlth, 2, 251-252

Brydia, L.E. & Willeboordse, F. (1968) Gas chromatographic analysis of isomeric diaminotoluenes. Analyt. Chem., 40, 110-113

Burnett, C., Lanman, B., Giovacchini, R., Wolcott, G., Scala, R. & Keplinger, M. (1975) Long-term toxicity studies on oxidation hair dyes. Fd Cosmet. Toxicol., 13, 353-357

Ciosek, A., Gesicka, E. & Kesy-Dabrowska, I. (1969) Occupational exposure of the workers employed in manufacturing polyurethane foams. Med. Pracy, 20, 417-424

Giles, A.L., Jr, Chung, C.W. & Kommineni, C. (1976) Dermal carcinogenicity study by mouse-skin painting with 2,4-toluenediamine alone or in representative hair dye formulations. J. Toxicol. environm. Hlth, 1, 433-440

Glinsukon, T., Benjamin, T., Grantham, P.H., Weisburger, E.K. & Roller, P.P. (1975) Enzymatic *N*-acetylation of 2,4-toluenediamine by liver cytosols from various species. Xenobiotica, 5, 475-483

Glinsukon, T., Benjamin, T., Grantham, P.H., Lewis, N.L. & Weisburger, E.K. (1976) *N*-Acetylation as a route of 2,4-toluenediamine metabolism by hamster liver cytosol. Biochem. Pharmacol., 25, 95-97

Grantham, P.H., Glinsukon, T., Mohan, L.C., Benjamin, T., Roller, P.P., Mitchell, F.E. & Weisburger, E.K. (1975) Metabolism of 2,4-toluenediamine in the rat (Abstract No. 143). Toxicol. appl. Pharmacol., 33, 179

Grasselli, J.G., ed. (1973) CRC Atlas of Spectral Data and Physical Constants for Organic Compounds, Cleveland, Ohio, Chemical Rubber Co., p. B-952

Hiasa, Y. (1970) *m*-Toluylenediamine carcinogenesis in rat liver. J. Nara med. Ass., 21, 1-19

IARC (1976) Information Bulletin on the Survey of Chemicals Being Tested for Carcinogenicity, No. 6, Lyon, p. 184

Ito, N., Hiasa, Y., Konishi, Y. & Marugami, M. (1969) The development of carcinoma in liver of rats treated with *m*-toluylenediamine and the synergistic and antagonistic effects with other chemicals. Cancer Res., 29, 1137-1145

Japan Dyestuff Industry Association (1976) Statistics of Dyestuffs (1971-1975), Tokyo

Kotteman, C.M. (1966) Two-dimensional thin layer chromatographic procedure for the identification of dye intermediates in arylamine oxidation hair dyes. J. Ass. off. analyt. Chem., 49, 954-959

Kreshkov, A.P., Aldarova, N.S. & Izyneev, A.A. (1966) Determination of certain amines and their mixtures by potentiometric titration in nonaqueous solutions. Tr. Buryat. Kompleks. Nauch.-Issled. Inst., Akad. Nauk SSSR, Sib. Otd., No. 20, 233-238

Lepri, L., Desideri, P.G. & Coas, V. (1974) Chromatographic and electrophoretic behaviour of primary aromatic amines on anion-exchange thin layers. J. Chromat., 90, 331-339

Levin, V., Nippoldt, B.W. & Rebertus, R.L. (1967) Spectrophotometric determination of primary aromatic amines with thiotrithiazyl chloride - application to determination of toluene-2,4-diisocyanate in air. Analyt. Chem., 39, 581-584

Lowe, A. (1970) The chemistry of isocyanates. Proc. roy. Soc. Med., 63, 367-368

Mathias, A. (1966) Analysis of diaminotoluene isomer mixtures by nuclear magnetic resonance spectrometry. Analyt. Chem., 38, 1931-1932

Pienta, R.J., Poiley, J.A. & Lebherz, W.B., III (1977) Morphological transformation of early passage golden Syrian hamster embryo cells derived from cryopreserved primary cultures as a reliable *in vitro* bioassay for identifying diverse carcinogens. Int. J. Cancer, 19, 642-655

Prager, B., Jacobson, P., Schmidt, P. & Stern, D., eds (1930) Beilsteins Handbuch der Organischen Chemie, 4th ed., Vol. 13, Syst. No. 1778, Berlin, Springer-Verlag, p. 124

Shah, M.J., Pienta, R.J., Lebherz, W.B., III & Andrews, A.W. (1977) Comparative studies of bacterial mutation and hamster cell transformation induced by 2,4-toluenediamine (Abstract No. 89). In: Weinhouse, S., Foti, M. & Bergbauer, P.A., eds, Proceedings of the 68th Annual Meeting of the American Association of Cancer Research, Denver, Co., 1977, Vol. 18, Baltimore, Williams & Wilkins, p. 23

Shvyrkova, L.A. & Burakova, T.P. (1974) Biamperometric titration of amines. Tr. Mosk. Khim.-Tekhnol. Inst., 81, 100-101

The Society of Dyers and Colourists (1971a) Colour Index, 3rd ed., Vol. 4, Yorkshire, UK, pp. 4020, 4154, 4281, 4285, 4294, 4479, 4480, 4490, 4643, 4850

The Society of Dyers and Colourists (1971b) Colour Index, 3rd ed., Vol. 1, Yorkshire, UK, pp. 1623, 1683

The Society of Dyers and Colourists (1971c) Colour Index, 3rd ed., Vol. 2, Yorkshire, UK, pp. 2002, 2403, 2426, 2428

The Society of Dyers and Colourists (1971d) Colour Index, 3rd ed., Vol. 3, Yorkshire, UK, pp. 3263, 3684, 3689, 3702

The Society of Dyers and Colourists (1975) Colour Index, revised 3rd ed., Vol. 5, Yorkshire, UK, pp. 5079, 5091, 5126, 5128, 5132, 5133, 5197, 5209, 5282, 5287, 5288, 5292, 5294, 5295, 5297

Umeda, M. (1955) Production of rat sarcoma by injections of propylene glycol solution of *m*-toluylenediamine. Gann, 46, 597-603

US International Trade Commission (1976a) Imports of Benzenoid Chemicals and Products, 1974, USITC Publication 762, Washington DC, US Government Printing Office, p. 28

US International Trade Commission (1976b) Synthetic Organic Chemicals, US Production and Sales, 1974, USITC Publication 776, Washington DC, US Government Printing Office, pp. 22, 38, 44

US International Trade Commission (1977a) Synthetic Organic Chemicals, US Production and Sales, 1975, USITC Publication 804, Washington DC, US Government Printing Office, pp. 22, 36, 42, 49, 60, 62, 65, 74

US International Trade Commission (1977b) Imports of Benzenoid Chemicals and Products, 1975, USITC Publication 806, Washington DC, US Government Printing Office, p. 27

US Tariff Commission (1919) Report on Dyes and Related Coal-Tar Chemicals, 1918, revised ed., Washington DC, US Government Printing Office, p. 29

Wall, F.E. (1972) Bleaches, hair colorings and dye removers. In: Balsam, M.S. & Sagarin, E., eds, Cosmetics Science and Technology, 2nd ed., Vol. 2, New York, Wiley-Interscience, pp. 308, 309, 310

Waring, R.H. & Pheasant, A.E. (1976) Some phenolic metabolites of 2,4-diaminotoluene in the rabbit, rat and guinea-pig. Xenobiotica, 6, 257-262

Weast, R.C., ed. (1976) CRC Handbook of Chemistry and Physics, 57th ed., Cleveland, Ohio, Chemical Rubber Co., pp. C-523, D-198

Willeboordse, F., Quick, Q. & Bishop, E.T. (1968) Direct gas chromatographic analysis of isomeric diaminotoluenes. Analyt. Chem., 40, 1455-1458

2,5-DIAMINOTOLUENE (SULPHATE)

1. Chemical and Physical Data

2,5-Diaminotoluene

1.1 Synonyms and trade names

Colour Index No.: 76042

Chem. Abstr. Services Reg. No.: 95-70-5

Chem. Abstr. Name: 2-Methyl-1,4-benzenediamine

4-Amino-2-methylaniline; 2-methyl-*para*-phenylenediamine; toluene-2,5-diamine; *para*-toluenediamine; toluylene-2,5-diamine; *para*-toluylenediamine; *para*-tolylenediamine; *para,meta*-tolylenediamine

1.2 Chemical formula and molecular weight

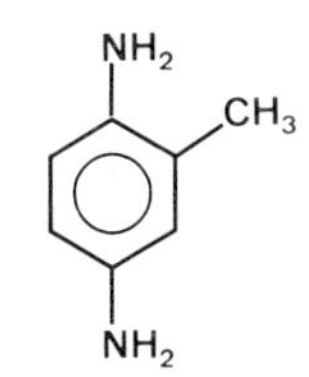

$C_7H_{10}N_2$ Mol. wt: 122.2

1.3 Chemical and physical properties of the pure substance

From Weast (1976), unless otherwise specified

(a) Description: Colourless plates

(b) Boiling-point: 273-274°C

(c) Melting-point: 64°C

(d) Spectroscopy data: Mass spectral data have been tabulated (Grasselli, 1973).

(e) Solubility: Soluble in water, ethanol and ether; slightly soluble in cold benzene but soluble in hot benzene

(f) Stability: See 'General Remarks on the Substances Considered', p. 27.

1.4 Technical products and impurities

2,5-Diaminotoluene is available in Japan with a minimum purity of 95% and containing nitroaminotoluene as an impurity.

2,5-Diaminotoluene sulphate

1.1 Synonyms and trade names

Colour Index No.: 76043

Colour Index Name: Oxidation Base 4

Chem. Abstr. Services Reg. No.: 6369-59-1

Chem. Abstr. Name: 2-Methyl-1,4-benzenediamine sulfate

2-Methyl-*para*-phenylenediamine sulphate; toluene-2,5-diamine sulphate; *para*-toluenediamine sulphate; toluylene 2,5-diamine sulphate; *para*-toluylenediamine sulphate; *para*-tolylenediamine sulphate; *para,meta*-tolylenediamine sulphate

Fouramine STD

1.2 Empirical formula and molecular weight

$C_7H_{10}N_2.H_2SO_4$ Mol. wt: 220.2

1.3 Chemical and physical properties of the pure substance

(a) Description: Off-white powder (technical product) (Ashland Chemical Co., 1975)

(b) Solubility: Soluble in water and ethanol (The Society of Dyers and Colourists, 1971a)

1.4 Technical products and impurities

2,5-Diaminotoluene sulphate is available in the US as a commercial grade with the following typical specifications: purity, 95% min.; residue on ignition, 0.2% max.; loss on drying, 0.5% max.; and iron content, 25 mg/kg max. (Ashland Chemical Co., 1975).

2. Production, Use, Occurrence and Analysis

For background information on this section, see preamble, p. 15.

2.1 Production and use

(a) Production

2,5-Diaminotoluene was prepared by Nietzki in 1877 by reductive cleavage of 4-amino-2,3'-dimethylazobenzene (*ortho*-aminoazotoluene[1]) with tin and hydrochloric acid (Prager *et al.*, 1930); reductive cleavage may also be carried out with zinc dust and hydrochloric acid (Thirtle, 1968). It can be prepared by the electrolytic reduction of 2,5-dinitrotoluene (Tallec & Gueguen, 1966) or by condensing 2-amino-1-methylbenzene and toluene-4-sulphonyl chloride to 4-toluenesulphono-2-toluidide, which is then coupled with diazotized aminobenzenesulphonic acid and reduced (Wlodarski, 1971).

2,5-Diaminotoluene was first produced commercially in the US in 1920 (US Tariff Commission, 1921), but commercial production has not been reported since 1966 (US Tariff Commission, 1968). 2,5-Diaminotoluene sulphate was first produced commercially in the US in 1927 (US Tariff Commission, 1928). Commercial production of an undisclosed amount (see preamble, p. 15) was last reported in 1970 (US Tariff Commission, 1972); however, it is believed still to be produced by two companies.

US imports through principal US customs districts were reported to be 1300 kg 2,5-diaminotoluene and 29.4 thousand kg 2,5-diaminotoluene sulphate in 1974 (US International Trade Commission, 1976) and 3400 kg 2,5-diaminotoluene in 1975 (US International Trade Commission, 1977a).

2,5-Diaminotoluene and its sulphate have been produced commercially in Japan since 1955. One Japanese company produced an estimated 850 kg 2,5-diaminotoluene and sulphate combined in 1976. Negligible amounts of these chemicals are imported and none exported.

---

[1]See IARC (1975).

No data on their production in Europe were available to the Working Group.

(b) Use

2,5-Diaminotoluene can be used as an intermediate in the production of two dyes, C.I. Basic Red 2 and C.I. Acid Brown 103, which are believed to be of commercial world significance (The Society of Dyers and Colourists, 1971a, 1975). C.I. Basic Red 2 is produced commercially in the US by one company (US International Trade Commission, 1977b); it is used to dye cotton, wool, silk, leather and paper and in pigments, spirit inks, as a biological stain, and as a solvent dye. C.I. Acid Brown 103 is used to dye leather (The Society of Dyers and Colourists, 1971b).

2,5-Diaminotoluene sulphate is used to dye furs deep brown (The Society of Dyers and Colourists, 1971c).

2,5-Diaminotoluene and its sulphate are used in hair-dye formulations to produce black, drab and warm-brown, blond and grey shades on the hair. 2,5-Diaminotoluene sulphate is also used in bleach-toner formulations for silver (at levels of 0.1%), smoke (0.04%) and platinum-blond (0.08%) shades (Wall, 1972).

In Japan, both 2,5-diaminotoluene and its sulphate are used in hair dyes.

The use of 2,5-diaminotoluene in hair-dye formulations will be forbidden in Italy after 1977 (Dal Falco, 1976).

2.2 Occurrence

2,5-Diaminotoluene and its sulphate are not known to occur as natural products. No data on their occurrence in the environment were available to the Working Group.

2.3 Analysis

The Association of Official Analytical Chemists has published an Official Final Action for the determination of 2,5-toluenediamine in hair dyes by two methods: (1) gravimetric determination of its diacetyl derivative and (2) an iodometric titration method (Horwitz, 1970).

A simple colorimetric test for detection of aromatic amines, including 2,5-diaminotoluene, in hair dyes involves hydrochloric acid and vanillin in isopropyl alcohol (Lange, 1966).

Nuclear magnetic resonance spectrometry has been used to determine the relative amounts of diaminotoluene isomers in isomeric mixtures, with an error of 2.5% or less (Mathias, 1966).

Paper chromatography, using three solvent systems and three development systems, has been used to separate and identify 169 primary aromatic amines, including 2,5-diaminotoluene (Cee & Gasparic, 1966).

Thin-layer chromatography has been used to separate and identify aromatic amines, including 2,5-diaminotoluene: on silica gel buffered with sodium acetate, using visualisation with *para*-dimethylaminocinnam-aldehyde (Bassl *et al.*, 1967), and on polystyrene-based anion exchangers or on microcrystalline cellulose and Cellex D anion exchangers (Lepri *et al.*, 1974). Hair dyes have been analysed for aromatic amines, including 2,5-diaminotoluene, by extraction with methanol, separation by thin-layer chromatography and visualisation with *para*-dimethylaminobenzaldehyde (Legatowa, 1973). Two-dimensional thin-layer chromatography has been used to separate and identify dye intermediates, including 2,5-diaminotoluene, in arylamine oxidation hair dyes (Kottemann, 1966).

Gas chromatography has been used to determine isomeric diaminotoluenes, including 2,5-diaminotoluene, in their mixtures, using a flame ionization detector (Boufford, 1968) or a thermal conductivity detector (accuracy, ±1.9%) (Willeboordse *et al.*, 1968). Gas chromatography of isomeric diaminotoluenes, including 2,5-diaminotoluene, as their *N*-trifluoroacetyl derivatives, has been used for quality control in their production, with an accuracy of ±0.03 (Brydia & Willeboordse, 1968).

3. Biological Data Relevant to the Evaluation of Carcinogenic Risk to Man

3.1 Carcinogenicity and related studies in animals[1]

(a) Skin application

Mouse: Groups of 28 male and 28 female Swiss-Webster mice, 4-7 weeks old, were treated either with hair-dye formulations containing 3% 2,5-diaminotoluene sulphate, 1.5% *para*-phenylenediamine and either 0.2% or 0.6% 2,4-diaminotoluene or with a hair-dye base without the dye components. The dye preparation was mixed with an equal volume of 6% hydrogen peroxide before application. Each week, 0.05 ml of the test solution was applied to shaved skin in the intrascapular region, and the treatment was discontinued after various times depending on toxicity. An untreated control group was also included (Giles *et al.*, 1976) [A large number of animals were unaccounted for in the final analysis of tumour incidence, thus making the published data inadequate for the evaluation of the carcinogenicity of this chemical (see also Bridges & Green, 1976)].

Two groups of 50 male and 50 female 6-8-week old Swiss-Webster mice received weekly or fortnightly applications of 0.05 ml of a freshly prepared 1:1 mixture of hair-dye preparations, containing 3% 2,5-toluenediamine sulphate and 1.5% *para*-phenylenediamine with either 0.2% 2,4-toluenediamine, 0.38% 2,4-diaminoanisole sulphate or 0.17% *meta*-phenylenediamine, and 6% hydrogen peroxide. The mixture was painted on shaved skin in the midscapular region for 18 months, at which time the surviving animals were killed. The incidence of lung tumours did not differ significantly from that in untreated control mice (Burnett *et al.*, 1975) [The Working Group noted the high incidence of tumours in the mice treated with the dye base alone, which made up the greatest part of the hair-dye formulation].

---

[1]The Working Group was aware of carcinogenicity studies in progress in mice and rats involving oral administration of 2,5-diaminotoluene sulphate in the diet (IARC, 1976).

Rat: Three experimental preparations in a carboxymethyl cellulose gel were tested: formulation 1, containing 4% 2,5-diaminotoluene (calculated as free base but used as sulphate); formulation 2, containing 3% 2,5-diaminotoluene and 0.75% 2,4-diaminoanisole; and formulation 3, used as control without added dye intermediates. Each formulation was mixed with an equal volume of 6% hydrogen peroxide immediately before use, and 0.5 g of the mixture was applied to the dorsal skin. Three groups, each of 50 male and 50 female Sprague-Dawley rats, 12 weeks old, were treated twice weekly for 2 years with either formulation 1 or 2 or left untreated. A fourth group of 25 male and 25 female rats were treated with formulation 3. No statistically significant differences were observed in tumour incidence between the experimental and control groups (Kinkel & Holzmann, 1973) [The Working Group noted the absence of information about survival times and the high incidence of tumours in the controls].

3.2 Other relevant biological data

(a) Experimental systems

The oral LD_{50} of 2,5-diaminotoluene sulphate in oil-in-water emulsion in rats is 98 mg/kg bw; the i.p. LD_{50} of the compound in dimethylsulphoxide in rats is 49 mg/kg bw (Burnett *et al.*, 1977).

Groups of 2-4 Sprague-Dawley rats were given 2,5-diaminotoluene sulphate either as a single s.c. injection of 0.75-24 mg/kg bw, or those doses repeated daily for 3-5 days, or as a single i.p. injection of 4-32 mg/kg bw. Even with the highest dose (32 mg/kg bw), no toxicologically significant increases in methaemoglobin formation were reported, although a few Heinz bodies were found 2-3 days after several s.c. injections (Kinkel & Holzmann, 1973).

2,5-Diaminotoluene sulphate is absorbed through the skin of dogs and excreted in the urine. About 40 mg of the compound were absorbed from 50 ml of a lauryl sulphate-based gel containing 1.4 g 2,5-diaminotoluene in 3 hours. When 6% hydrogen peroxide was added to the gel immediately before use, the amount of 2,5-diaminotoluene sulphate absorbed was less than 3 mg (Kiese *et al.*, 1968).

Doses of 0.5 ml of hair-dye formulations containing 4% 2,5-diaminotoluene sulphate or 3% 2,5-diaminotoluene and 0.75% 2,4-diaminoanisole sulphates (percentages calculated as free base), in carboxymethyl cellulose gel, mixed 1:1 with 6% hydrogen peroxide, were applied to the shaved dorsal skin of Sprague-Dawley rats twice weekly for 2 years. No differences were observed in body weight gain, food intake, mean lifespan, mortality, haematological parameters or liver function (bromosulphthalein excretion or serum transaminase levels), and autopsies revealed no pathological changes (Kinkel & Holzmann, 1973).

A hair-dye formulation, containing among its constituents 3% 2,5-diaminotoluene sulphate (and including 4% 2,4-diaminoanisole sulphate and 2% *para*-phenylenediamine), was tested for teratogenicity in a group of 20 mated Charles River CD female rats. The formulation was applied topically to a shaved site in the dorsoscapular region at a dose of 2 ml/kg bw on days 1, 4, 7, 10, 13, 16 and 19 of gestation; just prior to use it was mixed with an equal volume of 6% hydrogen peroxide. The mothers were killed on day 20 of gestation. On comparison with 3 control groups, skeletal changes were seen in 6/169 live foetuses ($P<0.05 \rightarrow P<0.01$). In a group of 20 female rats treated with a formulation containing 6% 2,5-diaminotoluene sulphate, no increase in abnormalities was found in foetuses when compared with controls (Burnett *et al.*, 1976).

2,5-Diaminotoluene induces reverse mutations in *Salmonella typhimurium* TA 1538 after addition of a rat liver post-mitochondrial supernatant fraction obtained from animals pretreated with polychlorinated biphenyl (Aroclor 1254) (Ames *et al.*, 1975).

No dominant lethals were induced in Charles River rats injected intraperitoneally three times weekly for 8 weeks with 20 mg/kg bw 2,5-diaminotoluene before mating (no positive control substance was tested) (Burnett *et al.*, 1977).

This compound did not induce micronucleated cells in bone marrow when 120 mg/kg bw were administered orally to two groups of 5 male and 5 female rats in two doses separated by an interval of 24 hours (Hossack & Richardson, 1977).

(b) Man

Ten mg 2,5-diaminotoluene sulphate (equivalent to 5.54 mg 2,5-diaminotoluene) injected subcutaneously into 6 adult volunteers were excreted in the urine as the *N,N'*-diacetyl derivative, which accounted for approximately half of the injected dose, over 48 hours. During the process of hair dyeing with a dye formulation in which a total of 2.5 g 2,5-diaminotoluene sulphate were applied in conjunction with 6% hydrogen peroxide added just before application, it was estimated that 4.6 mg 2,5-diaminotoluene sulphate had been absorbed through the skin of 5 volunteers (Kiese & Rauscher, 1968) [See also 'General Remarks on the Substances Considered'].

In two case reports, patients who developed aplastic anaemia were noted to have been 'using a new hair dye' containing 2,5-diaminotoluene or its sulphate prior to the onset of symptoms. In one of these cases, the use was stated to have begun one month prior to onset (Toghill & Wilcox, 1976), while in the other case the duration of use was unspecified (Hamilton & Sheridan, 1976).

3.3 Case reports and epidemiological studies

No data were available to the Working Group.

4. Comments on Data Reported and Evaluation

4.1 Animal data

2,5-Diaminotoluene sulphate in commercial hair-dye formulations, or in experimental mixtures similar to those used as hair dyes, was inadequately tested by skin application on mice and rats. No data on 2,5-diaminotoluene or 2,5-diaminotoluene sulphate alone were available. No evaluation of their carcinogenicity can be made.

4.2 Human data

No case reports or epidemiological studies were available to the Working Group (see, however, 'General Remarks on the Substances Considered', p. 29).

5. References

Ames, B.N., Kammen, H.O. & Yamasaki, E. (1975) Hair dyes are mutagenic: identification of a variety of mutagenic ingredients. Proc. nat. Acad. Sci. (Wash.), 72, 2423-2427

Ashland Chemical Co. (1975) Technical Data, 2,5-Toluenediamine Sulfate, Fine Chemicals Department, Columbus, Ohio

Bassl, A., Heckemann, H.J. & Baumann, E. (1967) Thin-layer chromatography of primary aromatic amines. I. J. Prakt. Chem., 36, 265-273

Boúfford, C.E. (1968) Determination of isomeric diaminotoluenes by direct gas liquid chromatography. J. Gas Chromat., 6, 438-440

Bridges, B.A. & Green, M.H.L. (1976) Carcinogenicity of hair dyes by skin painting in mice. J. Toxicol. environm. Hlth, 2, 251-252

Brydia, L.E. & Willeboordse, F. (1968) Gas chromatographic analysis of isomeric diaminotoluenes. Analyt. Chem., 40, 110-113

Burnett, C., Lanman, B., Giovacchini, R., Wolcott, G., Scala, R. & Keplinger, M. (1975) Long-term toxicity studies on oxidation hair dyes. Fd Cosmet. Toxicol., 13, 353-357

Burnett, C., Goldenthal, E.I., Harris, S.B., Wazeter, F.X., Strausburg, J., Kapp, R. & Voelker, R. (1976) Teratology and percutaneous toxicity studies on hair dyes. J. Toxicol. environm. Hlth, 1, 1027-1040

Burnett, C., Loehr, R. & Corbett, J. (1977) Dominant lethal mutagenicity study on hair dyes. J. Toxicol. environm. Hlth, 2, 657-662

Cee, A. & Gasparic, J. (1966) Identification of organic compounds. LXIII. Paper chromatographic separation and identification of primary aromatic amines in aqueous solvent systems. Mikrochim. acta, 1-2, 295-309

Dal Falco (Il Ministro per la Sanita) (1976) Decreto ministeriale, 18 giugno 1976, Art. 1. Gazzetta Ufficiale Della Repubblica Italiana, N. 166, 5025

Giles, A.L., Jr, Chung, C.W. & Kommineni, C. (1976) Dermal carcinogenicity study by mouse-skin painting with 2,4-toluenediamine alone or in representative hair dye formulations. J. Toxicol. environm. Hlth, 1, 433-440

Grasselli, J.G., ed. (1973) CRC Atlas of Spectral Data and Physical Constants for Organic Compounds, Cleveland, Ohio, Chemical Rubber Co., p. B-952

Hamilton, S. & Sheridan, J. (1976) Aplastic anaemia and hair dye. Brit. med. J., ii, 834

Horwitz, W., ed. (1970) Official Methods of Analysis of the Association of Official Analytical Chemists, 11th ed., Washington DC, Association of Official Analytical Chemists, pp. 613-614

Hossack, D.J.N. & Richardson, J.C. (1977) Examination of the potential mutagenicity of hair dye constituents using the micronucleus test. Experientia, 33, 377-378

IARC (1975) IARC Monographs on the Evaluation of Carcinogenic Risk of Chemicals to Man, 8, Some Aromatic Azo Compounds, Lyon, pp. 61-74

IARC (1976) Information Bulletin on the Survey of Chemicals Being Tested for Carcinogenicity, No. 6, Lyon, p. 288

Kiese, M. & Rauscher, E. (1968) The absorption of *p*-toluenediamine through human skin in hair dyeing. Toxicol. appl. Pharmacol., 13, 325-331

Kiese, M., Rachor, M. & Rauscher, E. (1968) The absorption of some phenylenediamines through the skin of dogs. Toxicol. appl. Pharmacol., 12, 495-507

Kinkel, H.J. & Holzmann, S. (1973) Study of long-term percutaneous toxicity and carcinogenicity of hair dyes (oxidizing dyes) in rats. Fd Cosmet. Toxicol., 11, 641-648

Kottemann, C.M. (1966) Two-dimensional thin layer chromatographic procedure for the identification of dye intermediates in arylamine oxidation hair dyes. J. Ass. off. analyt. Chem., 49, 954-959

Lange, F.W. (1966) Fast method for detection of *p*-phenylenediamine in hair dyes. Seifen-Ole-Fette-Wachse, 92, 751-753

Legatowa, B. (1973) Determination of aromatic amines and aminophenols in hair dyes. Rocz. Panstw. Zakl. Hig., 24, 393-402

Lepri, L., Desideri, P.G. & Coas, V. (1974) Chromatographic and electrophoretic behaviour of primary aromatic amines on anion-exchange thin layers. J. Chromat., 90, 331-339

Mathias, A. (1966) Analysis of diaminotoluene isomer mixtures by nuclear magnetic resonance spectrometry. Analyt. Chem., 38, 1931-1932

Prager, B., Jacobson, P., Schmidt, P. & Stern, D., eds (1930) Beilsteins Handbuch der Organischen Chemie, 4th ed., Vol. 13, Syst. No. 1778, Berlin, Springer-Verlag, p. 144

The Society of Dyers and Colourists (1971a) Colour Index, 3rd ed., Vol. 4, Yorkshire, UK, pp. 4007, 4099, 4451, 4457, 4643, 4851

The Society of Dyers and Colourists (1971b) Colour Index, 3rd ed., Vol. 1, Yorkshire, UK, pp. 1451, 1633

The Society of Dyers and Colourists (1971c) Colour Index, 3rd ed., Vol. 3, Yorkshire, UK, p. 3260

The Society of Dyers and Colourists (1975) Colour Index, revised 3rd ed., Vol. 5, Yorkshire, UK, pp. 5046, 5079, 5209

Tallec, A. & Gueguen, M.-J. (1966) Réduction sélective, à potentiel contrôle, de quelques dinitrobenzènes substitués par un groupement alcoyle. C.R. Acad. Sci. (Paris), 262, 484-487

Thirtle, J.R. (1968) Phenylenediamines. In: Kirk, R.E. & Othmer, D.F., eds, Encyclopedia of Chemical Technology, 2nd ed., Vol. 15, New York, John Wiley and Sons, p. 217

Toghill, P.J. & Wilcox, R.G. (1976) Aplastic anaemia and hair dye. Brit. med. J., i, 502-503

US International Trade Commission (1976) Imports of Benzenoid Chemicals and Products, 1974, USITC Publication 762, Washington DC, US Government Printing Office, p. 28

US International Trade Commission (1977a) Imports of Benzenoid Chemicals and Products, 1975, USITC Publication 806, Washington DC, US Government Printing Office, p. 27

US International Trade Commission (1977b) Synthetic Organic Chemicals, US Production and Sales, 1975, USITC Publication 804, Washington DC, US Government Printing Office, p. 60

US Tariff Commission (1921) Census of Dyes and Coal-Tar Chemicals, 1920, Tariff Information Series No. 23, Washington DC, US Government Printing Office, p. 28

US Tariff Commission (1928) Census of Dyes and of Other Synthetic Organic Chemicals, 1927, Tariff Information Series No. 37, Washington DC, US Government Printing Office, p. 33

US Tariff Commission (1968) Synthetic Organic Chemicals, US Production and Sales, 1966, TC Publication 248, Washington DC, US Government Printing Office, p. 90

US Tariff Commission (1972) Synthetic Organic Chemicals, US Production and Sales, 1970, TC Publication 479, Washington DC, US Government Printing Office, p. 50

Wall, F.E. (1972) Bleaches, hair colorings, and dye removers. In: Balsam, M.S. & Sagarin, E., eds, Cosmetics Science and Technology, 2nd ed., Vol. 2, New York, Wiley-Interscience, pp. 308,317

Weast, R.C., ed. (1976) CRC Handbook of Chemistry and Physics, 57th ed., Cleveland, Ohio, Chemical Rubber Co., p. C-523

Willeboordse, F., Quick, Q. & Bishop, E.T. (1968) Direct gas chromatographic analysis of isomeric diaminotoluenes. Analyt. Chem., 40, 1455-1458

Wlodarski, L. (1971) 2,5-Diaminotoluene. Polish Patent 63,097, 20 August

meta-PHENYLENEDIAMINE (HYDROCHLORIDE)

1. Chemical and Physical Data

meta-Phenylenediamine

1.1 Synonyms and trade names

Colour Index No.: 76025

Colour Index Name: Developer 11

Chem. Abstr. Services Reg. No.: 108-45-2

Chem. Abstr. Name: 1,3-Benzenediamine

3-Aminoaniline; *meta*-aminoaniline; *meta*-benzenediamine; 1,3-diaminobenzene; *meta*-diaminobenzene; 1,3-phenylenediamine

Developer C; Developer H; Developer M; Direct Brown BR; Direct Brown GG

1.2 Chemical formula and molecular weight

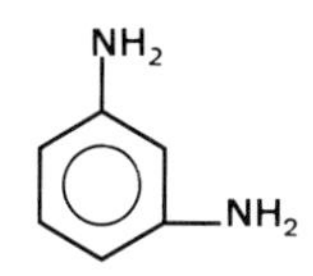

$C_6H_8N_2$ Mol. wt: 108.1

1.3 Chemical and physical properties of the pure substance

From Grasselli (1973), unless otherwise specified

(a) Description: White crystals (Windholz, 1976)

(b) Boiling-point: 282-284°C

(c) Melting-point: 63-64°C

(d) Refractive index: n_D^{58} 1.6339

(e) Spectroscopy data: λ_{max} 240 nm (E_1^1 = 708); 293 nm (E_1^1 = 263) in cyclohexane; fluorescence spectra have been measured (Inoue *et al.*, 1970).

(f) Solubility: Soluble in water, ethanol, ether and benzene (Grasselli, 1973); soluble in methanol, chloroform, acetone, dimethyl formamide, methyl ethyl ketone and dioxane (Windholz, 1976)

(g) Volatility: Vapour pressure is 1 mm at 99.8°C and 10 mm at 147°C (Weast, 1976).

(h) Stability: See 'General Remarks on the Substances Considered', p. 27.

1.4 Technical products and impurities

meta-Phenylenediamine is available in the US as a technical grade with the following typical specifications: purity, 99.5% min. (as determined by gas chromatography); water-insoluble matter, 0.1% max.; moisture content, 0.1% max.; dinitrobenzene, 0.1% max.; *para*-phenylenediamine (1,4-diaminobenzene), 500 mg/kg max.; *ortho*-phenylenediamine (1,2-diaminobenzene), 200 mg/kg max.; freezing-point, 62.7°C min. (E.I. du Pont de Nemours & Co., Inc., 1976).

In Japan, *meta*-phenylenediamine is available with a minimum purity of 98.5%, a minimum melting-point of 61.5°C and containing dinitrobenzene and phenylenediamine isomers as impurities.

meta-Phenylenediamine hydrochloride

1.1 Synonyms and trade names

Chem. Abstr. Services Reg. No.: 541-69-5

Chem. Abstr. Name: 1,3-Benzenediamine hydrochloride

3-Aminoaniline dihydrochloride; *meta*-aminoaniline dihydrochloride; *meta*-benzenediamine dihydrochloride; 1,3-diaminobenzene dihydrochloride; *meta*-diaminobenzene dihydrochloride; 1,3-phenylenediamine dihydrochloride

1.2 Empirical formula and molecular weight

$C_6H_8N_2.2HCl$ Mol. wt: 181.0

1.3 Chemical and physical properties of the pure substance

From Windholz (1976)

(a) Description: White crystalline powder

(b) Solubility: Soluble in water and ethanol

1.4 Technical products and impurities

No data were available to the Working Group.

2. Production, Use, Occurrence and Analysis

For background information on this section, see preamble, p. 15.

2.1 Production and use

(a) Production

meta-Phenylenediamine was prepared by Hofmann in 1861 by reducing 1,3-dinitrobenzene with iron and acetic acid (Prager *et al.*, 1930). It is produced commercially by reducing 1,3-dinitrobenzene with iron and hydrochloric acid or with iron, ammonium polysulphide and water gas (Thirtle, 1968).

meta-Phenylenediamine has been produced commercially in the US for over fifty years (US Tariff Commission, 1919); in 1975, only one US company reported an undisclosed amount (see preamble, p. 15) (US International Trade Commission, 1977a). No evidence was found that the hydrochloride has ever been produced commercially in the US.

US imports of *meta*-phenylenediamine through principal US customs districts amounted to 146.7 thousand kg in 1972 (US Tariff Commission, 1973), 90.1 thousand kg in 1974 (US International Trade Commission, 1976) and 35.2 thousand kg in 1975 (US International Trade Commission, 1977b).

It is believed that at least two companies produce *meta*-phenylenediamine in western Europe. Between 10-100 thousand kg are produced annually in the UK and less than 50 thousand kg are imported. It is believed that 100-1000 thousand kg/year are imported into Switzerland.

meta-Phenylenediamine has been produced commercially in Japan for more than thirty years, and three Japanese companies currently produce it; production amounted to 296 thousand kg in 1974 and 171 thousand kg in 1975. In 1975, Japan imported about 60 thousand kg *meta*-phenylenediamine from the Federal Republic of Germany and the US. There are no Japanese exports of this chemical (Japan Dyestuff Industry Association, 1976). Small amounts of *meta*-phenylenediamine hydrochloride are sold by minor dye companies in that country.

(b) Use

meta-Phenylenediamine is used for the production of various dyes, as a component of hair-dye formulations, as a curing agent for epoxy resins, for the production of heat-resistant fibres and in many other minor applications.

meta-Phenylenediamine can be used for the production of over 140 dyes, 37 of which are believed to have commercial world significance (The Society of Dyers and Colourists, 1971a). Only twelve of these are produced commercially in the US: C.I. Direct Black 38 (US production in 1975 was 984.3 thousand kg), Direct Black 22 (149.4 thousand kg), Basic Orange 2 (120.3 thousand kg), Basic Brown 1 (9.1 thousand kg), Basic Brown 2, Direct Brown 1, Direct Brown 44, Direct Black 9, Direct Black 19, Mordant Brown 1, Mordant Brown 12 and Solvent Orange 3 (US International Trade Commission, 1977a). These dyes are used to colour various textile fibres and other materials, e.g., leather, paper, polishes, spirit inks, etc. (The Society of Dyers and Colourists, 1971b,c,d).

meta-Phenylenediamine is also used as a direct dye developer to obtain black, blue and brown shades (The Society of Dyers and Colourists, 1971c); and it is included in hair-dye formulations to produce brown, golden-blond, blue and grey shades on the hair (Wall, 1972).

It is also employed as a curing agent to harden epoxy resins used in casting and plastic tooling, laminating applications and in adhesives (Weschler, 1965). It is added to isophthaloyl chloride to produce poly-*meta*-phenylene isophthalamide resin, a polyamide fibre (Nomex®) used in high-temperature applications, e.g., protective clothing, electrical insulation and other miscellaneous uses, such as ironing-board covers, fire hoses and aircraft upholstery. It is estimated that US production of Nomex® will reach approximately 4.5 million kg per year in 1977.

meta-Phenylenediamine is also used in the manufacture of rubber-curing agents, ion-exchange resins, decolourizing resins, formaldehyde condensates, resinous polyamides, polymer blocks, textile fibres, urethanes, petroleum additives, rubber additives and corrosion inhibitors; it finds use in photography and as an analytical reagent for bromine and gold (Windholz, 1976).

In Switzerland, it is used as a dyestuff intermediate. Its use in hair-dye formulations will be forbidden in Italy after 1977 (Dal Falco, 1976).

In Japan, *meta*-phenylenediamine is used in the production of heat-resistant fibres (80%) and dyes (20%).

meta-Phenylenediamine hydrochloride is used chiefly as an analytical reagent for nitrite (Windholz, 1976). It is also used in hair-dye formulations to produce golden-blond shades on the hair (Wall, 1972).

2.2 Occurrence

meta-Phenylenediamine and *meta*-phenylenediamine hydrochloride are not known to occur as natural products. No data on their occurrence in the environment were available to the Working Group.

2.3 Analysis

Aromatic amines, including *meta*-phenylenediamine, have been determined by non-aqueous titration (Mincheva, 1971). Primary amines, including this compound, have been determined in amounts greater than 1 mg by atomic absorption spectrophotometric determination of the copper-amine complex (Mitsui & Fujumura, 1974). A method for detecting μg amounts of aniline derivatives, including *meta*-phenylenediamine, has been suggested, in which a complex with 2,4,7-trinitro-9-fluorenone is identified by nuclear magnetic resonance or mass spectra (Hutzinger, 1969).

It has been identified in epoxy resin systems by mass spectrometry (D'Oyly-Watkins & Winsor, 1970). Ultra-violet spectrophotometry has been used to analyse this compound in air by aspiration through ethanol and absorbance measurement (Simonov *et al.*, 1971). *meta*-Phenylenediamine has been determined colorimetrically in hardened epoxy coatings, with an error of ±10%, by measuring the coloured chloranil-amine complex (Chapurin *et al.*, 1971), in solution by reaction with thiotrithiazyl chloride for levels of 0.2-5 mg/l (Levin *et al.*, 1967), by oxidation with peroxydisulphate (Gupta & Srivastava, 1971), by reaction with sodium chlorite for levels of 70-700 mg/l (Popa *et al.*, 1966) and by reaction with a cupric chloride-triphenylphosphine complex for levels of 5-100 mg/l (Hashmi *et al.*, 1969). Isomeric phenylenediamines have been detected in their mixtures as coloured derivatives with diazotized sulphanilic acid (Legradi, 1967).

Paper chromatography, using six different solvent systems and fifteen spray reagents and ultra-violet light, has been used to separate *meta*-phenylenediamine from other aminobenzene derivatives (Reio, 1970).

Thin-layer chromatography on silica gel has been used to separate and identify *meta*-phenylenediamine in aqueous extracts of epoxy resins (Kazarinova *et al.*, 1974), in mixtures of amines used as epoxy resin hardeners (Gedemer, 1969), in emulsified hair dyes by extraction with dilute acetic acid after break down of the emulsion by freezing (Zelazna & Legatowa, 1971) and in oxidation hair dyes by a two-dimensional technique (Kottemann, 1966). Aromatic amines, including *meta*-phenylenediamine, have been separated and identified by thin-layer chromatography using silica gel-calcium oxalate with two solvent systems and detection by exposure to nitrogen oxides (Srivastava & Dua, 1975). Carboxymethyl cellulose and alginic acid have been used as adsorbents for both thin-layer chromatographic and electrophoretic techniques (Cozzi *et al.*, 1969). Polystyrene-based anion exchangers, microcrystalline cellulose and Cellex D anion exchangers have also been used as adsorbents (Lepri *et al.*, 1974).

Column chromatography on an anion-exchange resin has been used to separate *meta*-phenylenediamine from various benzene derivatives (Funasaka *et al.*, 1972) and on a sulphonated cation-exchange resin for separation of phenylenediamine isomers (Funasaka *et al.*, 1969). Gel permeation chromatography on a styrene-divinylbenzene copolymer gel has also been used to separate aromatic amines, including this compound (Protivová & Pospíšil, 1974).

Gas chromatography has been used to separate isomeric phenylenediamines (Knight, 1971), and *meta*-phenylenediamine has been identified in cured epoxy resins by pyrolysis gas chromatography (Inagaki & Hayashi, 1968).

3. Biological Data Relevant to the Evaluation of Carcinogenic Risk to Man

3.1 Carcinogenicity and related studies in animals

(a) Skin application

Mouse: Two groups of 50 male and 50 female 6-8-week old Swiss-Webster mice received weekly or fortnightly applications of 0.05 ml of a freshly prepared 1:1 mixture of a hair-dye formulation, containing among its constituents 0.17% *meta*-phenylenediamine, and 6% hydrogen peroxide. The mixture was painted on shaved skin in the midscapular region for 18 months, at which time surviving animals were killed. The incidence of lung tumours did not differ significantly from that in untreated control animals (Burnett *et al.*, 1975) [The Working Group noted the high incidence of tumours in the mice treated with the dye base alone, which made up the greatest part of the hair-dye formulation].

(b) Subcutaneous and/or intramuscular administration

Rat: Four groups, each of 5 Wistar-King rats (sex not specified), were treated either with *meta*-phenylenediamine at a dose of 9 or 18 mg/kg bw or with *meta*-phenylenediamine hydrochloride at a dose of 12 or 24 mg/kg bw. The compounds were each dissolved in 0.5 ml water and injected subcutaneously into the back on alternate days, for 5 months in the higher dose groups and for 11 months in the lower dose groups. In two control groups, 5 rats were injected with 0.5 ml distilled water for either 5 months or 11 months. Fibrosarcomas were produced in one rat in the group treated with

9 mg/kg bw *meta*-phenylenediamine and in one rat treated with 24 mg/kg bw *meta*-phenylenediamine hydrochloride. No tumours were found in any of the other groups nor in the control groups (observation period not specified) (Saruta *et al.*, 1962) [The Working Group noted the inadequacy of the experiment].

3.2 Other relevant biological data

(a) Experimental systems

The s.c. LD_{50} of *meta*-phenylenediamine in DD mice is 90 mg/kg bw and that of the hydrochloride, 120 mg/kg bw (Saruta *et al.*, 1962). The i.p. LD_{50} of *meta*-phenylenediamine in dimethylsulphoxide in rats is 283 mg/kg bw; its oral LD_{50} in oil-in-water emulsion in rats is 650 mg/kg bw (Burnett *et al.*, 1977).

Toxic effects included changes in the central nervous system, decreased detoxifying activity of the liver and dermatitis at the site of application (Burnett *et al.*, 1975, 1976; Saruta *et al.*, 1962; Tainter *et al.*, 1929).

When 1.5 g *meta*-phenylenediamine hydrochloride in 50 ml of a lauryl sulphate-based gel were applied to the dorsal skin of dogs, approximately 60 mg were absorbed. This was determined by comparing the blood concentration of *meta*-phenylenediamine and the methaemoglobin levels with those obtained after the i.v. injection of 6 mg/kg bw *meta*-phenylenediamine (Kiese *et al.*, 1968).

A commercially available hair-dye preparation, containing 1.5% *meta*-phenylenediamine and 1% *ortho*-phenylenediamine, among other constituents, was applied to the dorsal skin of 6 male and 6 female adult white rabbits at a dose of 1 ml/kg bw twice weekly for 13 weeks. The formulation was mixed 1:1 with 6% hydrogen peroxide just prior to use. Blood and urine analyses, body weight gain or organ weights were similar to those in controls, except for a statistically significant increase in methaemoglobin in males ($P<0.05$) (Burnett *et al.*, 1976).

The teratogenicity of the same dye formulation was tested in a group of 20 mated Charles River CD female rats. It was applied topically to a shaved site in the dorsoscapular region at a dose of 2 ml/kg bw on days

1, 4, 7, 10, 13, 16 and 19 of gestation; just prior to use it was mixed with an equal volume of 6% hydrogen peroxide. The mothers were killed on day 20 of gestation, and no evidence of teratogenicity was observed (Burnett *et al.*, 1976).

meta-Phenylenediamine induced reverse mutations in *Salmonella typhimurium* TA 1538 in the presence of a rat liver post-mitochondrial supernatant fraction from animals pretreated either with polychlorinated biphenyl (Aroclor 1254) or with phenobarbital (Ames *et al.*, 1975; Garner & Nutman, 1977).

No dominant lethals were induced in Charles River rats injected intraperitoneally three times weekly for 8 weeks with 20 mg/kg bw *meta*-phenylenediamine before mating (no positive control substance was tested) (Burnett *et al.*, 1977).

(b) Man

meta-Phenylenediamine is excreted rapidly, unchanged, with little absorption (Hanżlik, 1922).

Workers who came into contact with *meta*-phenylenediamine during its production have been examined (Orlov, 1974). Their ages ranged from 30-50 years, and the duration of potential exposure was 5-10 years. Some workers (13.4%) complained of dysuria. The scratch test with *meta*-phenylenediamine allergen was positive in 8% of the workers studied; they also displayed eosinophiluria, and 0.3-40 μg *meta*-phenylenediamine were found per 100 ml urine. Cystoscopy demonstrated oedema of mucosa, polypous swellings and infiltration of the area of the triangle and cervix of the urinary bladder; the eosinophilic character of these formations was confirmed cytologically.

3.3 Case reports and epidemiological studies

No data were available to the Working Group.

4. Comments on Data Reported and Evaluation

4.1 Animal data

meta-Phenylenediamine as a constituent of a hair-dye formulation was inadequately tested in mice by skin painting. *meta*-Phenylenediamine and its hydrochloride were inadequately tested in rats by subcutaneous injection. No evaluation of the carcinogenicity of this compound or of its hydrochloride can be made.

4.2 Human data

No case reports or epidemiological studies were available to the Working Group (see, however, 'General Remarks on the Substances Considered', p. 29).

5. References

Ames, B.N., Kammen, H.O. & Yamasaki, E. (1975) Hair dyes are mutagenic: identification of a variety of mutagenic ingredients. Proc. nat. Acad. Sci. (Wash.), 72, 2423-2427

Burnett, C., Lanman, B., Giovacchini, R., Wolcott, G., Scala, R. & Keplinger, M. (1975) Long-term toxicity studies on oxidation hair dyes. Fd Cosmet. Toxicol, 13, 353-357

Burnett, C., Goldenthal, E.I., Harris, S.B., Wazeter, F.X., Strausburg, J., Kapp, R. & Voelker, R. (1976) Teratology and percutaneous toxicity studies on hair dyes. J. Toxicol. environm. Hlth, 1, 1027-1040

Burnett, C., Loehr, R. & Corbett, J. (1977) Dominant lethal mutagenicity study on hair dyes. J. Toxicol. environm. Hlth, 2, 657-662

Chapurin, V.I., Shaposhnik, S.S. & Kodner, M.S. (1971) Rapid method for detecting small amounts of *m*-phenylenediamine. Lakokrasoch. Mater. Ikh Primen., 4, 61-62

Cozzi, D., Desideri, P.G., Lepri, L. & Coas, V. (1969) Thin-layer chromatographic and electrophoretic behaviour of primary aromatic amines on weak ion exchangers. J. Chromat., 43, 463-472

Dal Falco (Il Ministro per la Sanita) (1976) Decreto ministeriale, 18 giugno 1976, Art. 1. Gazzetta Ufficiale Della Repubblica Italiana, N. 166, 5025

D'Oyly-Watkins, C. & Winsor, D.E. (1970) Identification of crosslinking agents in some epoxy resin systems by time-of-flight mass spectrometry. Dyn. Mass Spectrom., 1, 175-181

Funasaka, W., Fujimura, K. & Kuriyama, S. (1969) Ligand-exchange chromatography. I. Separation of phenylenediamine isomers by ligand-exchange chromatography. Bunseki Kagaku, 18, 19-24

Funasaka, W., Hanai, T., Fujimura, K. & Ando, T. (1972) Non-aqueous solvent chromatography. II. Separation of benzene derivatives in the anion-exchange and *n*-butyl alcohol system. J. Chromat., 72, 187-191

Garner, R.C. & Nutman, C.A. (1977) Testing of some azo dyes and their reduction products for mutagenicity using *Salmonella typhimurium* TA 1538. Mutation Res. (in press)

Gedemer, T.J. (1969) Characterization of amine hardeners and amine-cured epoxy resin pyrolyzates by thin-layer chromatography. In: Proceedings of the 27th Annual Technical Conference, Society of Plastics Engineering, Technical Paper, Vol. 15, Stamford, Conn., Society of Plastic Engineering Inc., pp. 471-474

Grasselli, J.G., ed. (1973) CRC Atlas of Spectral Data and Physical Constants for Organic Compounds, Cleveland, Ohio, Chemical Rubber Co., p. B-223

Gupta, R.C. & Srivastava, S.P. (1971) Oxidation of aromatic amines by peroxodisulphate ion. II. Identification of aromatic amines on the basis of absorption maxima of coloured oxidation products. Z. Analyt. Chem., 257, 275-277

Hanzlik, P.J. (1922) The pharmacology of some phenylenediamines (continued). J. industr. Hyg., 4, 448-462

Hashmi, M.H., Ajmal, A.I. & Adil, A.S. (1969) Spectrophotometric determination of *m*-phenylenediamine. Mikrochim. acta, 4, 778-781

Hutzinger, O. (1969) Electron acceptor complexes for chromogenic detection and mass spectrometric identification of phenol and aniline derivatives, related fungicides, and metabolites. Analyt. Chem., 41, 1662-1665

Inagaki, M. & Hayashi, M. (1968) Identification of hardeners in cured epoxy resins by pyrolysis gas chromatography. Nagoya-Shi Kogyo Kenkyusho Kenkyu Hokoku, 40, 23-27

Inoue, H., Asaumi, E., Hinohara, T., Sekiguchi, S. & Matsui, K. (1970) Fluorescent whitening agents. III. Fluorescence of aminostilbenes. Kogyo Kagaku Zasshi, 73, 187-194

Japan Dyestuff Industry Association (1976) Statistics of Dyestuffs, (1971-1975), Tokyo

Kazarinova, N.F., Dukhovnaya, I.S. & Myannik, L.E. (1974) Determination of *m*-phenylenediamine in aqueous media. Gig. i Sanit., 11, 61-63

Kiese, M., Rachor, M. & Rauscher, E. (1968) The absorption of some phenylenediamines through the skin of dogs. Toxicol. appl. Pharmacol., 12, 495-507

Knight, J.A. (1971) Gas chromatographic analysis of γ-irradiated aniline for aminoaromatic products. J. Chromat., 56, 201-208

Kottemann, C.M. (1966) Two-dimensional thin layer chromatographic procedure for the identification of dye intermediates in arylamine oxidation hair dyes. J. Ass. off. analyt. Chem., 49, 954-959

Legradi, L. (1967) Detection of coexisting isomeric phenylenediamines, aminophenols, and dihydric phenols. Mikrochim. acta, 4, 608-625

Lepri, L., Desideri, P.G. & Coas, V. (1974) Chromatographic and electrophoretic behaviour of primary aromatic amines on anion-exchange thin layers. J. Chromat., 90, 331-339

Levin, V., Nippoldt, B.W. & Rebertus, R.L. (1967) Spectrophotometric determination of primary aromatic amines with thiotrithiazyl chloride. Application to determination of toluene-2,4-diisocyanate in air. Analyt. Chem., 39, 581-584

Mincheva, E. (1971) Nonaqueous titration of aromatic amines. Khim. Ind. (Sofia), 43, 327-328

Mitsui, T. & Fujimura, Y. (1974) Indirect determination of primary amines by atomic absorption spectrophotometry. Bunseki Kagaku, 23, 1309-1314

Orlov, N.S. (1974) Allergic cystitis of chemical etiology. Urol. i Nefrol., 4, 33-36

E.I. du Pont de Nemours & Co., Inc. (1976) Sales Specification: *m*-Phenylenediamine Technical, Dyes and Chemicals Division, Wilmington, Delaware

Popa, G., Radulescu-Jercan, E. & Albert, F.M. (1966) Photometric determination of some aromatic amines. Rev. Roumaine Chim., 11, 1449-1452

Prager, B., Jacobson, P., Schmidt, P. & Stern, D., eds (1930) Beilsteins Handbuch der Organischen Chemie, 4th ed., Vol. 13, Syst. No. 1756, Berlin, Springer-Verlag, p. 33

Protivová, J. & Pospíšil, J. (1974) Antioxidants and stabilizers. XLVII. Behaviour of amine antioxidants and antiozonants and model compounds in gel permeation chromatography. J. Chromat., 88, 99-107

Reio, L. (1970) Third supplement for the paper chromatographic separation and identification of phenol derivatives and related compounds of biochemical interest using a 'reference system'. J. Chromat., 47, 60-85

Saruta, N., Yamaguchi, S. & Matsuoka, T. (1962) Sarcoma produced by subdermal administration of metaphenylenediamine and metaphenylenediamine hydrochloride. Kyushu J. med. Sci., 13, 175-180

Simonov, V.A., Bartenev, V.D. & Mikhailova, I.A. (1971) Spectrophotometric determination of aniline and *m*-phenylenediamine present together in the air. Gig. i Sanit., 36, 55-57

The Society of Dyers and Colourists (1971a) Colour Index, 3rd ed., Vol. 4, Yorkshire, UK, pp. 4643, 4820

The Society of Dyers and Colourists (1971b) Colour Index, 3rd ed., Vol. 1, Yorkshire, UK, pp. 1623, 1683

The Society of Dyers and Colourists (1971c) Colour Index, 3rd ed., Vol. 2, Yorkshire, UK, pp. 2002, 2345, 2363, 2428, 2432, 2433, 2439

The Society of Dyers and Colourists (1971d) Colour Index, 3rd ed., Vol. 3, Yorkshire, UK, pp. 3155, 3158, 3580

Srivastava, S.P. & Dua, V.K. (1975) TLC [thin-layer chromatography] separation of closely related amines. Fresenius' Z. analyt. Chem., 276, 382

Tainter, M.L., James, M. & Vandeventer, W. (1929) Comparative edemic actions of *ortho-*, *meta-* and *para*-phenylenediamines in different species. Arch. int. Pharmacodyn., 36, 152-162

Thirtle, J.R. (1968) Phenylenediamines and toluenediamines. In: Kirk, R.E. & Othmer, D.F., eds, Encyclopedia of Chemical Technology, 2nd ed., Vol. 15, New York, John Wiley and Sons, pp. 216-217, 222-223

US International Trade Commission (1976) Imports of Benzenoid Chemicals and Products, 1974, USITC Publication 762, Washington DC, US Government Printing Office, p. 25

US International Trade Commission (1977a) Synthetic Organic Chemicals, US Production and Sales, 1975, USITC Publication 804, Washington DC, US Government Printing Office, pp. 40, 49, 51, 60, 62, 64, 65, 72, 73

US International Trade Commission (1977b) Imports of Benzenoid Chemicals and Products, 1975, USITC Publication 806, Washington DC, US Government Printing Office, p. 24

US Tariff Commission (1919) Report on Dyes and Related Coal-Tar Chemicals, 1918, revised ed., Washington DC, US Government Printing Office

US Tariff Commission (1973) Imports of Benzenoid Chemicals and Products, 1972, TC Publication 601, Washington DC, US Government Printing Office, p. 21

Wall, F.E. (1972) Bleaches, hair colorings, and dye removers. In: Balsam, M.S. & Sagarin, E., eds, Cosmetics Science and Technology, 2nd ed., Vol. 2, New York, Wiley-Interscience, pp. 308, 309, 310

Weast, R.C., ed. (1976) CRC Handbook of Chemistry and Physics, 57th ed., Cleveland, Ohio, Chemical Rubber Co., p. D-197

Weschler, J.R. (1965) Epoxy resins. In: Kirk, R.E. & Othmer, D.F., eds, Encyclopedia of Chemical Technology, 2nd ed., Vol. 8, New York, John Wiley and Sons, pp. 309-312

Windholz, M., ed. (1976) The Merck Index, 9th ed., Rahway, NJ, Merck & Co., pp. 947-948

Zelazna, K. & Legatowa, B. (1971) Identification of basic dyes in emulsified hair dyes by thin layer chromatography. Rocz. Panstw. Zakl. Hig., 22, 427-430

1. Chemical and Physical Data

para-Phenylenediamine

1.1 Synonyms and trade names

Colour Index No.: 76060

Colour Index Names: Developer 13; Oxidation Base 10

Chem. Abstr. Services Reg. No.: 106-50-3

Chem. Abstr. Name: 1,4-Benzenediamine

4-Aminoaniline; *para*-aminoaniline; *para*-benzenediamine; 1,4-diaminobenzene; *para*-diaminobenzene; 1,4-phenylenediamine

BASF Ursol D; Benzofur D; Developer PF; Durafur Black R; Fouramine D; Fourrine 1; Fourrine D; Fur Black 41867; Fur Brown 41866; Furro D; Fur Yellow; Futramine D; Nako H; Orsin; Pelagol D; Pelagol DR; Pelagol Grey D; Peltol D; Renal PF; Santoflex LC; Tertral D; Ursol D; Zoba Black D

1.2 Chemical formula and molecular weight

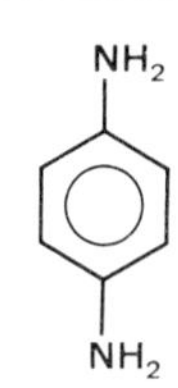

$C_6H_8N_2$ Mol. wt: 108.1

1.3 Chemical and physical properties of the pure substance

(a) Description: White crystals (Grasselli, 1973)

(b) Boiling-point: 267°C (Grasselli, 1973)

(c) Melting-point: 140°C (Grasselli, 1973)

(d) Spectroscopy data: λ_{max} 246 nm (E_1^1 = 788), 315 nm (E_1^1 = 184) (Grasselli, 1973); mass (Grasselli, 1973), fluorescence (Inoue *et al.*, 1970) and far infra-red spectra (Pommez *et al.*, 1967) have been recorded.

(e) Solubility: Slightly soluble in water; soluble in ethanol, ether, benzene, chloroform (Grasselli, 1973) and acetone (Boutwell & Bosch, 1959)

(f) Volatility: Vapour pressure is <1 mm at 21°C (technical product) (E.I. du Pont de Nemours & Co., Inc., 1977).

(g) Stability: See 'General Remarks on the Substances Considered', p. 27.

1.4 Technical products and impurities

One technical grade of *para*-phenylenediamine available in the US has the following specifications: purity, 99.2% min.; moisture content, 0.1% max.; *ortho*-phenylenediamine (1,2-diaminobenzene) content, 0.1% max.; and iron content, 50 mg/kg max. (E.I. du Pont de Nemours & Co., Inc., 1973).

In the UK, *para*-phenylenediamine has a minimum purity of 99% and contains traces of 1-amino-4-nitrobenzene and 4,4'-diaminoazobenzene.

In Japan, *para*-phenylenediamine is available as a commercial grade with a minimum purity of 99.5%, minimum melting-point of 139°C and containing 1-amino-4-nitrobenzene and 4-aminophenol as impurities.

para-Phenylenediamine hydrochloride

1.1 Synonyms and trade names

Colour Index No.: 76061

Colour Index Name: Oxidation Base 10A

Chem. Abstr. Services Reg. No.: 624-18-0

Chem. Abstr. Name: 1,4-Benzenediamine dihydrochloride

4-Aminoaniline dihydrochloride; *para*-aminoaniline dihydrochloride; *para*-benzenediamine dihydrochloride; 1,4-diaminobenzene dihydrochloride;

para-diaminobenzene dihydrochloride; 1,4-phenylenediamine dihydrochloride; *para*-phenylenediamine dihydrochloride

Durafur Black RC; Fourrine 64; Fourrine DS; Pelagol CD; Pelagol Grey CD

1.2 Empirical formula and molecular weight

$C_6H_8N_2.2HCl$ Mol. wt: 181.0

1.3 Chemical and physical properties of the pure substance

(a) Description: White crystals (Windholz, 1976)

(b) Solubility: Freely soluble in water; slightly soluble in ethanol and ether (Windholz, 1976)

1.4 Technical products and impurities

No data were available to the Working Group.

2. Production, Use, Occurrence and Analysis

For background information on this section, see preamble, p. 15.

2.1 Production and use

(a) Production

para-Phenylenediamine was prepared by Rinne & Zincke in 1874 by reducing 1,4-dinitrobenzene with tin and hydrochloric acid (Prager *et al.*, 1930). It is produced commercially by reducing 1-amino-4-nitrobenzene with (1) iron and hydrochloric acid, or (2) iron, ammonium polysulphide and hydrogen or (3) iron and ferrous chloride (Thirtle, 1968).

para-Phenylenediamine has been produced commercially in the US for over fifty years (US Tariff Commission, 1919). Commercial production of the hydrochloride in the US was first reported in 1971 (US Tariff Commission, 1973a). In 1975, undisclosed amounts (see preamble, p. 15) of *para*-phenylenediamine were reported by two companies and of the hydrochloride by one company (US International Trade Commission, 1977).

US imports of *para*-phenylenediamine through principal US customs districts amounted to 75.5 thousand kg in 1972 (US Tariff Commission, 1973b), 12.0 thousand kg in 1973 (US Tariff Commission, 1974) and 29.2 thousand kg in 1974 (US International Trade Commission, 1976).

It is believed that at least four companies produce *para*-phenylenediamine in western Europe. About 10-100 thousand kg are produced annually in the UK, with imports amounting to less than 100 thousand kg. It is believed that 10-100 thousand kg/year *para*-phenylenediamine are imported into Switzerland.

para-Phenylenediamine was first produced commercially in Japan in 1937; four companies currently produce it. Japanese production of *para*-phenylenediamine in 1971-1975 was 80 thousand kg, 60 thousand kg, 114 thousand kg, 113 thousand kg and 55 thousand kg per year, respectively. There are no imports or exports of this chemical (Japan Dyestuff Industry Association, 1976).

(b) Use

para-Phenylenediamine has commercial application in dyestuff manufacture, in hair-dye formulations, as a monomer for polyparaphenylene terephthalamide (Fiber B), in photographic developers and in a variety of antioxidants.

It is used as a dye developer to obtain black and brown shades (The Society of Dyers and Colourists, 1971a), and both this compound and its hydrochloride are used in dyeing furs black, grey, dull violet or reddish-brown shades, and for printing on cellulosic textile materials (The Society of Dyers and Colourists, 1971b).

para-Phenylenediamine is used as an intermediate in the production of five dyes which are believed to have commercial world significance: C.I. Direct Orange 27, Disperse Yellow 9, Solvent Orange 53, Sulphur Brown 23, and Leuco Sulphur Brown 23 (The Society of Dyers and Colourists, 1971c); however, none of these dyes are produced commercially in the US. It is also an unisolated intermediate in the production of ten dyes derived from 4-aminoacetanilide, 4-aminoformanilide, 4-aminooxanilic acid or 4-nitroaniline (The Society of Dyers and Colourists, 1971c).

para-Phenylenediamine is used in hair-dye formulations and can produce a variety of shades on the hair depending on the formulation. Typical levels in hair-dye products range from 0.20% in golden-blond dyes to 3.75% in black hair dyes: since most hair-dye formulations are proprietary, exact concentrations are not available. *para*-Phenylenediamine is also used as a dye intermediate in colour shampoos and bleach-toners. The hydrochloride is used in hair-dye formulations to produce a warm brown shade on the hair (Wall, 1972).

para-Phenylenediamine is used with terephthaloyl chloride to produce polyparaphenylene terephthalamide (Fiber B), used primarily as tire cord. It is estimated that annual US use of Fiber B may reach 11 million kg in 1977.

para-Phenylenediamine and its derivatives are also used as developers for black-and-white and colour photography. The derivatives are important antioxidants in synthetic and natural rubbers, petroleum products, cellulose ethers and alfalfa meal (Thirtle, 1968). *para*-Phenylenediamine has also reportedly been used in vulcanization acceleration and in photochemical measurements (Windholz, 1976).

The hydrochloride has been used as an analytical reagent in the testing of blood, hydrogen sulphide, amyl alcohol and milk (Windholz, 1976).

In Switzerland, *para*-phenylenediamine is used as a dyestuff and industrial chemical intermediate.

In Japan, it is used as a component of hair dyes (60%) and as a dye intermediate (40%).

According to the US Occupational Safety and Health Administration's health standards for air contaminants, an employee's exposure to *para*-phenylenediamine should not exceed 0.1 mg/m^3 in the working atmosphere in any eight-hour work shift of a forty-hour work week (US Occupational Safety and Health Administration, 1976). The maximum allowable concentration in the Federal Republic of Germany is also 0.1 mg/m^3; this is also the level accepted in Japan as the maximum workplace tolerance concentration.

2.2 Occurrence

para-Phenylenediamine and *para*-phenylenediamine hydrochloride are not known to occur as natural products. No data on their occurrence in the environment were available to the Working Group.

2.3 Analysis

The Association of Official Analytical Chemists has published an Official Final Action for the determination of *para*-phenylenediamine in hair dyes by gravimetric determination as its diacetyl derivative or by an iodimetric titration method (Horwitz, 1970). Titrimetric methods have been described (Ignaczak & Dziegiec, 1975; Ratnikova *et al.*, 1974) and chronopotentiometry has been used for mM concentrations (Bamberger & Strohl, 1969).

Atomic absorption spectrophotometry of a copper complex (detection limit, about 1 mg) has been used for the analysis of aniline derivatives, including *para*-phenylenediamine (Mitsui & Fujimura, 1974), and nuclear magnetic resonance and mass spectrometry of the 2,4,7-trinitro-9-fluorenone derivative have been used for detecting μg amounts (Hutzinger, 1969).

Colorimetric methods have been used to analyse aromatic amines, including *para*-phenylenediamine, by its reaction with: peroxydisulphate (Gupta & Srivastava, 1971), ruthenium trichloride-triphenylphosphine (for levels of 5-35 mg/l) (Hashmi *et al.*, 1969), thiotrithiazyl chloride (for levels of 1-20 mg/l) (Levin *et al.*, 1967), sodium chlorite (for levels of 70-1700 mg/l) (Popa *et al.*, 1966) or 2,6-xylenol (for levels of 2-10 mg/l with an accuracy of ±2%) (Corbett, 1975) or by coupling with diazotized sulphanilic acid and other compounds (Legradi, 1967). A spot test for the detection of *para*-phenylenediamine in hair dyes uses a hydrochloric acid-vanillin-isopropanol reagent (Lange, 1966).

Paper chromatography has been used to separate and identify aminobenzene derivatives, including *para*-phenylenediamine (Reio, 1970) and to detect free *para*-phenylenediamine in the presence of pyrocatechol in dyed furs (Galatik, 1972) and in dye intermediates by the action of bromine vapours (Matrka & Kroupa, 1971).

Thin-layer chromatography and electrophoresis have been used to separate amines, including *para*-phenylenediamine, on various adsorbents and with various vizualisation methods (Bassl *et al.*, 1967; Cozzi *et al.*, 1969; Drost & Reith, 1967; Lepri *et al.*, 1974; Srivastava & Dua, 1975). Basic dyes, including *para*-phenylenediamine, have been identified in hair dyes by thin-layer chromatography (Kottemann, 1966; Legatowa, 1973; Zelazna & Legatowa, 1971).

Column chromatography on an anion-exchange resin (Funasaka *et al.*, 1972), ligand-exchange chromatography (Funasaka *et al.*, 1969) and gel-permeation chromatography on polystyrene have been used to analyse *para*-phenylenediamine and other aromatic amines (Protivová & Pospíšil, 1974).

Gas chromatography has been used to separate the isomeric phenylenediamines (Knight, 1971); it has also been used to determine *para*-phenylenediamine in hair dyes (Pinter & Kramer, 1967). The *N*-substituted 2,5-dimethylpyrrole derivative has been used to determine aromatic amines, including *para*-phenylenediamine, in amounts as low as 1 ng, using gas chromatography (Walle, 1968).

3. Biological Data Relevant to the Evaluation of Carcinogenic Risk to Man

3.1 Carcinogenicity and related studies in animals[1]

(a) Oral administration

Rat: *para*-Phenylenediamine was administered orally in daily doses of 0.06, 0.3 and 10 mg/kg bw for 8 months to groups: of 10 rats (5 males and 5 females) for the 0.06 and 0.3 mg/kg bw doses and of 5 rats (sex not specified) for the 10 mg/kg bw dose. Four animals were given daily doses of 30 mg/kg bw, but three of them died within a 4-month period. The strain

[1]The Working Group was aware of carcinogenicity tests in progress on *para*-phenylenediamine hydrochloride in mice and rats, involving oral administration in the diet, and on *para*-phenylenediamine in mice and rabbits, involving skin application (IARC, 1976).

of rats, survival times and the solvent used were not mentioned. No tumours were observed in any of the treated animals within the 8-month period nor in 5 controls (Saruta *et al.*, 1958) [The Working Group noted the inadequacy of the experiment].

(b) Skin application

Mouse: A group of 30 adult, albino mice were treated twice weekly for 20 weeks with one drop of a 5% solution of *para*-phenylenediamine in acetone applied to the backs of the animals. None of the 19 which survived at 20 weeks had skin tumours (Boutwell & Bosch, 1959) [The Working Group noted the short duration of this experiment].

Six groups of 50 male and 50 female 6-8-week old Swiss-Webster mice received weekly or fortnightly applications of 0.05 ml of a freshly prepared 1:1 mixture of three different hair-dye formulations, all containing among their constituents 1.5% *para*-phenylenediamine and 3% 2,5-diaminotoluene sulphate and either 0.2% 2,4-diaminotoluene, 0.38% 2,4-diaminoanisole sulphate or 0.17 *meta*-phenylenediamine, and 6% hydrogen peroxide. The mixtures were painted on shaved skin in the midscapular region for 18 months, at which time the surviving animals were killed. The incidence of lung tumours did not differ significantly from that in untreated control animals (Burnett *et al.*, 1975) [The Working Group noted the high incidence of tumours in the mice treated with the dye base alone, which made up the greatest part of the hair-dye formulation].

Groups of 28 male and 28 female 4-7-week old Swiss-Webster mice received weekly applications on shaved skin in the intrascapular region of 0.05 ml of two different hair-dye formulations containing 1.5% *para*-phenylenediamine, 3% 2,5-diaminotoluene sulphate and either 0.2% or 0.6% 2,4-diaminotoluene among their constituents, or of a hair-dye base without the dye components; an untreated control group was also included. The hair dye was mixed with an equal volume of 6% hydrogen peroxide just before use (Giles *et al.*, 1976) [A large number of animals were unaccounted for in the final analysis of tumour incidence, thus making the published data inadequate for evaluation of the carcinogenicity of this chemical; see also Bridges & Green, 1976].

(c) Subcutaneous and/or intramuscular administration

Rat: *para*-Phenylenediamine was injected subcutaneously into the backs of rats divided into 2 groups, each of 5 animals (strain and sex not specified), at a daily dose of 12.5 or 20 mg/kg bw. Fibrosarcomas were produced at the site of injection in 2/5 rats that received 12.5 mg/kg bw after 7 months of administration. No s.c. tumours were observed in the group that received 20 mg/kg bw nor in 5 controls (Saruta *et al.*, 1958) [The Working Group noted the inadequacy of the experiment].

3.2 Other relevant biological data

(a) Experimental systems

The oral LD_{50} of *para*-phenylenediamine is 250 mg/kg bw in rabbits and 100 mg/kg bw in cats. The s.c. LD_{50} is 170 mg/kg bw in rats, 200 mg/kg bw in rabbits and 100 mg/kg bw in dogs (Spector, 1956). The i.p. LD_{50} of *para*-phenylenediamine in dimethylsulphoxide in rats is 37 mg/kg bw; the oral LD_{50} of the compound in oil-in-water emulsion in rats is 80 mg/kg bw (Burnett *et al.*, 1977).

From studies of the intracutaneous sensitization of guinea-pigs using *para*-phenylenediamine, hydroquinone, quinhydrone and benzoquinone, it has been suggested that benzoquinone formation plays an important role in the allergic action of *para*-phenylenediamine (Rajka & Blohm, 1970).

A characteristic oedema of the head and neck is produced by i.p. injection of *para*-phenylenediamine hydrochloride in rabbits (190 mg/kg bw) and cats (120 mg/kg bw) (Tainter & Hanzlik, 1924); the same effect has been produced in guinea-pigs (s.c. dose of 350 mg/kg bw) and rats (s.c. doses of 120-150 mg/kg bw) (Tainter *et al.*, 1929). More recent studies in rats demonstrate that s.c. administration of 3 mg of the hydrochloride induces skeletal muscle lesions: rhabdomyolysis with infiltration of myophages, necrosis with calcifications, accumulation of neutral lipids and dilatation of sarcoplasmic reticulum (Mascrès & Jasmin, 1974, 1975).

Four commercially available hair-dye formulations, containing 1, 2, 3 or 4% *para*-phenylenediamine and several aromatic amine derivatives among their constituents, were tested by skin application in groups of 6 male and 6 female adult white rabbits. A 1:1 mixture of each formulation with 6% hydrogen peroxide was applied topically twice weekly for 13 weeks at a dose of 1 ml/kg bw. Blood and urine parameters and organ weights of treated animals were not significantly different from those of controls (Burnett *et al.*, 1976).

para-Phenylenediamine hydrochloride was applied to the skin of dogs in gels and fluids, as they are used in hair dyeing; absorption was determined by comparing blood concentration with that obtained after i.v. injection of the diamine over a comparable period of time. It was calculated that 16 mg were absorbed when 50 ml of a lauryl sulphate-based gel containing 1.5 g *para*-phenylenediamine were applied for 3 hours and free access to air was allowed (the maximal blood concentration was 0.15 μg/ml). This amount rose to 110 mg if the treated area was immediately covered with aluminium foil (the maximal blood concentration was 0.5 μg/ml). If hydrogen peroxide was added to the gel immediately prior to application, no *para*-phenylenediamine was found in blood or urine, indicating that only less than 2 mg (minimum detectable) could have been absorbed (Kiese *et al.*, 1968).

Four commercially available hair-dye formulations, containing 1, 2, 3 or 4% *para*-phenylenediamine and several aromatic amine derivatives among their constituents, were tested for teratogenicity in groups of 20 mated Charles River CD female rats. Each formulation was applied topically to a shaved site in the dorsoscapular region at a dose of 2 ml/kg bw on days 1, 4, 7, 10, 13, 16 and 19 of gestation; just prior to its use, each formulation was mixed with an equal volume of 6% hydrogen peroxide. The mothers were killed on day 20 of gestation. No abnormal maternal or foetal effects were noted, except that in one group treated with a formulation containing 2% *para*-phenylenediamine, there were skeletal changes in 9/169 live foetuses (Burnett *et al.*, 1976).

N,N'-Diacetyl-*para*-phenylenediamine was identified as a urinary metabolite of *para*-phenylenediamine in dogs (Kiese *et al.*, 1968). *para*-Phenylenediamine is excreted in the urine of dogs (Ishidate & Hashimoto,

1959) and rats (Ishidate & Hashimoto, 1962; Stevenson *et al.*, 1942) after oral or s.c. administration of *para*-dimethylaminoazobenzene[1].

para-Phenylenediamine has been reported to induce reverse mutations in *Salmonella typhimurium* TA 1538 in the presence of a rat liver post-mitochondrial supernatant fraction from animals pretreated with phenobarbital (Garner & Nutman, 1977).

This compound did not induce micronucleated cells in bone marrow when 300 mg/kg bw were administered orally to two groups of 5 male and 5 female rats in two doses separated by an interval of 24 hours (Hossack & Richardson, 1977).

(b) Man

There have been many reports of hypersensitivity to *para*-phenylenediamine. The subject was reviewed by Mayer (1954), who pointed out that although *para*-phenylenediamine commonly provokes contact sensitization of the skin, it can also provoke atopically-induced skin allergies such as disseminated neurodermatitis. Other allergies, such as arthritis, asthma, conjunctivitis and gastrointestinal disturbances, have been noted after ectopic exposures, particularly in the dyeing of fur. The allergic reaction most often produced by hair dyes is contact dermatitis.

In a study of 'nylon stocking' dermatitis, 13 patients were hypersensitive to azodyes used in the manufacture of the stockings and to *para*-phenylenediamine; none of these subjects were hypersensitive to dyes used in the manufacture of the stockings that were derivatives of anthraquinone or acridine. Additionally, 3 out of 5 patients with known hypersensitivity to *para*-phenylenediamine, but with no previous history of 'stocking' dermatitis, were also hypersensitive to nylon stockings containing the allergenic dyes (Dobkevitch & Baer, 1947).

In Stockholm, 2,903 patients diagnosed as suffering from eczema during the period 1958-1961 were investigated by patch testing with a 2% aqueous

---

[1]See IARC (1975).

solution of *para*-phenylenediamine; 10.6% reacted positively (Modée & Skog, 1962). In patch tests made in Budapest on 691 patients with allergic dermatoses during a 17-month period, 6% of the patients responded to *para*-phenylenediamine (Korossy *et al.*, 1969). Positive reactions to patch tests with *para*-phenylenediamine may, however, indicate previous contact with local anaesthetics, sulphonamide, aniline dyes or rubber antioxidants, thus implying a more general 'para-group' allergy (Malten *et al.*, 1973).

para-Phenylenediamine poisoning was diagnosed in a 51-year-old woman who had regularly been dyeing her hair with a commercial preparation consisting of impure *para*-phenylenediamine and iron. Liver and spleen enlargement were observed, and the patient developed progressive neurological symptoms prior to her death 11 weeks after admission to hospital. Other symptoms of poisoning with *para*-phenylenediamine include vertigo, gastritis, diplopia, asthenia and exfoliative dermatitis (Davison, 1943).

Gastrointestinal and nervous symptoms were observed in a woman who had used a *para*-phenylenediamine hair-dye preparation (Close, 1932); and jaundice and subacute atrophy of the liver, resulting in death, were seen in a female hairdresser who had used *para*-phenylenediamine hair dyes over a period of 5 years (Israëls & Susman, 1934).

3.3 Case reports and epidemiological studies

No data were available to the Working Group.

4. Comments on Data Reported and Evaluation

4.1 Animal data

para-Phenylenediamine has been inadequately tested in mice by skin application and in rats by oral and subcutaneous administration. Studies in mice in which *para*-phenylenediamine as a constituent of hair-dye preparations was tested by skin application cannot be evaluated. No evaluation of the carcinogenicity of this compound can be made.

4.2 Human data

No case reports or epidemiological studies were available to the Working Group (see, however, 'General Remarks on the Substances Considered, p. 29).

5. References

Bamberger, R.L. & Strohl, J.H. (1969) Quantitative analysis of *p*-nitrophenol, hydroquinone, and *p*-phenylenediamine using thin-layer chronopotentiometry. Analyt. Chem., 41, 1450-1452

Bassl, A., Heckemann, H.J. & Baumann, E. (1967) Thin-layer chromatography of primary aromatic amines. I. J. Prakt. Chem., 36, 265-273

Boutwell, R.K. & Bosch, D.K. (1959) The tumor-promoting action of phenol and related compounds for mouse skin. Cancer Res., 19, 413-427

Bridges, B.A. & Green, M.H.L. (1976) Carcinogenicity of hair dyes by skin painting in mice. J. Toxicol. environm. Hlth, 2, 251-252

Burnett, C., Lanman, B., Giovacchini, R., Wolcott, G., Scala, R. & Keplinger, M. (1975) Long-term toxicity studies on oxidation hair dyes. Fd Cosmet. Toxicol., 13, 353-357

Burnett, C., Goldenthal, E.I., Harris, S.B., Wazeter, F.Z., Strausburg, J., Kapp, R. & Voelker, R. (1976) Teratology and percutaneous toxicity studies on hair dyes. J. Toxicol. environm. Hlth, 1, 1027-1040

Burnett, C., Loehr, R. & Corbett, J. (1977) Dominant lethal mutagenicity study on hair dyes. J. Toxicol. environm. Hlth, 2, 657-662

Close, W.J. (1932) A case of poisoning from hair dye (paraphenylenediamine). Med. J. Austr., 1, 53-54

Corbett, J.F. (1975) Application of oxidative coupling reactions to the assay of *p*-phenylenediamines and phenols. Analyt. Chem., 47, 308-313

Cozzi, D., Desideri, P.G., Lepri, L. & Coas, V. (1969) Thin-layer chromatographic and electrophoretic behaviour of primary aromatic amines on weak ion exchangers. J. Chromat., 43, 463-472

Davison, C. (1943) Paraphenylenediamine poisoning with changes in the central nervous system. Arch. Neurol. Psych., 49, 254-265

Dobkevitch, S. & Baer, R.L. (1947) Eczematous cross-hypersensitivity to azodyes in nylon stockings and to paraphenylendiamine. J. Invest. Dermatol., 9, 203-211

Drost, R.H. & Reith, J.F. (1967) Identification of toxic substances by means of Feldstein's extraction method, thin-layer chromatography, and UV spectrometry. I. Basic substances. Pharm. Weekbl., 102, 1379-1387

Funasaka, W., Fujimura, K. & Kuriyama, S. (1969) Ligand-exchange chromatography. I. Separation of phenylenediamine isomers by ligand-exhange chromatography. Bunseki Kagaku, 18, 19-24

Funasaka, W., Hanai, T., Fujimura, K. & Ando, T. (1972) Non-aqueous solvent chromatography. II. Separation of benzene derivatives in the anion-exchange and *n*-butyl alcohol system. J. Chromat., 72, 187-191

Galatik, J. (1972) Chromatographic determination of free *p*-phenylenediamine in the presence of pyrocatechol in furs dyed with oxidizing dyes. Kozarstvi, 22, 21-23

Garner, R.C. & Nutman, C.A. (1977) Testing of some azo dyes and their reduction products for mutagenicity using *Salmonella typhimurium* TA 1538. Mutation Res. (in press)

Giles, A.L., Jr, Chung, C.W. & Kommineni, C. (1976) Dermal carcinogenicity study by mouse-skin painting with 2,4-toluenediamine alone or in representative hair dye formulations. J. Toxicol. environm. Hlth, 1, 433-440

Grasselli, J.G., ed. (1973) CRC Atlas of Spectral Data and Physical Constants for Organic Compounds, Cleveland, Ohio, Chemical Rubber Co., p. B-223

Gupta, R.C. & Srivastava, S.P. (1971) Oxidation of aromatic amines by peroxodisulphate ion. II. Identification of aromatic amines on the basis of absorption maxima of coloured oxidation products. Z. analyt. Chem., 257, 275-277

Hashmi, M.H., Iftikhar, A.A., Rashid, A. & Qureshi, T. (1969) Spectrophotometric determination of *o*- and *p*-phenylenediamine. Mikrochim. acta, 1, 100-107

Horwitz, W., ed. (1970) Official Methods of Analysis of the Association of Official Analytical Chemists, 11th ed., Washington DC, Association of Official Analytical Chemists, p. 614

Hossack, D.J.N. & Richardson, J.C. (1977) Examination of the potential mutagenicity of hair dye constituents using the micronucleus test. Experientia, 33, 377-378

Hutzinger, O. (1969) Electron acceptor complexes for chromogenic detection and mass spectrometric identification of phenol and aniline derivatives, related fungicides, and metabolites. Analyt. Chem., 41, 1662-1665

IARC (1975) IARC Monographs on the Evaluation of Carcinogenic Risk of Chemicals to Man, 8, Some Aromatic Azo Compounds, Lyon, pp. 125-146

IARC (1976) Information Bulletin on the Survey of Chemicals Being Tested for Carcinogenicity, No. 6, Lyon, pp. 193, 252

Ignaczak, M. & Dziegiec, J. (1975) Use of ceric perchlorate in the determination of *p*-quinone, *p*-aminophenol, *p*-phenylenediamine, *p*-aminobenzoic acid, and sulfanilic acid. Chem. analyt. (Warsaw), 20, 229-232

Inoue, H., Asaumi, E., Hinohara, T., Sekiguchi, S. & Matsui, K. (1970) Fluorescent whitening agents. III. Fluorescence of aminostilbenes. Kogyo Kagaku Zasshi, 73, 187-194

Ishidate, M. & Hashimoto, Y. (1959) The metabolism of *p*-dimethylaminoazobenzene and related compounds. I. Metabolites of *p*-dimethylaminoazobenzene in dog urine. Chem. pharm. Bull., 7, 108-113

Ishidate, M. & Hashimoto, Y. (1962) Metabolism of 4-dimethylaminoazobenzene and related compounds. II. Metabolites of 4-dimethylaminoazobenzene and 4-aminoazobenzene in rat urine. Chem. pharm. Bull., 10, 125-133

Israëls, M.C.G. & Susman, W. (1934) Systemic poisoning by phenylenediamine with report of a fatal case. Lancet, i, 508-510

Japan Dyestuff Industry Association (1976) Statistics of Dyestuffs, (1971-1975), Tokyo

Kiese, M., Rachor, M. & Rauscher, E. (1968) The absorption of some phenylenediamines through the skin of dogs. Toxicol. appl. Pharmacol., 12, 495-507

Knight, J.A. (1971) Gas chromatographic analysis of γ-irradiated aniline for aminoaromatic products. J. Chromat., 56, 201-208

Korossy, S., Vincz, E., Doroszlay, J. & Munkácsi, A. (1969) Zur Revision des Allergenspektrums der Ungarn gebräuchlichen diagnostischen Standard-Reihe für Epikutantestung. Berufsdermatosen, 17, 252-263

Kottemann, C.M. (1966) Two-dimensional thin layer chromatographic procedure for the identification of dye intermediates in arylamine oxidation hair dyes. J. Ass. off. analyt. Chem., 49, 954-959

Lange, F.W. (1966) Fast method for detection of *p*-phenylenediamine in hair dyes. Seifen-Ole-Fette-Wachse, 92, 751-753

Legatowa, B. (1973) Determination of aromatic amines and aminophenols in hair dyes. Rocz. Panstw. Zakl. Hig., 24, 393-402

Legradi, L. (1967) Detection of coexisting isomeric phenylenediamines, aminophenols, and dihydric phenols. Mikrochim. acta, 4, 608-625

Lepri, L., Desideri, P.G. & Coas, V. (1974) Chromatographic and electrophoretic behaviour of primary aromatic amines on anion-exchange thin layers. J. Chromat., 90, 331-339

Levin, V., Nippoldt, B.W. & Rebertus, R.L. (1967) Spectrophotometric determination of primary aromatic amines with thiotrithiazyl chloride - application to determination of toluene-2,4-diisocyanate in air. Analyt. Chem., 39, 581-584

Malten, K.E., Kuiper, J.P. & van der Staak, W.B.J.M. (1973) Contact allergic investigations in 100 patients with ulcus cruris. Dermatologica, 147, 241-254

Mascrès, C. & Jasmin, G. (1974) Etude pathogénique des lésions musculaires induites par la *p*-phénylènediamine. Union méd. Can., 103, 672-677

Mascrès, C. & Jasmin, G. (1975) Altérations de la fibre musculaire induites chez le rat par la *p*-phénylènediamine. Path. Biol. (Paris), 23, 193-199

Matrka, M. & Kroupa, J. (1971) Analyse von Farbstoffen und von bei der Farbstofferzeugung anfallenden Zwischenprodukten. XIV. Sichtbarmachung aromatischer Diamine in der Papierchromatographie mit Hilfe von Bromdämpfen. Collec. Czech. Chem. Commun., 36, 2366-2371

Mayer, R.L. (1954) Group-sensitisation to compounds of quinone structure and its biochemical basis: role of these substances in cancer. Progr. Allergy, 4, 79-172

Mitsui, T. & Fujimura, Y. (1974) Indirect determination of primary amines by atomic absorption spectrophotometry. Bunseki Kagaku, 23, 1309-1314

Modée, J. & Skog, E. (1962) A comparison of results of patch testing in 1951 and in 1961. Acta dermato-venereol., 42, 280-289

Pinter, I. & Kramer, M. (1967) Gas chromatographic detection and determination of some aromatic diamines in hair dyes. Parfuem. Kosmet., 48, 126-128

Pommez, P., Lafaix, M. & Delorme, P. (1967) Far infrared absorption study of low-frequency oscillations in a series of *para*-disubstituted benzene derivatives. I. Spectra and assignment of frequencies observed. J. Chim. Phys., 64, 1450-1460

E.I. du Pont de Nemours & Co., Inc. (1973) Sales Specification: *p*-Phenylenediamine Technical, Wilmington, Delaware, Dyes and Chemicals Division

E.I. du Pont de Nemours & Co., Inc. (1977) US Department of Labor, Occupational Safety & Health Administration, Material Safety Data Sheet, Wilmington, Delaware

Popa, G., Radulescu-Jercan, E. & Albert, F.M. (1966) Photometric determination of some aromatic amines. Rev. Roumaine Chim., 11, 1449-1452

Prager, B., Jacobson, P., Schmidt, P. & Stern, D., eds (1930) Beilsteins Handbuch der Organischen Chemie, 4th ed., Vol. 13, Syst. No. 1765-1766, Berlin, Springer-Verlag, pp. 61-62

Protivová, J. & Pospíšil, J. (1974) Antioxidants and stabilizers. XLVII. Behaviour of amine antioxidants and antiozonants and model compounds in gel permeation chromatography. J. Chromat., 88, 99-107

Rajka, G. & Blohm, S.G. (1970) The allergenicity of paraphenylenediamine. II. Acta dermatovener (Stockholm), 50, 51-54

Ratnikova, T.V., Klochkov, V.I., Kharchevnikov, V.M., Devikina, L.I. & Lepilin, V.N. (1974) Analysis of some ingredients of rubbers. Zh. Prikl. Khim. (Leningrad), 47, 850-854

Reio, L. (1970) Third supplement for the paper chromatographic separation and identification of phenol derivatives and related compounds of biochemical interest using a 'reference system'. J. Chromat., 47, 60-85

Saruta, N., Yamaguchi, S. & Nakatomi, Y. (1958) Sarcoma produced by subdermal administration of paraphenylen-diamine. Kyushu J. med. Sci., 9, 94-101

The Society of Dyers and Colourists (1971a) Colour Index, 3rd ed., Vol. 2, Yorkshire, UK, p. 2002

The Society of Dyers and Colourists (1971b) Colour Index, 3rd ed., Vol. 3, Yorkshire, UK, p. 3262

The Society of Dyers and Colourists (1971c) Colour Index, 3rd ed., Vol. 4, Yorkshire, UK, pp. 4644, 4822

Spector, W.S., ed. (1956) Handbook of Toxicology, Vol. 1, Acute Toxicities of Solids, Liquids and Gases to Laboratory Animals, Philadelphia, W.B. Saunders Co., p. 232

Srivastava, S.P. & Dua, V.K. (1975) TLC [thin-layer chromatography] separation of closely related amines. Fresenius' Z. analyt. Chem., 276, 382

Stevenson, E.S., Dobriner, K. & Rhoads, C.P. (1942) The metabolism of dimethylaminoazobenzene (butter yellow) in rats. Cancer Res., 2, 160-167

Tainter, M.L. & Hanzlik, P.J. (1924) The mechanism of edema production by paraphenylenediamine. J. Pharmacol. exp. Ther., 24, 179-211

Tainter, M.L., James, M. & Vandeventer, W. (1929) Comparative edemic actions of *ortho*-, *meta*- and *para*- phenylenediamines in different species. Arch. int. Pharmacodyn., 36, 152-162

Thirtle, J.R. (1968) Phenylenediamines and toluenediamines. In: Kirk, R.E. & Othmer, D.F., eds, Encyclopedia of Chemical Technology, 2nd ed., Vol. 15, New York, John Wiley and Sons, pp. 216-224

US International Trade Commission (1976) Imports of Benzenoid Chemicals and Products, 1974, USITC Publication 762, Washington DC, US Government Printing Office, p. 25

US International Trade Commission (1977) Synthetic Organic Chemicals, US Production and Sales, 1975, USITC Publication 804, Washington DC, US Government Printing Office, p. 40

US Occupational Safety & Health Administration (1976) Air contaminants. US Code of Federal Regulations, Title 29, part 1910.1000, pp. 581-582, 585

US Tariff Commission (1919) Report on Dyes and Related Coal-Tar Chemicals, 1918, revised ed., Washington DC, US Government Printing Office

US Tariff Commission (1973a) Synthetic Organic Chemicals, US Production and Sales, 1971, TC Publication 614, Washington DC, US Government Printing Office, p. 47

US Tariff Commission (1973b) Imports of Benzenoid Chemicals and Products, 1972, TC Publication 601, Washington DC, US Government Printing Office, p. 21

US Tariff Commission (1974) Imports of Benzenoid Chemicals and Products, 1973, TC Publication 688, Washington DC, US Government Printing Office, p. 21

Wall, F.E. (1972) Bleaches, hair colorings, and dye removers. In: Balsam, M.S. & Sagarin, E., eds, Cosmetics Science and Technology, 2nd ed., Vol. 2, New York, Wiley-Interscience, pp. 308, 313, 317

Walle, T. (1968) Quantitative gas-chromatographic determination of primary amines in submicrogram quantities after condensation with 2,5-hexanedione. Acta pharm. suecica, 5, 353-366

Windholz, M., ed. (1976) The Merck Index, 9th ed., Rahway, NJ, Merck & Co., p. 948

Zelazna, K. & Legatowa, B. (1971) Identification of basic dyes in emulsified hair dyes by thin layer chromatography. Rocz. Panstw. Zakl. Hig., 22, 427-430

COLOURING AGENTS

ACRIDINE ORANGE

1. Chemical and Physical Data

1.1 Synonyms and trade names

Colour Index No.: 46005:1

Colour Index Name: Solvent Orange 15

Chem. Abstr. Services Reg. No.: 494-38-2

Chem. Abstr. Name: *N,N,N',N'*-Tetramethyl-3,6-acridinediamine

3,6-Bis(dimethylamino)acridine; 3,6-bis-dimethylaminoacridine; C.I. 46005B; 3,6-di(dimethylamino)acridine

Acridine Orange base; Acridine Orange NO; Brilliant Acridine Orange E; Euchrysine; Rhoduline Orange N; Rhoduline Orange NO; Waxoline Orange A

1.2 Chemical formula and molecular weight

$(CH_3)_2N$ … N … $N(CH_3)_2$

$C_{17}H_{19}N_3$ Mol. wt: 265.3

1.3 Chemical and physical properties of the pure substance

(a) Description: Orange-brown needles (Van Duuren *et al.*, 1969)

(b) Melting-point: 181-182°C (Van Duuren *et al.*, 1969)

(c) Spectroscopy data: λ_{max} 490 nm (E_1^1 = 1800); peak fluorescence at 530 nm (Porro & Morse, 1965); the nuclear magnetic resonance spectrum has been tabulated by Kokko & Goldstein (1963).

(d) Solubility: Soluble in ethanol, acetone (0.25%) and benzene; insoluble in mineral oil; soluble in stearic acid (25%) (The Society of Dyers and Colourists, 1971a). The zinc double chloride salt is soluble in water and ethanol (The Society of Dyers and Colourists, 1971b).

1.4 Technical products and impurities

No data were available to the Working Group.

2. Production, Use, Occurrence and Analysis

For background information on this section, see preamble, p. 15.

2.1 Production and use

(a) Production

Acridine orange was described in 1889 by Bender and can be prepared by nitrating 4,4'-methylenebis(*N,N*-dimethylaniline) to 4,4'-methylenebis-(*N,N*-dimethyl-3-nitroaniline), followed by reduction, cyclization and oxidation (The Society of Dyers and Colourists, 1971b). This method has been used commercially in Japan.

The commercial production of acridine orange in the US was first reported in 1925 (US Tariff Commission, 1926); however, no US manufacturers have recently reported commercial production of this chemical to the US International Trade Commission. Acridine orange is presumably produced during the commercial manufacture of its zinc double chloride salt, C.I. Basic Orange 14. Only one US company reported commercial production of an undisclosed amount (see preamble, p. 15) of this dye in 1973, the latest year in which it was reported (US International Trade Commission, 1975).

In Japan, C.I. Basic Orange 14 was produced commercially from 1953 to 1973; one company produced 1800 kg in 1972 but only 200 kg in 1973. Production ceased after 1973 due to declining demand. There are no imports.

As of 1975, there were no countries in the world in which acridine orange was known to be produced commercially (The Society of Dyers and Colourists, 1975).

(b) Use

Acridine orange has been used to colour fats, oils and waxes a bright orange (The Society of Dyers and Colourists, 1971a). It is presumably used as an unisolated intermediate in the manufacture of its zinc double chloride salt, C.I. Basic Orange 14, which can be used to dye silk, cotton, wool, bast fibres, leather and spirit inks and in the manufacture of pigments (The Society of Dyers and Colourists, 1971c).

2.2 Occurrence

Acridine orange is not known to occur as a natural product. No data on its occurrence in the environment were available to the Working Group.

2.2 Analysis

Cationic dyes, including acridine orange, have been determined (±2-3% accuracy) by titration with a commercial fluorescent brightener. An insoluble, nonfluorescent compound is formed, and the end-point is indicated by the appearance of fluorescence (Schiffner & Borrmeister, 1964). A chemiluminescent method of determining organic compounds, including acridine orange, has a sensitivity of 2 ng (Bowman & Alexander, 1966).

A paper chromatographic method for identifying basic dyes (Tajiri, 1969), a micro, dry-column chromatographic method for separation of mixtures of biological stains (Bauman, 1972), and a gel filtration chromatography method for the analysis and purification of biological stains (Horobin, 1971) can be used for analysis of the zinc double chloride salt of acridine orange.

3. Biological Data Relevant to the Evaluation of Carcinogenic Risk to Man

3.1 Carcinogenicity and related studies in animals

(a) Skin application

Mouse: Acridine orange (0.85 mg in 0.1 ml acetone) was applied to the skin of 20 female 8-week-old ICR/Ha Swiss mice thrice weekly for up to 455 days. Skin application of acridine orange alone resulted in the development in 3/20 mice of tumours involving the liver (1 hepatoma, 1 haemangioma and 1 reticulum-cell sarcoma), suggesting systemic absorption. Two additional groups, each of 20 or 40 mice, received acridine orange (single dose of 0.85 mg in 0.1 ml acetone) as initiator, followed 14 days later by thrice weekly applications of 0.1 ml acetone or 25 µg phorbol myristyl acetate in 0.1 ml acetone. Single applications of acridine orange neither initiated nor induced tumours of the skin. In order to investigate the promoting properties of acridine orange, a further group of 20 mice received a single dose of 150 µg 7,12-dimethylbenz[*a*])anthracene (DMBA) in 0.1 ml acetone applied to the skin as initiator, followed 14 days later by the acridine orange applications started at thrice weekly doses of 0.85 mg in 0.1 ml acetone. The incidence of skin tumours was markedly increased when acridine orange was applied repeatedly after a single dose of DMBA (6 papillomas and 6 carcinomas of the skin), the first papilloma being observed after 322 days; application of DMBA followed by thrice weekly doses of 0.1 ml acetone resulted in 3/20 mice with papillomas. No skin tumours occurred in 40 acetone-treated controls (Van Duuren *et al.*, 1969).

(b) Subcutaneous and/or intramuscular administration

Mouse: A group of 30 female 8-week-old ICR/Ha mice received weekly s.c. injections of acridine orange (0.26 mg in 0.1 ml tricaprylin) in the left axillary area for 442 days. A group of 100 control mice were left untreated, and a group of 30 mice received tricaprylin only. Of the mice given acridine orange, 1 had a fibrosarcoma and 1 a lymphoma at the site of injection. No local tumours were observed in the control groups (Van Duuren *et al.*, 1969).

Rat: A group of 20 female 6-week-old Sprague-Dawley rats were given weekly s.c. injections of acridine orange (0.5 ml in 0.1 ml tricaprylin) in the left axillary area for 550 days. A group of 20 rats were given tricaprylin alone, and a group of 30 received no treatment. Of the rats given acridine orange, 1 had a reticulum-cell sarcoma at the site of injection. No local tumours were observed in the two control groups (Van Duuren *et al.*, 1969).

(c) Other experimental systems

Mouse: A group of 27 10-14-week old Swiss/NIH mice of both sexes received single i.p. injections of 45 mg/kg bw acridine orange dissolved in water. A 5% solution of croton oil in liquid paraffin was applied with a glass rod to the skin 4 days after the i.p. injection of acridine orange and then twice weekly for 20 weeks, at which time the surviving animals were killed. Single i.p. injections of urethane (20 mg/mouse) were given to 25 mice as a positive control, and a group of 41 mice received the croton oil treatment only. Fifteen mice given acridine orange, 16 mice given urethane and 39 mice that did not receive a chemical were still alive at the termination of the experiment. Skin papillomas developed only in 6 mice given urethane, and not in mice in the two other groups. Lung adenomas were observed in 3/39 control mice, in 3/15 mice given acridine orange and in 14/16 urethane-treated mice (Trainin *et al.*, 1964).

3.2 Other relevant biological data

(a) Experimental systems

The s.c. LD_{50} of acridine orange in mice is 250 mg/kg bw (Rubbo, 1947).

This chemical is a powerful inhibitor of DNA, RNA and protein synthesis, forming non-covalent, tightly-bound complexes with DNA by intercalation of the bases in the double-stranded DNA molecule. Acridines are also light-sensitive compounds and form radicals upon exposure to visible light (Lerman, 1963; Zelenin & Liapunova, 1964).

No data on its embryotoxicity or teratogenicity were available to the Working Group.

Acridine orange induced frameshift mutations in replicating bacteriophage T4 (Orgel & Brenner, 1961) and reverse mutations in *Escherichia coli* strain S (an isolate obtained from *E. coli* strain B) (Zampieri & Greenberg, 1965) in the dark. Increased mutation rates were observed when the microorganism was irradiated with visible light while exposed to the chemical (Nakai & Saeki, 1964). Acridine orange inhibits the growth of an *E. coli pol* A^- mutant, compared with that of *pol* A^+ (Rosenkranz & Carr, 1971).

It induced reverse mutations in *Salmonella typhimurium* strains TA 100, TA 98 and TA 1537 in the presence of a rat liver postmitochondrial supernatant fraction from animals pretreated with polychlorinated biphenyl (Aroclor 1254) (McCann *et al.*, 1975). Acridine orange also induces mitotic gene conversion in *Saccharomyces cerevisiae* (Davies *et al.*, 1975; Fahrig, 1970).

(b) Man

No data were available to the Working Group.

3.3 Case reports and epidemiological studies

No data were available to the Working Group.

4. Comments on Data Reported and Evaluation

4.1 Animal data

Acridine orange has been tested in mice by skin application, in mice and rats by repeated subcutaneous injection and in mice by single intraperitoneal injection followed by skin application of croton oil. The available data are insufficient for an evaluation of the carcinogenicity of this compound to be made.

4.2 Human data

No case reports or epidemiological studies were available to the Working Group.

5. References

Bauman, A.J. (1972) Method and apparatus for micro dry column chromatography. US Patent 3,692,669, September 19 (to California Institute of Technology)

Bowman, R.L. & Alexander, N. (1966) Ozone-induced chemiluminescence of organic compounds. Science, 154, 1454-1456

Davies, P.J., Evans, W.E. & Parry, J.M. (1975) Mitotic recombination induced by chemical and physical agents in the yeast *Saccharomyces cerevisiae*. Mutation Res., 29, 301-314

Fahrig, R. (1970) Acridine-induced mitotic gene conversion (paramutation) in *Saccharomyces cerevisiae*: the effect of two different modes of binding to DNA. Mutation Res., 10, 509-514

Horobin, R.W. (1971) Analysis and purification of biological stains by gel filtration. Stain Technol., 46, 297-304

Kokko, J.P. & Goldstein, J.H. (1963) The nuclear magnetic resonance spectra of acridine and some of its derivatives. Spectrochim. acta, 19, 1119-1125

Lerman, L.S. (1963) The structure of the DNA-acridine complex. Proc. nat. Acad. Sci. (Wash.), 49, 94-102

McCann, J., Choi, E., Yamasaki, E. & Ames, B.N. (1975) Detection of carcinogens as mutagens in the *Salmonella*/microsome test: assay of 300 chemicals. Proc. nat. Acad. Sci. (Wash.), 72, 5135-5139

Nakai, S. & Saeki, T. (1964) Induction of mutation by photodynamic action in *Escherichia coli*. Genet. Res. (Camb.), 5, 158-161

Orgel, A. & Brenner, S. (1961) Mutagenesis of bacteriophage T_4 by acridines. J. molec. Biol., 3, 762-768

Porro, T.J. & Morse, H.T. (1965) Fluorescence and absorption spectra of biological dyes (II). Stain Technol., 40, 173-176

Rosenkranz, H.S. & Carr, H.S. (1971) Possible hazard in use of gentian violet. Brit. med. J., iii, 702-703

Rubbo, S.D. (1947) The influence of chemical constitution on toxicity. I. A general survey of the acridine series. Brit. J. exp. Path., 28, 1-11

Schiffner, R. & Borrmeister, B. (1964) Determination of cationic dyes. Faserforsch. Textiltech., 15, 211-214

The Society of Dyers and Colourists (1971a) Colour Index, 3rd ed., Vol. 3, Yorkshire, UK, pp. 3582-3583

The Society of Dyers and Colourists (1971b) Colour Index, 3rd ed., Vol. 4, Yorkshire, UK, p. 4431

The Society of Dyers and Colourists (1971c) Colour Index, 3rd ed., Vol. 1, Yorkshire, UK, p. 1625

The Society of Dyers and Colourists (1975) Colour Index, revised 3rd ed., Vol. 5, 1st rev., Yorkshire, UK, p. 5271

Tajiri, H. (1969) Paper chromatography of dyes. Senshoku Kogyo, 17, 513-519

Trainin, N., Kaye, A.M. & Berenblum, I. (1964) Influence of mutagens on the initiation of skin carcinogenesis. Biochem. Pharmacol., 13, 263-267

US International Trade Commission (1975) Synthetic Organic Chemicals, US Production and Sales, 1973, ITC Publication 728, Washington DC, US Government Printing Office, p. 68

US Tariff Commission (1926) Census of Dyes and other Synthetic Organic Chemicals, 1925, Tariff Information Series No. 34, Washington DC, US Government Printing Office, p. 64

Van Duuren, B.L., Sivak, A., Katz, C. & Melchionne, S. (1969) Tumorigenicity of acridine orange. Brit. J. Cancer, 23, 587-590

Zampieri, A. & Greenberg, J. (1965) Mutagenesis by acridine orange and proflavine in *Escherichia coli* strain S. Mutation Res., 2, 552-556

Zelenin, A.V. & Liapunova, E.A. (1964) Inhibition of protein synthesis by acridine orange. Nature (Lond.), 204, 45-46

BENZYL VIOLET 4B

1. Chemical and Physical Data

1.1 Synonyms and trade names

Colour Index No.: 42640

Colour Index Names: Acid Violet 49; Food Violet 2

Chem. Abstr. Services Reg. No.: 1694-09-3

Chem. Abstr. Name: *N*-(4-{[4-(Dimethylamino)phenyl][4-{ethyl[(3-sulfophenyl)methyl]amino}phenyl]methylene}-2,5-cyclohexadien-1-ylidene)-*N*-ethyl-3-sulfobenzenemethanaminium hydroxide inner salt, sodium salt

Acid violet 6B; C.I. Acid Violet 49, sodium salt; D and C Violet No. 1; FD and C Violet No. 1

Acid Fast Violet 5BN; Acid Violet; Acid Violet 4BNS; Acid Violet 5B; Acid Violet 6B; Acid Violet S; Acilan Violet S4BN; A. F. Violet No. 1; Aizen Acid Violet 5BH; Aizen Food Violet No. 1; Benzyl Violet; Benzyl Violet 3B; Calcocid Violet 4BNS; Cogilor Violet 411.12; Coomassie Violet; Dispersed Violet 12197; Eriosin Violet 3B; Fast Acid Violet 5BN; Formyl Violet S4BN; Hidacid Wool Violet 5B; Intracid Violet 4BNS; Kiton Violet 4BNS; Pergacid Violet 2B; Pergacid Violet 3B; Polaxal Violet 6B; Solar Violet 5BN; Tertracid Brilliant Violet 6B; 11386 Violet; Violet Acid 5B; Violet 6B; Violet No. 1; Wool Violet 4BN; Wool Violet 5BN

1.2 Chemical formula and molecular weight

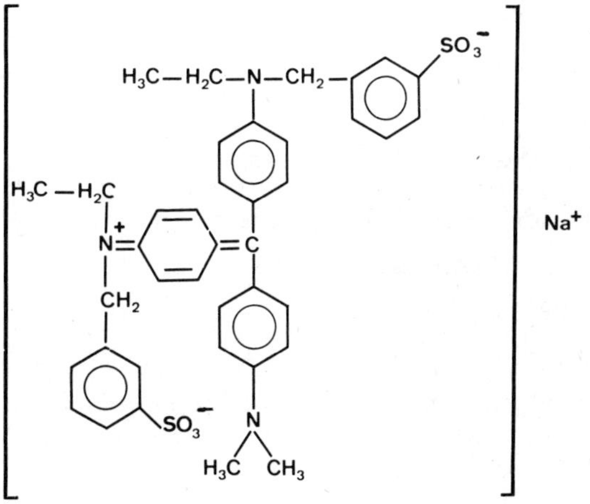

$C_{39}H_{40}N_3O_6S_2.Na$ Mol. wt: 733.9

1.3 Chemical and physical properties of the pure substance

(a) Description: Fine powder (Grasso *et al.*, 1974)

(b) Solubility: Soluble in water and ethanol; insoluble in vegetable oils (The Society of Dyers and Colourists, 1971a)

1.4 Technical products and impurities

The grade of benzyl violet 4B used as a food colour additive in Europe has the following specifications: total dye content, 85% min.; sum of volatile matter (at 135°C), chloride and sulphates, 15% max.; subsidiary dyes, 5% max.; intermediates, 0.5% max.; water-insoluble matter, 0.2% max.; ether-extractable matter, 0.2% max.; chromium, 20 mg/kg max.; lead, 10 mg/kg max.; and arsenic, 3 mg/kg max. (WHO, 1966).

2. Production, Use, Occurrence and Analysis

For background information on this section, see preamble, p. 15.

2.1 Production and use

(a) Production

The colouring properties of benzyl violet 4B were described by Schultz & Zierold in 1889 (The Society of Dyers and Colourists, 1971b). It is produced commercially by condensing 4-dimethylaminobenzaldehyde with α-(*N*-ethylanilino)-4-toluenesulphonic acid, followed by oxidation and conversion to the sodium salt (Zuckerman, 1964). It can also be prepared by condensing α-(*N*-ethylanilino)-3-toluenesulphonic acid with formaldehyde, followed by oxidation in the presence of *N*,*N*-dimethylaniline (The Society of Dyers and Colourists, 1971b).

Benzyl violet 4B has been produced commercially in the US for over 50 years (US Tariff Commission, 1927). In 1974, four US companies reported production of over 55 thousand kg (US International Trade Commission, 1976). US imports through principal US customs districts in 1972, the latest year in which it was reported, amounted to 800 kg (US Tariff Commission, 1973).

It is believed that 10-100 thousand kg/year benzyl violet 4B are produced in Switzerland.

It has been produced commercially in Japan for more than 30 years. Three Japanese companies currently produce it, and production amounted to 20.5 thousand kg in 1975. Japan exported 4.9 thousand kg in 1975; there are no imports.

(b) Use

Benzyl violet 4B is used to dye wool, silk, nylon, leather and anodized aluminium. It is also used as a biological stain, as a wood stain, in inks and in colouring paper (The Society of Dyers and Colourists, 1971b,c).

Benzyl violet 4B was used in the US as a colour additive for food, drugs and cosmetics until 10 April 1973, at which time the US Food and Drug Administration terminated its provisional listing as an additive and cancelled all certificates issued for it (US Food and Drug Administration, 1976). During the first nine months of 1967, 94.4% of total US sales of benzyl violet 4B was used in food, 4.4% in pharmaceutical products and 1.2% in cosmetics. Food uses were in meat inks, sweets and confections, pet foods, beverages, bakery goods, ice-cream, sherbet, dairy products, snack foods (Anon., 1968), gelatine desserts and puddings. Pharmaceutical products in which it was used were aqueous drug solutions, tablets and capsules; cosmetics uses were in lipsticks, rouges, hair rinses (Zuckerman, 1964) and temporary colour shampoos (Markland, 1966).

Benzyl violet 4B does not appear on any lists of colourants permitted for use in food in the European Economic Communities (The Society of Dyers and Colourists, 1975); however, some member states may its use as a food colour until 1978. It may be used in cosmetics, including those which may be in contact with mucous membranes (EEC, 1976). In Switzerland, benzyl violet 4B is used as a dyestuff; and in Japan, it is used to dye leather, paper and inks.

Benzyl violet 4B was considered by a FAO/WHO Working Group on Food Additives to be unsuitable for use in foods (WHO, 1977).

2.2 Occurrence

Benzyl violet 4B is not known to occur as a natural product. No data on its occurrence in the environment were available to the Working Group.

2.3 Analysis

A review of chromatographic and spectrophotometric methods for the identification of food dyes, including benzyl violet 4B, was published by the FAO (1963). The dye content of a sample of this colour may be determined by titration with titanous chloride (WHO, 1966).

Prior to the cancellation of the use of benzyl violet 4B in food, drugs and cosmetics, the Association of Official Analytical Chemists, published recommended methods for use in batch certifications to meet specifications defined by the US Food and Drug Administration (Horwitz, 1970).

Nuclear magnetic resonance spectroscopy has been evaluated as a means of identifying and differentiating food colours (Marmion, 1974). In one method for the determination of dyes, including benzyl violet 4B, in strawberry jam, they are first adsorbed from the sample onto a liquid anion exchanger, recovered and determined colorimetrically (Tonogai, 1973).

Paper electrophoresis has been used to separate dyes, including benzyl violet 4B, used in medicines (Dobrecky & De Carnevale Bonino, 1968). Paper chromatography has been used to separate these colours from food, drugs or cosmetics, using one solvent system (Dobrecky & De Carnevale Bonino, 1972), twelve solvent systems (Pearson, 1973a) or sixteen solvent systems (Dobrecky & De Carnevale Bonino, 1967). A photoelectric densitometer has been used to quantitate food dyes, including benzyl violet 4B, separated by paper chromatography (Sasaki & Iwata, 1972).

Thin-layer chromatography has been used to separate and identify colours, including benzyl violet 4B, used in food, pharmaceuticals or cosmetics on silica gel, using one or more solvent systems (Martelli & Proserpio, 1974; Pearson, 1973b), and on microcrystalline cellulose layers, using four solvent systems (Penner, 1968). Diethylaminoethyl-Sephadex may be used to isolate the dyes from food samples before thin-layer chromatographic separation (Takeshita *et al.*, 1972).

3. Biological Data Relevant to the Evaluation of Carcinogenic Risk to Man

3.1 Carcinogenicity and related studies in animals

(a) Oral administration

Mouse: Groups of 48 male and 50 female SPF-derived ASH-CS1 mice were fed diets containing 0 (control), 70, 700 or 3500 mg/kg of diet benzyl violet 4B (90% pure) for 80 weeks, at which time more than 60% of the animals in each group were still alive. The daily intake in the higher dose group was approximately 500 mg/kg bw. In the treated female groups, 1/43, 5/42 and 4/48 animals at the 3 dose levels, respectively, had thymic lymphosarcomas, compared with 1/42 mice of the untreated female controls; however, the increased incidence is not significant (P>0.05) (Grasso *et al.*, 1974).

Rat: Four groups of 15 male and 15 female 4-6-week old Wistar rats were fed diets containing 0 (control), 0.03, 0.3 or 3% benzyl violet 4B (food grade) for 75 weeks, at which time the experiment was terminated. The numbers of survivors at this time were 23 (control), 19, 26 and 19 rats in the four groups, respectively; tissues from 5 male and 5 female survivors in each of the groups were examined histologically. Squamous-cell carcinomas of the skin were observed in 1 female fed the lowest dose and in 1 male and 2 females fed the highest dose; of the latter, one had a liposarcoma and the other a lung carcinoma. One control male had a fibrosarcoma (Mannell *et al.*, 1962) [The Working Group noted the inadequate histological examination of tissues in this experiment].

Groups of 20 male or 20 female Sprague-Dawley rats, approximately 7-weeks old, were fed a diet containing 5% benzyl violet 4B (89.1% pure). The experiment was terminated at 364 days. Nine of 18 treated female rats that survived more than 70 days developed tumours: 5 mammary carcinomas were detected between 156 and 196 days and 4 ear-duct carcinomas between 138 and 345 days. In 10 untreated control females, 1 mammary fibroadenoma was seen at day 338; no tumours were observed in 15 treated and 10 control males that survived more than 70 days. The average survival times were 187 (males) and 210 (females) days in the treated groups and 228 (males) and 315 (females) days in control groups (Uematsu & Miyaji, 1973).

A group of 35 female 6-week-old Sprague-Dawley rats were fed a diet containing 1% benzyl violet 4B (91.9% pure) for 1 week, then 3% for 1 week and thereafter a diet containing 5% of the dye for 12 months. During this period, 11 rats developed mammary carcinomas, 4 had squamous-cell carcinomas, and 7 animals had both types of tumour. The majority of the squamous-cell carcinomas developed 'at the ear-duct but also appeared at the buccal and axillary regions of the skin'. The times of appearance of the first mammary carcinoma and the first squamous-cell carcinoma were 13 and 27 weeks, respectively. No skin tumours were observed in 35 controls (Ikeda *et al.*, 1974).

(b) Subcutaneous and/or intramuscular administration

Rat: In an abstract, it was reported that 18 young Osborne-Mendel rats of both sexes received weekly s.c. injections of 20 mg benzyl violet 4B (food grade) as a 2% aqueous solution for 2 years. Fibrosarcomas at the site of injection were seen in 14 rats. Saline- and glycerin-injected control animals had no tumours at the site of injection (Nelson & Davidow, 1957).

(c) Other experimental systems

Rat: A group of 20 male and 20 female 3-week-old Wistar rats had continual skin/fur contact with benzyl violet 4B (food grade) for 100 weeks: the tray of the cage was sprinkled with 3 g of the dye, and the rats were placed directly on the tray; the tray was cleaned twice weekly, and the same amount of dye was added after each cleaning. Malignant skin tumours were not observed in treated or control rats. The increased incidence of benign mammary tumours in the test group (8/14) compared with that in controls (1/14) was statistically significant ($P<0.05$) (Mannell *et al.*, 1964).

3.2 Other relevant biological data

The toxicology of many colouring materials, including benzyl violet 4B, has been reviewed by Drake (1975).

(a) Experimental systems

The acute oral LD_{50} of benzyl violet 4B in rats is greater than 2.0 g/kg bw (Lu & Lavalle, 1964).

Studies in rats and dogs with oral administration of benzyl violet 4B indicate that only small amounts of the dye (<5%) are absorbed and excreted, mainly in the faeces (Gaunt *et al.*, 1974; Hess & Fitzhugh, 1954, 1955).

Benzyl violet 4B exhibits high plasma protein binding (Gangolli *et al.*, 1972).

No data on the teratogenicity, embryotoxicity or mutagenicity of benzyl violet 4B were available to the Working Group.

(b) Man

No data were available to the Working Group.

3.3 Case reports and epidemiological studies

No data were available to the Working Group.

4. Comments on Data Reported and Evaluation[1]

4.1 Animal data

Benzyl violet 4B is carcinogenic in rats following its oral or subcutaneous administration: it produced mammary carcinomas and squamous-cell carcinomas of the skin after its oral administration to female rats and local fibrosarcomas following its subcutaneous injection in male and female rats. It also increased the incidence of benign mammary tumours in female rats following ambient exposure.

4.2 Human data

No case reports or epidemiological studies were available to the Working Group.

---

[1]See also the section, 'Animal Data in Relation to the Evaluation of Risk to Man', in the introduction to this volume, p. 13.

5. References

Anon. (1968) Guidelines for good manufacturing practice: use of certified FD&C colors in food. Food Techn., 22, 14-17

Dobrecky, J. & De Carnevale Bonino, R.C.D. (1967) Paper chromatography of dyes, using various solvents. SAFYBI, 7, 165-170

Dobrecky, J. & De Carnevale Bonino, R.C.D. (1968) Paper electrophoresis of some coloring agents used in medicines. SAFYBI, 8, 67-70

Dobrecky, J. & De Carnevale Bonino, R.C.D. (1972) Paper chromatography of coloring agents used in drugs and cosmetics. II. Rev. Farm. (Buenos Aires), 114, 21-22

Drake, J.J.-P. (1975) Food colours - harmless aesthetics or epicurean luxuries? Toxicology, 5, 3-42

EEC (1976) Législation. Journal Officiel des Communautés Européennes, No. L 262, p. 192

FAO (1963) Specifications for Identity and Purity of Food Additives, Vol. II, Food Colours, Rome, pp. 119-121, 128-129

Gangolli, S.D., Grasso, P., Golberg, L. & Hooson, J. (1972) Protein binding by food colourings in relation to the production of subcutaneous sarcomas. Fd Cosmet. Toxicol., 10, 449-462

Gaunt, I.F., Hardy, J., Kiss, I.S. & Gangolli, S.D. (1974) Short-term toxicity of violet 6B (FD & C Violet No. 1) in the rat. Fd Cosmet. Toxicol., 12, 11-19

Grasso, P., Hardy, J., Gaunt, I.F., Mason, P.L. & Lloyd, A.G. (1976) Long-term toxicity of violet 6B (FD & C Violet No. 1) in mice. Fd Cosmet. Toxicol., 12, 21-31

Hess, S.M. & Fitzhugh, O.G. (1954) Metabolism of coal-tar colors. II. Bile studies (Abstract 1201). Fed. Proc., 13, 365

Hess, S.M. & Fitzhugh, O.G. (1955) Absorption and excretion of certain triphenylmethane colors in rats and dogs. J. Pharmacol. exp. Ther., 114, 38-42

Horwitz, W., ed. (1970) Official Methods of Analysis of the Association of Official Analytical Chemists, 11th ed., Washington DC, Association of Official Analytical Chemicals, pp. 582-583, 588-589, 591-592, 600-601, 603

Ikeda, Y., Horiuchi, S., Imoto, A., Kodama, Y., Aida, Y. & Kobayashi, K. (1974) Induction of mammary gland and skin tumour in female rats by the feeding of benzyl violet 4B. Toxicology, 2, 275-284

Lu, F.C. & Lavalle, A. (1964) The acute toxicity of some synthetic colours used in drugs and foods. Canad. Pharm. J., 97, 30

Mannell, W.A., Grice, H.C. & Allmark, M.G. (1962) Chronic toxicity studies on food colours. V. Observations on the toxicity of brilliant blue FCF, guinea green B and benzyl violet 4B in rats. J. Pharm. Pharmacol., 14, 378-384

Mannell, W.A., Grice, H.C. & Dupuis, I. (1964) The effect on rats of long term exposure to guinea green B and benzyl violet 4B. Fd Cosmet. Toxicol., 2, 345-347

Markland, W.R. (1966) Hair preparations. In: Kirk, R.E. & Othmer, D.F., eds, Encyclopedia of Chemical Technology, 2nd ed., Vol. 10, New York, John Wiley and Sons, p. 793

Marmion, D.F. (1974) Applications of nuclear magnetic resonance spectroscopy to certifiable food colors. J. Ass. off. analyt. Chem., 57, 495-507

Martelli, A. & Proserpio, G. (1974) Identification of dyes in cosmetics by thin-layer chromatography. Relata Tech. Chim. Biol. Appl., 6, 157-164

Nelson, A.A. & Davidow, B. (1957) Injection site fibrosarcoma production in rats by food colours (Abstract 1571). Fed. Proc., 16, 367

Pearson, D. (1973a) Identification of EEC food colors. J. Ass. Publ. Analyt., 11, 127-134

Pearson, D. (1973b) Identification of UK and EEC food colours using standardized TLC [thin-layer chromatography] plates. J. Ass. Publ. Analyt., 11, 135-138

Penner, M.H. (1968) Thin-layer chromatography of certified coal tar color additives. J. pharm. Sci., 57, 2132-2135

Sasaki, H. & Iwata, T. (1972) Analytical studies of food dyes. IV. Direct densitometry of paper chromatograms of food and other dyes by transparent methods. Shokuhin Eiseigaku Zasshi, 13, 120-126

The Society of Dyers and Colourists (1971a) Colour Index, 3rd ed., Vol. 2, Yorkshire, UK, pp. 2778, 2808

The Society of Dyers and Colourists (1971b) Colour Index, 3rd ed., Vol. 4, Yorkshire, UK, p. 4397

The Society of Dyers and Colourists (1971c) Colour Index, 3rd ed., Vol. 1, Yorkshire, UK, p. 1270

The Society of Dyers and Colourists (1975) Colour Index, revised 3rd ed., Vol. 6, Yorkshire, UK, pp. 6235-6236

Takeshita, R., Yamashita, T. & Itoh, N. (1972) Separation and detection of water-soluble acid dyes on polyamide thin layers. J. Chromat., 73, 173-182

Tonogai, Y. (1973) Analysis of food additives. I. Determination of coal tar dyes in foods with liquid anion exchangers. Eisei Kagaku, 19, 231-235

Uematsu, K. & Miyaji, T. (1973) Induction of tumors in rats by oral administration of technical acid violet 6B. J. nat. Cancer Inst., 51, 1337-1338

US Food and Drug Administration (1976) Food and drugs. US Code of Federal Regulations, Title 21, parts 8.502, 8.510

US International Trade Commission (1976) Synthetic Organic Chemicals, US Production and Sales, 1974, USITC Publication 776, Washington DC, US Government Printing Office, pp. 52, 62, 76

US Tariff Commission (1927) Census of Dyes and Other Synthetic Organic Chemicals, 1926, Tariff Information Series No. 35, Washington DC, US Government Printing Office, p. 63

US Tariff Commission (1973) Imports of Benzenoid Chemicals and Products, 1972, TC Publication 601, Washington DC, US Government Printing Office, p. 38

WHO (1966) Specifications for identity and purity and toxicological evaluations of food colours. WHO/Food Add./66.25, pp. 107-108

WHO (1977) Evaluation of certain food additives. 21st Report of the Joint FAO/WHO Expert Committee on Food Additives. Wld Hlth Org. techn. Rep. Ser. (in press)

Zuckerman, S. (1964) Colors for foods, drugs, and cosmetics. In: Kirk, R.E. & Othmer, D.F., eds, Encyclopedia of Chemical Technology, 2nd ed., Vol. 5, New York, John Wiley and Sons, pp. 865, 867

Some confusion exists in the classification of literature for blue VRS: that concerning the dye patent blue V (C.I. 42051, Acid blue 3) has been classified under 'blue VRS'.

1. Chemical and Physical Data

1.1 Synonyms and trade names

Colour Index No.: 42045

Colour Index Names: Acid Blue 1; Food Blue 3

Chem. Abstr. Services Reg. No.: 129-17-9

Chem. Abstr. Name: *N*-(4-{[4-(Diethylamino)phenyl][2,4-disulfophenyl]methylene}-2,5-cyclohexadien-1-ylidene)-*N*-ethylethanaminium hydroxide inner salt, sodium salt

Acid blue-1; anhydro-4,4'-*bis*(diethylamino)triphenylmethanol-2',4"-disulphonic acid, monosodium salt; C.I. Acid Blue 1, sodium salt; (4-{[α-(*para*-diethylamino)phenyl]-2,4-disulphobenzylidene}-2,5-cyclohexadien-1-ylidene)diethylammonium hydroxide inner salt, sodium salt; sulphan blue

Acid Blue 1; Acid Blue V; Acid Brilliant Blue VF; Acid Brilliant Sky Blue Z; Acid Leather Blue V; Aizen Brilliant Acid Pure Blue VH; Alphazurine 2G; Amacid Blue V; Blue 1084; 1085 Blue; Brilliant Acid Blue A Export; Brilliant Acid Blue V Extra; Brilliant Acid Blue VS; Brilliant Blue GS; Bucacid Patent Blue VF; Carmin Blue VS; Carmine Blue VF; Cosmetic Green Blue R25396; Disulfine blue VN; Disulphine Blue VN 150; Dilsulphine VN: Edicol Supra Blue VR; Erio Brilliant Blue V; Erioglaucine; Erioglaucine supra; Fenazo Blue XF; Fenazo Blue XV; Hexacol Blue VRS; Hidacid Blue V; Intracid Pure Blue V; Kiton Pure Blue V; Kiton Pure Blue V.FQ; Leather Blue G; Lissamine Turquoise VN; Merantine Blue VF; Patent Blue; Patent Blue V; Patent Blue VF; Patent Blue VF-CF; Patent Blue VF Special; Patent Blue VS; Pontacyl Brilliant Blue V;

Sodium Blue VRS; Sodium Patent Blue V; Sulfacid Brilliant Blue 6J; Sumitomo Patent Pure Blue VX; Tertracid Carmine Blue V; Xylene Blue VS

1.2 Chemical formula and molecular weight

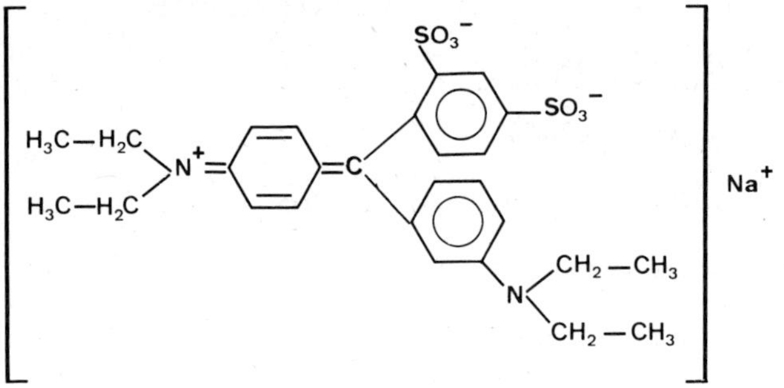

$C_{27}H_{31}N_2O_6S_2.Na$ Mol. wt: 566.7

1.3 Chemical and physical properties of the pure substance

(a) Description: Violet powder (Windholz, 1976)

(b) Solubility: Soluble in water (5%) and in ethanol (Windholz, 1976)

(c) Surface activity: A 2% aqueous solution showed a depression of surface tension of water of 38.5% at 37°C (Gangolli *et al.*, 1967).

1.4 Technical products and impurities

A grade of blue VRS produced in the UK has the following specifications: dye content, 82% min.; volatile matter (at 135°C), 6% max.; water-insoluble matter, 0.2% max.; subsidiary dye content, 5% max.; sodium chloride and sulphate, 10% max.; arsenic, 1 mg/kg; lead, 10 mg/kg; copper, 10 mg/kg; chromium, 20 mg/kg (Grasso & Golberg, 1966).

2. Production, Use, Occurrence and Analysis

For background information on this section, see preamble, p. 15.

2.1 Production and use

(a) Production

The colouring properties of blue VRS were first described in 1902 by Steiner. It can be prepared by condensing 4-formyl-1,3-benzenedisulphonic acid with *N*,*N*-diethylaniline, followed by oxidation and, finally, conversion to the sodium salt (The Society of Dyers and Colourists, 1971a).

The commercial production of blue VRS in the US was first reported in 1946 (US Tariff Commission, 1948). In 1975, only one US company reported an undisclosed amount (see preamble, p. 15) (US International Trade Commission, 1977a). US imports of blue VRS through principal US customs districts were about 2500 kg in 1974 (US International Trade Commission, 1976) and 1700 kg in 1975 (US International Trade Commission, 1977b).

The UK exports less than 10,000 kg blue VRS annually. It is not produced commercially in Japan.

(b) Use

Blue VRS is used to dye wool and silk a greenish blue (The Society of Dyers and Colourists, 1971b). It produces a bright blue hue on leather (The Society of Dyers and Colourists, 1971c) and has also been used for dyeing jute (Gantz *et al*., 1965).

Blue VRS is also used as a stain for paper; to colour phenol-formaldehyde resins, milled soaps, inks and methylated spirit wash; for the mass colouration of casein; and as a biological stain. Heavy metal salts of the free acid are used as pigments (The Society of Dyers and Colourists, 1971b), and the aluminium salt has been used to colour pharmaceuticals (The Society of Dyers and Colourists, 1971a).

Less than 10,000 kg blue VRS are used annually in the UK, 90% of which are found in cosmetic and household products.

Blue VRS was used in the past as a food dye (The Society of Dyers and Colourists, 1971b); however, it does not appear on any current lists of dyes permitted for use in food in the US, Japan or the European Economic Communities (The Society of Dyers and Colourists, 1975). It is provisionally accepted for use in cosmetics in which it does not come into contact with mucous membranes (EEC, 1976).

2.2 Occurrence

Blue VRS is not known to occur as a natural product. No data on its occurrence in the environment were available to the Working Group.

2.3 Analysis

The purity of batches of commercial dyes, including blue VRS, has been evaluated by thin-layer chromatography and nuclear magnetic resonance spectrometry (Hiranaka *et al.*, 1975). Thin-layer chromatography has also been used to separate and identify dyes, including blue VRS, used in foods, using two solvent series on diethylaminoethyl cellulose (Turner & Jones, 1971), three solvents on Sephadex G-25 (Parrish, 1968) or one solvent system on silica gel (Graham & Nya, 1969).

3. Biological Data Relevant to the Evaluation of Carcinogenic Risk to Man

3.1 Carcinogenicity and related studies in animals

(a) Subcutaneous and/or intramuscular administration

Rat: A group of 10 male and 10 female 5-week-old Wistar rats received weekly s.c. injections of 0.5 ml of a 4% solution of blue VRS (purity not given) in isotonic saline for 45 weeks (total dose, 900 mg). The experiment was terminated at 71 weeks, at which time 3 rats were still alive; animals that died were checked for cause of death, and grossly abnormal tissues from surviving rats were examined histologically. Two female rats developed rhabdomyosarcomas in the area of the injection site. In 20 control animals injected with isotonic saline alone, 8 of which were still alive at 71 weeks, no injection-site tumours were seen (Mannell & Grice, 1964).

Shell or Carworth Farm E albino rats of both sexes, weighing 100-150 g, were given twice weekly s.c. injections of either 0.5 ml of a 2% solution (20 rats) or 1 ml of a 1% aqueous solution (65 rats) of blue VRS (82% pure) for 55 or 72 weeks, respectively. The first tumours were detected at 45 and 50 weeks, when 19 and 60 rats were alive, respectively. Six out of 19 and 8/60 rats developed sarcomas at the site of repeated injections. Another 20 rats received twice weekly s.c. injections of 0.5 ml of a 2% aqueous solution of the dye for 65 weeks, but the injections were distributed among four s.c. sites in turn (dorsal to the scapulae and in the lumbar region). At 81 weeks, no sarcomas were observed in these rats at any of the injection sites (Grasso & Golberg, 1966; Grasso *et al.*, 1971).

A group of 66 male and 66 female Ash/CFE rats, weighing about 100 g, were given twice weekly s.c. injections of 0.5 ml of a 2% aqueous solution (buffered to pH 7.0) of blue VRS (82% pure) for different periods (24-117 injections per animal). Ninety-eight rats were still alive when the first tumours appeared at 47 weeks, and 21 developed injection-site sarcomas (Hooson *et al.*, 1973).

A group of 20 male and 20 female 10-week-old Wistar rats received weekly injections of 0.4 ml of a 4% solution of blue VRS (food grade) in 0.9% saline into the posterior thigh muscles of the right leg. Occasionally, if a rat became lame or disabled from the injections, they were given every two weeks. Five out of 19 males and 9/18 females developed rhabdomyosarcomas at the injection site. The first tumours appeared in males after 25 weeks and in females after 35 weeks (median latency periods were 30 and 50 weeks; average doses administered were 307 and 414 mg per animal, respectively). Metastases occurred in lymph nodes, lungs and liver. No tumours developed in control animals at the site of injections of 0.9% saline (Grice & Mannell, 1966; Grice *et al.*, 1966).

3.2 Other relevant biological data

The toxicological aspects of colouring materials, including blue VRS (Drake, 1975), and the toxicology and biochemical aspects of synthetic organic food colours, including blue VRS, have been reviewed (Dacre, 1969).

(a) Experimental systems

The oral LD_{50} of blue VRS in mice is more than 5 g/kg bw, and that for rats is more than 10 g/kg bw. The i.p. LD_{50} in mice is approximately 3 g/kg bw (Hall *et al.*, 1967).

Feeding of 1.5% and 3% blue VRS to rats for 90 days caused retardation of growth in males and occasional fatty change in the livers of females (Hall *et al.*, 1967).

This dye exhibits high plasma protein binding (Gangolli *et al.*, 1967, 1972).

No data on the embryotoxicity, teratogenicity or mutagenicity of blue VRS were available to the Working Group.

(b) Man

No data were available to the Working Group.

3.3 Case reports or epidemiological studies

No data were available to the Working Group.

4. Comments on Data Reported and Evaluation[1]

4.1 Animal data

Blue VRS is carcinogenic in rats following its subcutaneous or intramuscular injection; it produced sarcomas at the site of repeated injections.

4.2 Human data

No case reports or epidemiological studies were available to the Working Group.

---

[1]See also the section, 'Animal Data in Relation to the Evaluation of Risk to Man' in the introduction to this volume, p. 13.

5. References

Dacre, J.C. (1969) Synthetic organic food colours: toxicology and biochemical aspects. Fd Technol. N.Z., 4, 169-177

Drake, J.J.-P. (1975) Food colours - harmless aesthetics or epicurean luxuries? Toxicology, 5, 3-42

EEC (1976) Législation. Journal Officiel des Communautés Européennes, No. L 262, p. 199

Gangolli, S.D., Grasso, P. & Golberg, L. (1967) Physical factors determining the early local tissue reactions produced by food colourings and other compounds injected subcutaneously. Fd Cosmet. Toxicol., 5, 601-621

Gangolli, S.D., Grasso, P., Golberg, L. & Hooson, J. (1972) Protein binding by food colourings in relation to the production of subcutaneous sarcoma. Fd Cosmet. Toxicol., 10, 449-462

Gantz, G.M., Ellis, J.R., Duncan, J.J., Freas, J.G., Hall, R.C., Landberge, M.J., Moss, J., O'Neal, F., Pardey, W.N. & Vogel, T. (1965) Dyes: application and evaluation. In: Kirk, R.E. & Othmer, D.F., eds, Encyclopedia of Chemical Technology, 2nd ed., Vol. 7, New York, John Wiley and Sons, p. 525

Graham, R.J.T. & Nya, A.E. (1969) Thin-layer chromatography of synthetic food dyes. In: Proceedings of the 5th International Symposium on Chromatography and Electrophoresis, 1968, Ann Arbor, Michigan, Humphrey Scientific Publishers, Inc., pp. 486-490

Grasso, P. & Golberg, L. (1966) Early changes at the site of repeated subcutaneous injection of food colourings. Fd Cosmet. Toxicol., 4, 269-282

Grasso, P., Gangolli, S.D., Golberg, L. & Hooson, J. (1971) Physicochemical and other factors determining local sarcoma production by food additives. Fd Cosmet. Toxicol., 9, 463-478

Grice, H.C. & Mannell, W.A. (1966) Rhabdomyosarcomas induced in rats by intramuscular injections of blue VRS. J. nat. Cancer Inst., 37, 845-857

Grice, H.C., Dupuis, I., Dennery, M. & Mannell, W.A. (1966) Blue VRS-induced rhabdomyosarcomas (Abstract No. 25). Toxicol. appl. Pharmacol., 8, 342-343

Hall, D.E., Gaunt, I.F., Farmer, M. & Grasso, P. (1967) Acute (mouse and rat) and short-term (rat) toxicity studies on blue VRS. Fd Cosmet. Toxicol., 5, 165-170

Hiranaka, P.K., Kleinman, L.M., Sokoloski, E.A. & Fales, H.M. (1975) Chemical structure and purity of dyes used in lymphograms. Amer. J. Hosp. Pharm., 32, 928-930

Hooson, J., Grasso, P. & Gangolli, S.D. (1973) Injection site tumours and preceding pathological changes in rats treated subcutaneously with surfactants and carcinogens. Brit. J. Cancer, 27, 230-244

Mannell, W.A. & Grice, H.C. (1964) Chronic toxicity of brilliant blue FCF, blue VRS, and green S in rats. J. pharm. Pharmacol., 16, 56-59

Parrish, J.R. (1968) Chromatography of food dyes on Sephadex. J. Chromat., 33, 542-543

The Society of Dyers and Colourists (1971a) Colour Index, 3rd ed., Vol. 4, Yorkshire, UK, p. 4382

The Society of Dyers and Colourists (1971b) Colour Index, 3rd ed., Vol. 1, Yorkshire, UK, p. 1295

The Society of Dyers and Colourists (1971c) Colour Index, 3rd ed., Vol. 2, Yorkshire, UK, p. 2809

The Society of Dyers and Colourists (1975) Colour Index, revised 3rd ed., Vol. 6, Yorkshire, UK, pp. 6232-6236

Turner, T.D. & Jones, B.E. (1971) The identification of blue triphenylmethane food dyes by thin-layer chromatography. J. pharm. Pharmacol., 23, 806-807

US International Trade Commission (1976) Imports of Benzenoid Chemicals and Products, 1974, USITC Publication 762, Washington DC, US Government Printing Office, p. 43

US International Trade Commission (1977a) Synthetic Organic Chemicals, US Production and Sales, 1975, USITC Publication 804, Washington DC, US Government Printing Office, p. 57

US International Trade Commission (1977b) Imports of Benzenoid Chemicals and Products, 1975, USITC Publication 806, Washington DC, US Government Printing Office, p. 40

US Tariff Commission (1948) Synthetic Organic Chemicals, US Production and Sales, 1946, Report No. 159, Second Series, Washington DC, US Government Printing Office, p. 80

Windholz, M., ed. (1976) The Merck Index, 9th ed., Rahway, NJ, Merck & Co., p. 1165

BRILLIANT BLUE FCF
DIAMMONIUM AND DISODIUM SALTS

The common name, brilliant blue FCF, has been applied to both the diammonium and disodium salts. Some confusion exists in the classification of the literature for these compounds: although most is classified under the diammonium salt, the disodium salt is the one of primary commercial importance. The disodium salt is used in foods, and the diammonium salt appears to have had limited usage only in drugs and cosmetics.

1. Chemical and Physical Data

Brilliant blue FCF, diammonium salt

1.1 Synonyms and trade names

Colour Index No.: 42090

Colour Index Names: Acid Blue 9; Food Blue 2

Chem. Abstr. Services Reg. No.: 2650-18-2

Chem. Abstr. Name: *N*-Ethyl-*N*-(4-[(4-{ethyl[(3-sulfophenyl)methyl]-amino}phenyl)(2-sulfophenyl)methylene]-2,5-cyclohexadien-1-ylidene)-3-sulfobenzenemethanaminium hydroxide inner salt, diammonium salt

Acid blue 9; C.I. Acid Blue 9, diammonium salt; D and C Blue No. 4; ethyl({4-[*para*-ethyl(*meta*-sulphobenzyl)amino]-α-(*ortho*-sulphophenyl)benzylidene}-2,5-cyclohexadien-1-ylidene)-*meta*-sulphobenzylammonium hydroxide inner salt, diammonium salt

1.2 Chemical formula and molecular weight

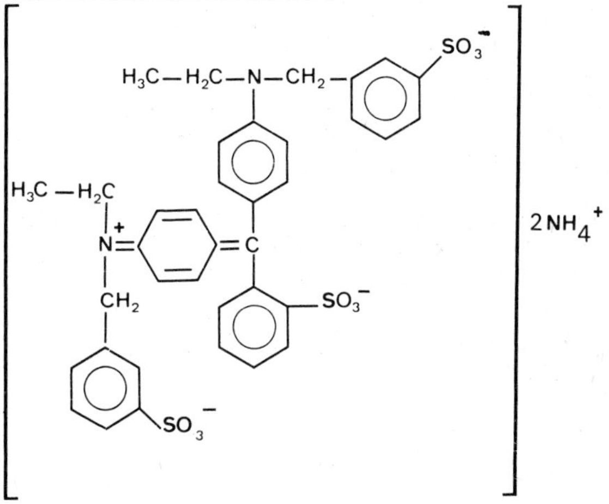

$C_{37}H_{34}N_2O_9S_3 \cdot 2NH_4$ Mol. wt: 782.8

1.3 Chemical and physical properties of the pure substance

Solubility: Soluble in water (5% at 20°C and 98°C) and in ethanol (The Society of Dyers and Colourists, 1971a,b)

1.4 Technical products and impurities

Drug and cosmetic grades of brilliant blue FCF diammonium salt used in the US must meet the following specifications: pure dye, 82.0% min.; volatile matter (at 135°C), 10.0% max.; ammonium chlorides and sulphates, 5.0% max.; water-insoluble matter, 1.0% max.; mixed oxides, 1.0% max.; and ether extracts, 0.5% max. (US Food and Drug Administration, 1976).

Brilliant blue FCF, disodium salt

1.1 Synonyms and trade names

Colour Index No.: 42090

Colour Index Names: Acid Blue 9; Food Blue 2

Chem. Abstr. Services Reg. No.: 3844-45-9

Chem. Abstr. Name: *N*-Ethyl-*N*-(4-[(4-{ethyl[(3-sulfophenyl)methyl]-amino}phenyl)(2-sulfophenyl)methylene]-2,5-cyclohexadien-1-ylidene)-3-sulfobenzenemethanaminium hydroxide inner salt, disodium salt

C.I. Acid Blue 9, disodium salt; D and C Blue No. 1; D and C Blue No. 4; ethyl(4-{*para*[ethyl(*meta*-sulphobenzyl)amino]-α-(*ortho*-sulphophenyl)benzylidene}-2,5-cyclohexadien-1-ylidene)(*meta*-sulpho-benzyl)ammonium hydroxide inner salt, disodium salt; FD and C Blue 1; FD and C Blue No. 1; FDC Blue No. 1

Acid Sky Blue A; Acilan Turquoise Blue AE; A. F. Blue No. 1; Aizen Brilliant Blue FCF; Aizen Food Blue No. 1; Alphazurine; Alphazurine FG; Alphazurine FGND; Amacid Blue FG; Amacid Blue FG Conc; 1206 Blue; 11388 Blue; Blue Dye Number 1 food additive; Brilliant Blue; Brilliant Blue FCF; Brilliant Blue Lake; Bucacid Azure Blue; Calcocid Blue EG; Calcocid Blue 2G; Canacert Brilliant Blue FCF; Cogilor Blue 512.12; Cosmetic Blue Lake; Dispersed Blue 12195; Disulphine Lake Blue EG; Dolkwal Brilliant Blue; Edicol Blue Cl 2; Edicol Supra Blue E6; Erioglaucine; Erioglaucine A;

Erioglaucine E; Erioglaucine G; Eriosky Blue; Fenazo Blue XI; Fenazo Blue XR; Food Blue 1; Hexacol Brilliant Blue A; Hidacid Azure Blue; Intracid Pure Blue L; Kiton Blue AR; Kiton Pure Blue L; Maple Brilliant Blue FCF; Merantine Blue EG; Neptune Blue BRA; Neptune Blue BRA Concentration; Patent Blue AE; Patent Blue 2Y; Peacock Blue X-1756; Usacert Blue No. 1; Xylene Blue VSG

1.2 Empirical formula and molecular weight

$C_{37}H_{34}N_2O_9S_3.2Na$ Mol. wt: 792.8

1.3 Chemical and physical properties of the pure substance

(a) Description: Reddish-violet powder or granules with a metallic lustre (Japanese Union of Food Additives Associations, 1974)

(b) Spectroscopy data: λ_{max} 630 nm (Japanese Union of Food Additives Associations, 1974); the nuclear magnetic resonance spectrum has been published (Marmion, 1974).

(c) Solubility: Soluble in water (5% at 20°C and 98°C) and in ethanol; insoluble in vegetable oils (The Society of Dyers and Colourists, 1971a,b)

(d) Surface activity: A 3% aqueous solution showed a depression of surface tension of water of 35.0% at 37°C (Gangolli *et al.*, 1967).

1.4 Technical products and impurities

The grade of brilliant blue FCF disodium salt intended for use as a food colour additive in the US must meet the following specifications: total colour, 85.0% min.; sum of volatile matter (at 135°C), sodium chlorides and sulphates, 15.0% max.; subsidiary colours (isomers of brilliant blue FCF, disodium salt), 6.0% max.; leuco base, 5.0% max.; sum of *ortho*-, *meta*- and *para*-sulphobenzaldehydes, 1.5% max.; *N*-ethyl-*N*-(*meta*-sulphobenzyl)sulphanilic acid, 0.3% max.; water-insoluble matter 0.2% max.; chromium, 50 mg/kg max.; lead, 10 mg/kg max.; and arsenic, 3 mg/kg max. (US Food and Drug Administration, 1976).

The food grade of brilliant blue FCF, disodium salt that is available in western Europe conforms to these same specifications.

In Japan, the grade of brilliant blue FCF, disodium salt used in food meets the following specifications: pure colouring matter, 92±2%; subsidiary colours (isomers of brilliant blue FCF, disodium salt), 6.0% max.; loss on drying, 4.0% max.; inorganic salt, 3.0% max.; water-insoluble matter, 0.1% max.; iron, 100 mg/kg max.; chromium, 25 mg/kg max.; lead, 10 mg/kg max.; copper, 5 mg/kg max.; zinc, 5 mg/kg max.; arsenic (as As_2O_3), 2 mg/kg max.; λ_{max} = 630±2 nm; and pH = 5.0-7.0.

2. Production, Use, Occurrence and Analysis

For background information on this section, see preamble, p. 15.

2.1 Production and use

(a) Production

The colouring properties of brilliant blue FCF, diammonium and disodium salts were first described by Sandmeyer in 1896 (The Society of Dyers and Colourists, 1971c). The disodium salt is produced commercially in the US and Japan by condensing benzaldehyde-2-sulphonic acid with benzylethylaniline sulphonic acid followed by oxidation and conversion to the disodium salt (The Society of Dyers and Colourists, 1971c; Zuckerman, 1964). This method was also used in the US for production of the diammonium salt.

Commercial production of brilliant blue FCF, diammonium salt was last reported in the US in 1965 (US Tariff Commission, 1967). This salt has never been produced commercially in Japan.

Brilliant blue FCF, disodium salt has been produced commercially in the US for over fifty years (US Tariff Commission, 1922). In 1975, three US companies produced 622 thousand kg of the general dye grade, and another four companies produced 56 thousand kg of the food, drug and cosmetic grade (US International Trade Commission, 1977).

An estimated 300-500 thousand kg of unspecified salts of brilliant blue FCF are produced annually in western Europe. Less than 10 thousand kg of the disodium salt are imported and exported annually to the UK.

Brilliant blue FCF, disodium salt has been produced commercially in Japan since 1941. Four Japanese companies now manufacture it, and production during the years 1973-1975 amounted to 32 thousand kg, 26 thousand kg and 18 thousand kg per year, respectively. Small amounts are exported to eastern Asia and the US; none is imported.

(b) Use

Brilliant blue FCF, diammonium salt

Brilliant blue FCF, diammonium salt can be used to dye wool, silk, nylon, paper, leather, soap and woodstains; it is also used as an indicator and as a biological stain (The Society of Dyers and Colourists, 1971a). It was used in the US to dye aqueous drug solutions and in hair rinses (Zuckerman, 1964). It has never had any commercial application in Japan.

Brilliant blue FCF, diammonium salt is listed in the US, subject to certification (see section 1.4, p. 172), for use in drugs and cosmetics (US Food and Drug Administration, 1976). However, its use as food colour is not permitted in the US or Japan, nor is it included in the list of colourants permitted for use in foods in the European Economic Communities (The Society of Dyers and Colourists, 1975). Brilliant blue FCF, diammonium salt is permitted for use in the European Economic Communities in cosmetics which come into contact with mucous membranes (EEC, 1976).

Brilliant blue FCF, disodium salt

Brilliant blue FCF, disodium salt can be used to dye wool, silk, nylon, paper, leather, soap and woodstains; it is also used as an indicator, as a biological stain (The Society of Dyers and Colourists, 1971a,b) and as a dye in toilet-bowl cleaners. The barium salt of the acid dye, C.I. Pigment Blue 24, is used in printing inks, rubber and polyvinyl chloride resin, and also in the production of green pigments (The Society of Dyers and Colourists, 1971d), but separate production figures are not available.

Brilliant blue FCF, disodium salt is also used in the US as a colour additive for foods, drugs and cosmetics. In 1967, approximately 92% of the food grade was used in food, 7% in drugs and 1% in cosmetics (Anon., 1968). It occurs as a food colour in beverages, sweets and confections, bakery goods, pudding powders, ice-cream, sherbet, dairy products, pet food, cereals, sausages, maraschino cherries, snack foods and meat inks (Anon., 1968). It is used to colour aqueous drugs, tablets and capsules, and its aluminium lake is used in ointments (Zuckerman, 1964).

The cosmetics in which it can occur include bath salts and hair rinses (Markland, 1966), and its aluminium lake is used in lipsticks, rouges, face powders and talcums (Zuckerman, 1964).

Less than 10 thousand kg of the food grade of brilliant blue FCF, disodium salt are used annually in the UK in food, drugs and cosmetics. In Switzerland, it is used as a dyestuff.

In Japan, brilliant blue FCF, disodium salt is used in inks and toilet-bowl cleaners (70%) and as a food colour (30%).

Brilliant blue FCF, disodium salt is provisionally listed in the US subject to certification (see section 1.4, p. 173), for use in food, drugs and cosmetics (US Food and Drug Administration, 1976). In Japan, it is approved for use as a food colour in all but the following foods: beans, fresh fish and shellfish, fresh vegetables, roasted soya bean flour, *Laminariales* (kelp), meats, cured meats, marmalade, soya bean paste, noodles, porphyra, soaked fish meats, soya sauce, sponge cakes and tea (Japanese Union of Food Additives Associations, 1974). Brilliant blue FCF, disodium salt does not appear on the list of colourants permitted for use in foods in the European Economic Communities (The Society of Dyers and Colourists, 1975); however, member states may permit its use in foods until 1978. It is permitted for use in cosmetics which come into contact with mucous membranes (EEC, 1976).

The Joint FAO/WHO Expert Committee on Food Additives has set an acceptable daily intake for man of 0-12.5 mg/kg bw brilliant blue FCF, disodium salt (WHO, 1970). The Codex Alimentarius Commission has limited its use (singly or in combination with other colours) in tinned green peas to 100 mg/kg and in tinned apple sauce to 200 mg/kg (Codex Alimentarius Commission, 1973).

2.2 Occurrence

Brilliant blue FCF, diammonium and disodium salts are not known to occur as natural products. No data on their occurrence in the environment were available to the Working Group.

2.3 Analysis

Brilliant blue FCF, diammonium salt

Blue food dyes, including brilliant blue FCF, diammonium salt, have been separated by thin-layer chromatography on diethylaminoethyl cellulose using two solvent systems (Turner & Jones, 1971). Silica gel G or cellulose CC41 coatings have also been used (Chapman & Oakland, 1968). Brilliant blue FCF, diammonium salt has been separated from light green SF yellowish on microcrystalline cellulose powder thin-layer plates (Hayes *et al.*, 1972). Methods of extraction of food dyes, including brilliant blue FCF, diammonium salt, from various kinds of foods include the use of polyamide columns for purification of the dyes prior to thin-layer chromatographic analysis (Gilhooley *et al.*, 1972).

Brilliant blue FCF, disodium salt

Chromatographic and spectrophotometric methods for the identification of food dyes have been reviewed (FAO, 1963). The Association of Official Analytical Chemists has published recommended methods for use in batch certifications of brilliant blue FCF, disodium salt, so that it meet specifications defined by the US Food and Drug Administration (Horwitz, 1975), and an Official First Action for its detection in foods by a solvent extraction method (Horwitz, 1970) or by column chromatography with spectrophotometric identification (Horwitz, 1975).

Titrimetric methods can be used for determining the dye content in samples of brilliant blue FCF, disodium salt either alone (FAO, 1963; Rao & Dutt, 1970) or in combination with spectrophotometric identification (Nielsen, 1968).

Nuclear magnetic resonance spectroscopy has been evaluated as a means of identifying and differentiating food colours, including brilliant blue FCF, disodium salt (Marmion, 1974), and for confirming the identity of dyes considered for use in lymphograms, with purity determination by thin-layer chromatography (Hiranaka *et al.*, 1975).

A liquid anion-exchange system has been used to isolate colourants, including brilliant blue FCF, disodium salt, in strawberry jam prior to spectrophotometric determination (Tonogai, 1973).

Paper chromatography, using sixteen solvent systems, has been used to separate dyes, including brilliant blue FCF, disodium salt, used in foods and drugs (Dobrecky & De Carnevale Bonino, 1967a) and in carbonated beverages (Arino Espada, 1970). Adsorption on defatted wool has been employed to isolate the dyes prior to separation by paper chromatography, elution and spectrophotometric determination (Takagi, 1968). Two-dimensional paper chromatography has been used to separate dyes, including brilliant blue FCF, disodium salt, used in foods, drugs and cosmetics (Dobrecky & De Carnevale Bonino, 1967b); and paper electrophoresis has been used for those found in medicines (Dobrecky & De Carnevale Bonino, 1968).

Food colours, including brilliant blue FCF, disodium salt, have been separated and detected by thin-layer chromatography on Avicel using three solvent systems (Wang & Tsai, 1973) and on polyamide using one solvent system (Takeshita *et al.*, 1972).

Baker's yeast has been used to isolate colours, including brilliant blue FCF, disodium salt, in confectionery prior to thin-layer chromatography (Onozaki & Minami, 1971); and diethylaminoethyl-Sephadex has been used for colours in foods (Takeshita & Sakagami, 1969). Spectrodensitometry has been used to quantitate pharmaceutical dyes on thin-layer plates (Penner, 1968).

Separation and detection of dyes, including brilliant blue FCF, disodium salt, in cosmetics and detergents in the presence of large amounts of surface-active agents has been carried out with thin-layer chromatography (Matsumoto *et al.*, 1972).

3. Biological Data Relevant to the Evaluation of Carcinogenic Risk to Man

3.1 Carcinogenicity and related studies in animals

(a) Oral administration

Mouse: Groups of 48 male (20-31 g) and 50 female (19-23 g) ASH/CS1 mice were fed diets containing 0 (control), 0.015, 0.15 or 1.5% brilliant blue FCF, disodium salt (purity, 85%) for 80 weeks, providing intakes of approximately 0, 20, 200 or 2000 mg/kg bw/day. At 80 weeks, 16 male controls, 8, 12 and 9 treated males, 17 female controls and 13, 16 and 22 treated females had died, in the different groups, respectively. No significant increase in tumour incidence was observed in treated mice compared with that in controls. One squamous-cell carcinoma of the stomach was observed in a female treated with the highest dose; 7 kidney tumours (6 adenomas and 1 adenocarcinoma) were observed among 30 males examined that had received the 0.15% dose, compared with 1 kidney adenoma in 44 controls examined ($P<0.05$) (Rowland *et al.*, 1977).

Rat: Four groups of 15 male and 15 female Wistar rats, 4-6-weeks old, were fed diets containing 0 (control), 0.03, 0.3 or 3% brilliant blue FCF, disodium salt (food grade) for 75 weeks, at which time the experiment was terminated. The numbers of survivors at this time were 23 (control), 19, 20 and 19 rats in the four groups, respectively; tissues from 5 male and 5 female survivors in each group were examined histologically. One lymphosarcoma occurred in a male fed the 0.3% dose level (Mannell *et al.*, 1962) [The Working Group noted the inadequate histological examination of tissues in this experiment].

Groups of 12 male and 12 female 3-week old Osborne-Mendel rats received 0 (control), 0.5, 1.0, 2.0 or 5% brilliant blue FCF, disodium salt in their diet for 2 years. No significant increase in mortality or pathological changes were observed in these animals when compared with untreated control rats (Hansen *et al.*, 1966) [The Working Group noted the inadequacy of the experiment].

(b) Subcutaneous and/or intramuscular administration

Rat: Groups of 48 and 27 rats (sex and strain unspecified) received twice weekly s.c. injections of 1 ml of an aqueous solution of brilliant blue FCF, disodium salt containing 7.4 or 20 mg of the dye (37% pure). The average period of treatment was 20.5 months, and the average total dose per animal was 1.22 or 3.5 g, respectively, calculated on the basis of pure dye. Injection-site fibrosarcomas developed in 10 and 20 animals. Three of 4 rats given twice weekly s.c. injections of 20 mg brilliant blue FCF, disodium salt (37% pure) for 9 months and observed for a further 22 months developed injection-site sarcomas. In groups of 31 and 14 rats that served as controls and that were injected subcutaneously with 0.9% and 3.4% saline for more than 8 months (maximum, 29 or 27 months), no injection-site tumours occurred (Gross, 1961).

Groups of 25 and 40 rats (sex and strain unspecified) received twice weekly s.c. injections of 1 ml of an aqueous solution of brilliant blue FCF, disodium salt containing 7.4 or 20 mg of the dye (>90% pure). The average periods of treatment were 17 and 12.5 months, and the average total dose per animal was 1.07 and 2.5 g, respectively, calculated on the basis of pure dye. Injection-site fibrosarcomas developed in 18 and 25 rats. All of 7 rats given twice weekly s.c. injections of 20 mg brilliant blue FCF, disodium salt (>90% pure) for 9 months and observed for a further 10.5 months developed injection-site sarcomas. The controls were the same as those used in the experiment with 37% pure brilliant blue FCF, disodium salt described above (Gross, 1961).

Ten male and 10 female 5-week-old Wistar rats received weekly s.c. injections of 0.5 ml of a 4% solution of brilliant blue FCF, disodium salt (food grade) in isotonic saline for 45 weeks (total dose, 900 mg). The experiment was terminated at 71 weeks, at which time 9 of the animals were still alive. No fibrosarcomas were observed (Mannell & Grice, 1964).

Nine male and 9 female Osborne-Mendel rats, about 4-weeks old, received weekly s.c. injections of 30 mg brilliant blue FCF, disodium salt as a 3% aqueous solution for 2 years; 5 rats were still alive at 18 months. A total of 16 rats developed injection-site sarcomas, and none occurred in 18 saline-injected control animals (Hansen *et al.*, 1966).

3.2 Other relevant biological data

The toxicology of colouring materials, including brilliant blue FCF, disodium salt, has been reviewed (Drake, 1975).

(a) Experimental systems

The s.c. LD_{50} of brilliant blue FCF, disodium salt in mice is 4.6 g/kg bw (Gross, 1961). The oral LD_{50} of an unspecified salt of brilliant blue FCF in rats was more than 2.0 g/kg bw (Lu & Lavalle, 1964).

Studies in rats, rabbits and dogs involving oral administration of brilliant blue FCF, disodium salt indicate that only small amounts of the dye (<5%) are absorbed and that it is excreted mainly in the faeces (Hess & Fitzhugh, 1953, 1955). Following its i.v. injection in rats, over 90% of the dye was excreted in the bile within 4 hours (Iga *et al.*, 1971a).

Five and 3 beagle dogs, 6-7 months old, received diets containing 2% and 1% brilliant blue FCF, disodium salt, respectively, for 1 year, at which time the experiment was terminated. No clinical signs and no gross or microscopic pathology were attributed to ingestion of the dye (Hansen *et al.*, 1966).

Brilliant blue FCF, disodium salt exhibits high plasma protein binding (Gangolli *et al.*, 1967, 1972; Iga *et al.*, 1971a,b).

No adequate data on the embryotoxicity, teratogenicity or mutagenicity of these colours were available to the Working Group.

(b) Man

No data were available to the Working Group.

3.3 Case reports and epidemiological studies

No data were available to the Working Group.

4. Comments on Data Reported and Evaluation[1]

4.1 Animal data

Brilliant blue FCF, disodium salt is carcinogenic in rats after its subcutaneous injection: it produced fibrosarcomas following repeated injections. It also produced an increased incidence of kidney tumours in mice after its oral administration.

4.2 Human data

No case reports or epidemiological studies were available to the Working Group.

[1]See also the section, 'Animal Data in Relation to the Evaluation of Risk to Man' in the introduction to this volume, p. 13.

5. References

Anon. (1968) Guidelines for good manufacturing practice: use of certified FD&C colors in food. Food Techn., 22, 14-17

Arino Espada, M. (1970) Chromatographic identification of dyes used in carbonated beverages. Inform. Quím. Analyt., 24, 63-67

Chapman, W.B. & Oakland, D. (1968) Differentiation of blue coloring matters in food and drugs with particular reference to blue VRS and patent blue V. J. Ass. Publ. Analyt., 6, 124-128

Codex Alimentarius Commission (1973) List of Additives Evaluated for Their Safety-In-Use In Food, 1st Series, Rome, Joint FAO/WHO Food Standards Programme, CAC/FAL 1-1973, pp. 16, 66

Drake, J.J.-P. (1975) Food colours - harmless aesthetics or epicurean luxuries? Toxicology, 5, 3-42

Dobrecky, J. & De Carnevale Bonino, R.C.D. (1967a) Paper chromatography of dyes, using various solvents. SAFYBI, 7, 165-170

Dobrecky, J. & De Carnevale Bonino, R.C.D. (1967b) Paper chromatography separation of coloring agents used legally in food, drugs, and cosmetics in the USA. Rev. Asoc. Bioquím. Argent., 32, 16-19

Dobrecky, J. & De Carnevale Bonino, R.C.D. (1968) Paper electrophoresis of some coloring agents used in medicines. SAFYBI, 8, 67-70

EEC (1976) Législation. Journal Officiel des Communautés Européennes, No. L 262, p. 191

FAO (1963) Specifications for Identity and Purity of Food Additives, Vol. II, Food Colors, Rome, pp. 107-108, 119-121, 128-129

Gangolli, S.D., Grasso, P. & Golberg, L. (1967) Physical factors determining the early local tissue reactions produced by food colourings and other compounds injected subcutaneously. Fd Cosmet. Toxicol., 5, 601-621

Gangolli, S.D., Grasso, P., Golberg, L. & Hooson, J. (1972) Protein binding by food colourings in relation to the production of subcutaneous sarcoma. Fd Cosmet. Toxicol., 10, 449-462

Gilhooley, R.A., Hoodless, R.A., Pitman, K.G. & Thomson, J. (1972) Separation and identification of food colours. IV. Extraction of synthetic water-soluble food colours. J. Chromat., 72, 325-331

Gross, E. (1961) Über die Erzeugung von Sarkomen durch die besonders gereinigten Triphenylmethanfarbstoffe Lichtgrün SF und Patentblau AE bei der wiederholten subcutanen Injektion an der Ratte. Z. Krebsforsch., 64, 287-304

Hansen, W.H., Fitzhugh, O.G., Nelson, A.A. & Davis, K.J. (1966) Chronic toxicity of two food colors, brilliant blue FCF and indigotine. Toxicol. appl. Pharmacol., 8, 29-36

Hayes, W.P., Nyaku, N.Y. & Burns, D.T. (1972) Separation and identification of food dyes. III. Improved resolution of selected dye pairs. J. Chromat., 71, 585-587

Hess, S.M. & Fitzhugh, O.G. (1953) Metabolism of coal-tar dyes. I. Triphenylmethane dyes (Abstract No. 1090). Fed. Proc., 12, 330-331

Hess, S.M. & Fitzhugh, O.G. (1955) Absorption and excretion of certain triphenylmethane colors in rats and dogs. J. Pharmacol. exp. Ther., 114, 38-42

Hiranaka, P.K., Kleinman, L.M., Sokoloski, E.A. & Fales, H.M. (1975) Chemical structure and purity of dyes used in lymphograms. Amer. J. Hosp. Pharm., 32, 928-930

Horwitz, W., ed. (1970) Official Methods of Analysis of the Association of Official Analytical Chemists, 11th ed., Washington DC, Association of Official Analytical Chemists, pp. 576-578

Horwitz, W., ed. (1975) Official Methods of Analysis of the Association of Official Analytical Chemists, 12th ed., Washington DC, Association of Official Analytical Chemists, pp. 629-632, 636, 638-639

Iga, T., Awazu, S. & Nogami, H. (1971a) Pharmacokinetic study of biliary excretion. II. Comparison of excretion behavior in triphenylmethane dyes. Chem. pharm. Bull., 19, 273-281

Iga, T., Awazu, S., Hanano, M. & Nogami, H. (1971b) Pharmacokinetic studies of biliary excretion. IV. The relationship between the biliary excretion behavior and the elimination from plasma of azo dyes and triphenylmethane dyes in rat. Chem. pharm. Bull., 19, 2609-2616

Japanese Union of Food Additives Associations (1974) Restrictions for the use of food additives. The Japanese Standards of Food Additives, 3rd ed., Tokyo, pp. 25, 206-207

Lu, F.C. & Lavalle, A. (1964) The acute toxicity of some synthetic colours used in drugs and foods. Canad. pharm. J., 97, 30

Mannell, W.A. & Grice, H.C. (1964) Chronic toxicity of brilliant blue FCF, blue VRS, and green S in rats. J. Pharm. Pharmacol., 16, 56-59

Mannell, W.A., Grice, H.C. & Allmark, M.G. (1962) Chronic toxicity studies on food colours. V. Observations on the toxicity of brilliant blue FCF, guinea green B and benzyl violet 4B in rats. J. Pharm. Pharmacol., 14, 378-384

Markland, W.R. (1966) Hair preparations. In: Kirk, R.E. & Othmer, D.F., eds, Encyclopedia of Chemical Technology, 2nd ed., Vol. 10, New York, John Wiley and Sons, p. 793

Marmion, D.M. (1974) Applications of nuclear magnetic resonance spectroscopy to certifiable food colors. J. Ass. off. analyt. Chem., 57, 495-507

Matsumoto, I., Takashiba, K. & Homma, Y. (1972) Quality assurance of cosmetics. VI. Detection of micro amounts of coal-tar dyes in cosmetics containing anionic surface-active agents. Eisei Kagaku, 18, 278-280

Nielsen, L.S. (1968) Color caeruleus DAK (brilliant blue FFC): comments on the monograph in DAK 63. Arch. Pharm. Chemi, 75, 1155-1158

Onozaki, H. & Minami, K. (1971) Separate determinations of coal-tar dyes in candies by use of bakers' yeasts as an adsorbent. Nippon Shokuhin Kogyo Gakkai-Shi, 18, 346-350

Penner, M.H. (1968) Thin-layer chromatography of certified coal tar color additives. J. pharm. Sci., 57, 2132-2135

Rao, N.V. & Dutt, V.V.S.E. (1970) Cerimetric determination of triphenylmethane dyes. Fresenius' Z. analyt. Chem., 251, 187-188

Rowland, I.R., Gaunt, I.F., Hardy, J., Kiss, I.S. & Butterworth, K.R. (1977) Long-term toxicity of brilliant blue FCF in mice. Fd Cosmet. Toxicol. (in press)

The Society of Dyers and Colourists (1971a) Colour Index, 3rd ed., Vol. 1, Yorkshire, UK, p. 1299

The Society of Dyers and Colourists (1971b) Colour Index, 3rd ed., Vol. 2, Yorkshire, UK, pp. 2778, 2809

The Society of Dyers and Colourists (1971c) Colour Index, 3rd ed., Vol. 4, Yorkshire, UK, p. 4385

The Society of Dyers and Colourists (1971d) Colour Index, 3rd ed., Vol. 3, Yorkshire, UK, p. 3349

The Society of Dyers and Colourists (1975) Colour Index, revised 3rd ed., Vol. 6, Yorkshire, UK, pp. 6232, 6235-6236

Takagi, S. (1968) Determination of coal-tar dyes in prepared foods. I. Method for some beverages. Shokuhin Eiseigaku Zasshi, 9, 399-404

Takeshita, R. & Sakagami, Y. (1969) Food additives. XII. Detection of some water-soluble acid dyes in foods. Eisei Kagaku, 15, 72-76

Takeshita, R., Yamashita, T. & Itoh, N. (1972) Separation and detection of water-soluble acid dyes on polyamide thin layers. J. Chromat., 73, 173-182

Tonogai, Y. (1973) Analysis of food additives. I. Determination of coal tar dyes in foods with liquid anion exchangers. Eisei Kagaku, 19, 231-235

Turner, T.D. & Jones, B.E. (1971) The identification of blue triphenylmethane food dyes by thin-layer chromatography. J. pharm. Pharmacol., 23, 806-807

US Food and Drug Administration (1976) Food and drugs. US Code of Federal Regulations, Title 21, parts 8.206, 8.501, 8.4021, 9.80, 9.240

US International Trade Commission (1977) Synthetic Organic Chemicals, US Production and Sales, 1975, USITC Publication 804, Washington DC, US Government Printing Office, pp. 48, 52, 57, 70

US Tariff Commission (1922) Census of Dyes and Other Synthetic Organic Chemicals, 1921, Tariff Information Series No. 26, Washington DC, US Government Printing Office, p. 66

US Tariff Commission (1967) Synthetic Organic Chemicals, US Production and Sales, 1965, TC Publication 206, Washington DC, US Government Printing Office, p. 104

Wang, R.T. & Tsai, Y.H. (1973) Avicel thin-layer chromatography of food dyes. Tai-Wah K'o Hsueh, 27, 32-35

WHO (1970) Toxicological evaluation of some food colours, emulsifiers, stabilizers, anti-caking agents and certain other substances. WHO/Food Add/70.36, pp. 24-27

Zuckerman, S. (1964) Colors for foods, drugs, and cosmetics. In: Kirk, R.E. & Othmer, D.F., eds, Encyclopedia of Chemical Technology, 2nd ed., Vol. 5, New York, John Wiley and Sons, pp. 865-868, 872-873

FAST GREEN FCF

1. Chemical and Physical Data

1.1 Synonyms and trade names

Colour Index No.: 42053

Colour Index Name: Food Green 3

Chem. Abstr. Services Reg. No.: 2353-45-9

Chem. Abstr. Name: *N*-Ethyl-*N*-(4-[(4-{ethyl[(3-sulfophenyl)methyl]-amino}phenyl)(4-hydroxy-2-sulfophenyl)methylene]-2,5-cyclohexadien-1-ylidene)-3-sulfobenzenemethanaminium hydroxide inner salt, disodium salt

Disodium salt of 4-{[4-(*N*-ethyl-*para*-sulphobenzylamino)phenyl][4-hydroxy-2-sulphoniumphenyl]methylene} 1-(*N*-ethyl-*N*-*para*-sulphobenzyl)-$\Delta^{2,5}$-cyclohexadienimine; ethyl[4-(α-{*para*-[ethyl(*meta*-sulphobenzyl)-amino]phenyl}-4-hydroxy-3-sulphobenzylidene)-2,5-cyclohexadien-1-ylidene](*meta*-sulphobenzyl)ammonium hydroxide inner salt, disodium salt; FD and C Green No. 3

Aizen Food Green No. 3; Green 1724; Solid Green FCF

1.2 Chemical formula and molecular weight

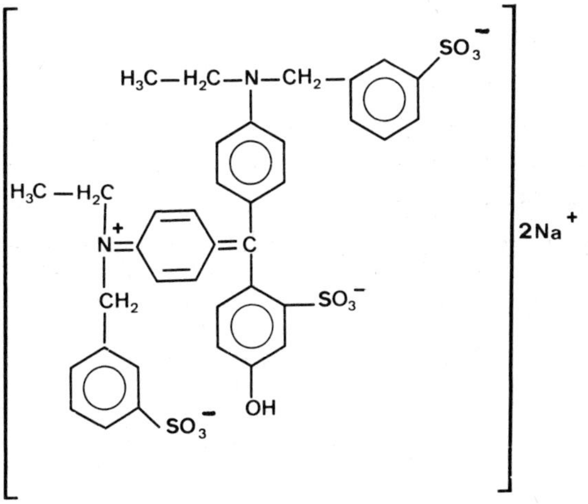

$C_{37}H_{34}N_2O_{10}S_3.2Na$ Mol. wt: 808.9

1.3 Chemical and physical properties of the pure substance

(a) Description: Dark-green powder or granules with a metallic lustre (Japanese Union of Food Additives Associations, 1974)

(b) Spectroscopy data: λ_{max} 628 nm (Japanese Union of Food Additives Associations, 1974)

(c) Solubility: Soluble in water and in ethanol; insoluble in vegetable oils (The Society of Dyers and Colourists, 1971a)

(d) Surface activity: A 3% aqueous solution showed a depression of surface tension of water of 31.1% at 37°C (Gangolli *et al.*, 1967).

1.4 Technical products and impurities

Fast green FCF is available in the US in a grade certified by the US Food and Drug Administration as suitable for use as a colour additive in food, drugs and cosmetics; it must be at least 84.5% pure dye (determined by titration with titanium trichloride) and contain not more than: 10.0% volatile matter (at 135°C), 5.0% subsidiary dyes, 5.0% sodium chloride and sodium sulphate, 1.0% mixed oxides, 0.5% water-insoluble matter, 0.4% ether extracts, 1.4 mg/kg arsenic (as As_2O_3), 100 mg/kg lead and trace quantities of heavy metals other than arsenic and lead (US Food and Drug Administration, 1976). Fast green FCF available in western Europe conforms to these same specifications.

In Japan, fast green FCF is available as a colour additive grade that contains a minimum of 85.0% pure dye (determined by titanium trichloride titration) and not more than: 10% volatile matter (at 135°C), 6% subsidiary dyes, 5% inorganic chloride and sulphate, 0.3% water-insoluble matter (Japanese Union of Food Additives Associations, 1974), 500 mg/kg iron, 200 mg/kg zinc, 50 mg/kg chromium, 20 mg/kg lead and 2 mg/kg arsenic (as As_2O_3).

3. Production, Use, Occurrence and Analysis

For background information on this section, see preamble, p. 15.

2.1 Production and use

(a) Production

Fast green FCF can be prepared by condensing 4-hydroxybenzaldehyde with *N*-ethyl-*N*-phenylbenzylamine, followed by sulphonation, oxidation and isolation as the sodium salt (The Society of Dyers and Colourists, 1971b). It is produced commercially by the condensation of 4-hydroxybenzaldehyde-2-sulphonic acid with α-(*N*-ethylanilino)-4-toluenesulphonic acid (Zuckerman, 1964).

This colour has been produced commercially in the US since 1927 (US Tariff Commission, 1928); only one US company reported an undisclosed amount (see preamble, p. 15) in 1975 (US International Trade Commission, 1977).

Less than 10 thousand kg fast green FCF are imported and exported annually in the UK.

It was first produced commercially in Japan in 1945, but there has been no commercial production in recent years. Japan imports less than 1000 kg annually, and there are no exports.

(b) Use

In the US, about 88% of the 1400 kg fast green FCF sold in the first nine months of 1967 were used in food, 11% in pharmaceuticals and about 1% in cosmetics (Anon., 1968). The products in which it is used are beverages, sweets and confections, maraschino cherries, pudding powers, bakery goods, ice-cream, sherbet and dairy products (Anon., 1968). It is used to dye pharmaceutical capsules (Zuckerman, 1964) and as a colour ingredient in temporary colour shampoos, at concentrations of about 0.5-2% (Markland, 1966). Fast green FCF is also used as a histological stain in a specific test for DNA (Parker, 1966).

Less than 10 thousand kg fast green FCF are used annually in the UK; over 90% of this is in cosmetics. In Japan, it is used as a food colour only.

Fast green FCF is provisionally listed in the US for use in foods, drugs and cosmetics and is subject to certification that it conform to the specifications set forth by the US Food and Drug Administration (see section 1.4, p. 188) (US Food and Drug Administration, 1976).

In Japan, this colour is not permitted for use in beans, fresh fish and shellfish, fresh vegetables, roasted soya bean flour, *Laminariales* (kelp), fresh and cured meat, marmalade, soya bean paste, noodles, soaked fish meat, soya sauce, sponge cake or tea (Japanese Union of Food Additives Associations, 1974).

It is not among the colourants permitted for use in foods in the European Economic Communities (The Society of Dyers and Colourists, 1975); however, it may be used in cosmetics which come into contact with mucous membranes (EEC, 1976).

The Joint FAO/WHO Expert Committee on Food Additives has assigned an acceptable daily intake for man of 0-12.5 mg/kg bw (WHO, 1970). Fast green FCF has been specifically approved for use in tinned apple sauce and tinned pears at a maximum level of 200 mg/kg, singly or in combination with other colours (Codex Alimentarius Commission, 1973).

2.2 Occurrence

Fast green FCF is not known to occur as a natural product. No data on its occurrence in the environment were available to the Working Group.

2.3 Analysis

Chromatographic and spectrophotometric methods for the identification of food dyes, including fast green FCF, have been reviewed (FAO, 1963).

The Association of Official Analytical Chemists has published recommended methods for use in batch certifications of fast green FCF so that it meet specifications defined by the US Food and Drug Administration (Horwitz, 1975), and an Official First Action for the detection of synthetic organic colour additives, including fast green FCF, in foods by selective solvent extractions (Horwitz, 1970) or by column chromatography and spectrophotometric identification (Horwitz, 1975).

The dye content in a sample of fast green FCF may be determined by titration (FAO, 1963). Nuclear magnetic resonance spectroscopy has been evaluated as a means of identifying and differentiating food colours (Marmion, 1974).

Food dyes, including fast green FCF, have been separated by paper chromatography and quantitated by photoelectric densitometry (Sasaki & Iwata, 1972). Adsorption onto defatted wool has been used to isolate dyes used in beverages, including fast green FCF, prior to paper chromatography (Takagi, 1968). Two-dimensional paper chromatography has also been used to separate dyes, including fast green FCF, used in food, drugs and cosmetics (Dobrecky & De Carnevale Bonino, 1967).

Dyes used in foods, including fast green FCF, have been separated by thin-layer chromatography on diethylaminoethyl cellulose using two solvent series (Turner & Jones, 1971), on polyamide-silica gel and polyamide-diatomaceous earth (Hsu *et al.*, 1971) and on microcrystalline cellulose (Hayes *et al.*, 1972). Diethylaminoethyl-Sephadex has been used to isolate the dye from food products prior to thin-layer chromatography (Takeshita & Sakagami, 1969; Takeshita *et al.*, 1972), and baker's yeast has been used for confectionery products (Onozaki & Minami, 1971). Thin-layer chromatography on silica gel has also been used to separate and identify dyes in cosmetics (Martelli & Proserpio, 1974) and in the presence of large amounts of anionic surface-active agents (Matsumoto *et al.*, 1972). Separation on microcrystalline cellulose using four solvent systems has been used with spectrodensitometric quantitation (Penner, 1968).

3. Biological Data Relevant to the Evaluation of Carcinogenic Risk to Man

3.1 Carcinogenicity and related studies in animals

(a) Oral administration

Mouse: Groups of 50 male and 50 female C_3HeB/FeJ mice were fed diets containing 1% or 2% fast green FCF for 2 years; 100 males and 100 females served as controls. After 78 weeks, 56 controls and 27 and 17 mice in the two treated groups, respectively, were still alive. The numbers of animals with tumours were not different in the three groups (Hansen *et al.*, 1966) [The Working Group noted the inadequacy of the experiment].

Rat: Five groups of 4-week-old Osborne-Mendel rats, 50 per group and divided evenly by sex, were fed diets containing 0 (control), 0.5, 1.0, 2.0 or 5.0% fast green FCF for 2 years. Mortality was not higher in treated groups. The numbers of animals with malignant tumours and the types of tumours were not different among all groups when compared with controls (Hansen *et al.*, 1966) [The Working Group noted the inadequacy of the experiment; only 12 animals fed 5% dye were examined microscopically in detail].

(b) Subcutaneous and/or intramuscular administration

Rat: Eighteen 3-week-old Osborne-Mendel rats of both sexes received weekly s.c. injections of approximately 30 mg (1 ml of a 3% aqueous solution) fast green FCF for life. Four animals were still alive after 18 months. Fifteen animals developed injection-site fibrosarcomas, which appeared after about 40 weeks of treatment. No injection-site tumours were seen in 64 control rats injected with saline (51 alive at 18 months and 19 at 2 years) (Hansen *et al.*, 1966; Nelson & Hagan, 1953).

Two groups of 16 female Fischer rats, approximately 1-month old, were given repeated s.c. injections of 0.5 ml of a 3% or 6% aqueous solution of fast green FCF (purity, 92%; solution adjusted to pH 7.1). Initially, injections of 6% were given twice weekly to one group and injections of 3% twice weekly to the other. After 17 weeks, treatment in both groups was with 3% twice weekly for 9 weeks. During the following 22 weeks both groups were injected usually once a week, occasionally twice a week. Thirteen of 16 rats started at the 6% level and 10/14 of the other group developed fibrosarcomas at the injection site; the first tumours appeared after 7 months. No injection-site tumours occurred in 10 control rats injected with distilled water. Five rats received a single s.c. injection of 0.5 ml of a 3% aqueous solution of the dye, and no local tumours developed within the 16-month observation period (Hesselbach & O'Gara, 1960).

3.2 Other relevant biological data

(a) Experimental systems

The oral LD_{50} of fast green FCF in rats is more than 2.0 g/kg bw (Lu & Lavalle, 1964).

Studies in rats, rabbits and dogs involving oral administration of fast green FCF indicate that only small amounts of the dye (<5%) are absorbed and that it is mainly excreted in the faeces (Hess & Fitzhugh, 1953, 1954, 1955). Following its i.v. injection in rats, over 90% of the dye was excreted in the bile within 4 hours (Iga *et al.*, 1971).

Groups of 2 male and 2 female beagle dogs, 6-9-months old, were fed diets containing 0, 1.0 or 2.0% fast green FCF (reaching daily amounts of 512-781 mg/kg bw and 258-288 mg/kg bw at the 2% and 1% level, respectively) for 2 years, at which time the experiment was terminated. No toxic effects were recorded (Hansen *et al.*, 1966).

Fast green FCF exhibits high plasma protein binding (Gangolli *et al.*, 1967, 1972; Iga *et al.*, 1971).

No adequate data on the embryotoxicity, teratogenicity or mutagenicity of this colour were available to the Working Group.

(b) Man

No data were available to the Working Group.

3.3 Case reports and epidemiological studies

No data were available to the Working Group.

4. Comments on Data Reported and Evaluation[1]

4.1 Animal data

Fast green FCF has been tested in mice and rats by oral administration and in rats by subcutaneous injection. It was carcinogenic in rats, producing sarcomas at the site of repeated subcutaneous injections.

4.2 Human data

No case reports or epidemiological studies were available to the Working Group.

---

[1]See also the section, 'Animal Data in Relation to the Evaluation of Risk to Man' in the introduction to this volume, p. 13.

5. References

Anon. (1968) Guidelines for good manufacturing practice: use of certified FD&C colors in food. Food Techn., 22, 14-17

Codex Alimentarius Commission (1973) List of Additives Evaluated for Their Safety-In-Use In Food, 1st Series, Rome, Joint FAO/WHO Food Standards Programme, CAC/FAL 1-1973, pp. 16, 67

Dobrecky, J. & De Carnevale Bonino, R.C.D. (1967) Paper chromatographic separation of coloring agents used legally in food, drugs, and cosmetics in the USA. Rev. Asoc. Bioquím. Argent., 32, 16-19

EEC (1976) Législation. Journal Officiel des Communautés Européennes, No. L 262, p. 191

FAO (1963) Specifications for Identity and Purity of Food Additives, Vol. II, Food Colors, Rome, pp. 99-100, 119-121, 128-129

Gangolli, S.D., Grasso, P. & Golberg, L. (1967) Physical factors determining the early local tissue reactions produced by food colourings and other compounds injected subcutaneously. Fd Cosmet. Toxicol., 5, 601-621

Gangolli, S.D., Grasso, P., Golberg, L. & Hooson, J. (1972) Protein binding by food colourings in relation to the production of subcutaneous sarcoma. Fd Cosmet. Toxicol., 10, 449-462

Hansen, W.H., Long, E.L., Davis, K.J., Nelson, A.A. & Fitzhugh, O.G. (1966) Chronic toxicity of three food colourings: guinea green B, light green SF yellowish and fast green FCF in rats, dogs and mice. Fd Cosmet. Toxicol., 4, 389-410

Hayes, W.P., Nyaku, N.Y. & Burns, D.T. (1972) Separation and identification of food dyes. III. Improved resolution of selected dye pairs. J. Chromat., 71, 585-587

Hess, S.M. & Fitzhugh, O.G. (1953) Metabolism of coal-tar dyes. I. Triphenylmethane dyes (Abstract No. 1090). Fed. Proc., 12, 330-331

Hess, S.M. & Fitzhugh, O.G. (1954) Metabolism of coal-tar colors. II. Bile studies (Abstract No. 1201). Fed. Proc., 13, 365

Hess, S.M. & Fitzhugh, O.G. (1955) Absorption and excretion of certain triphenylmethane colors in rats and dogs. J. Pharmacol. exp. Ther., 114, 38-42

Hesselbach, M.L. & O'Gara, R.W. (1960) Fast green- and light green-induced tumors: induction, morphology, and effect on host. J. nat. Cancer Inst., 24, 769-793

Horwitz, W., ed. (1970) Official Methods of Analysis of the Association of Official Analytical Chemists, 11th ed., Washington DC, Association of Official Analytical Chemists, pp. 575-578

Horwitz, W., ed. (1975) Official Methods of Analysis of the Association of Official Analytical Chemists, 12th ed., Washington DC, Association of Official Analytical Chemists, pp. 629-632, 636-638

Hsu, H.-Y., Chiang, H.-C. & Yang, P.-C. (1971) Polyamine mixed layer chromatography of food dyes. T'ai-Wan Yao Hsueh Tsa Chih, 23, 34-36

Iga, T., Awazu, S. & Nogami, H. (1971) Pharmacokinetic study of biliary excretion. II. Comparison of excretion behavior in triphenylmethane dyes. Chem. pharm. Bull., 19, 273-281

Japanese Union of Food Additives Associations (1974) The Japanese Standards of Food Additives, 3rd ed., Tokyo, pp. 25, 213-214

Lu, F.C. & Lavalle, A. (1964) The acute toxicity of some synthetic colours used in drugs and foods. Canad. pharm. J., 97, 30

Markland, W.R. (1966) Hair preparations. In: Kirk, R.E. & Othmer, D.F., eds, Encyclopedia of Chemical Technology, 2nd ed., Vol. 10, New York, John Wiley and Sons, p. 793

Marmion, D.M. (1974) Applications of nuclear magnetic resonance spectroscopy to certifiable food colors. J. Ass. off. analyt. Chem., 57, 495-507

Martelli, A. & Proserpio, G. (1974) Identification of dyes in cosmetics by thin-layer chromatography. Relata Tech. Chim. Biol. Appl., 6, 157-164

Matsumoto, I., Takashiba, K. & Yasuko, H. (1972) Quality assurance of cosmetics. VI. Detection of micro amounts of coal-tar dyes in cosmetics containing anionic surface-active agents. Eisei Kagaku, 18, 278-280

Nelson, A.A. & Hagan, E.C. (1953) Production of fibrosarcomas in rats at site of subcutaneous injection of various food dyes (Abstract No. 1307). Fed. Proc., 12, 397-398

Onozaki, H. & Minami, K. (1971) Separate determinations of coal-tar dyes in candies by use of bakers' yeasts as an adsorbent. Nippon Shokuhin Kogyo Gakkai-Shi, 18, 346-350

Penner, M.H. (1968) Thin-layer chromatography of certified coal tar color additives. J. pharm. Sci., 57, 2132-2135

Parker, R.C. (1966) Methods of Tissue Culture, 3rd ed., New York, Hoeber Medical Division, Harper & Row, pp. 307-309

Sasaki, H. & Iwata, T. (1972) Analytical studies of food dyes. IV. Direct densitometry of paper chromatograms of food and other dyes by transparent methods. Shokuhin Eiseigaku Zasshi, 13, 120-126

The Society of Dyers and Colourists (1971a) Colour Index, 3rd ed., Vol. 2, Yorkshire, UK, p. 2779

The Society of Dyers and Colourists (1971b) Colour Index, 3rd ed., Vol. 4, Yorkshire, UK, p. 4383

The Society of Dyers and Colourists (1975) Colour Index, revised 3rd ed., Vol. 6, Yorkshire, UK, pp. 6235-6236

Takagi, S. (1968) Determination of coal-tar dyes in prepared foods. I. Method for some beverages. Shokuhin Eiseigaku Zasshi, 9, 399-404

Takeshita, R. & Sakagami, Y. (1969) Food additives. XII. Detection of some water-soluble acid dyes in foods. Eisei Kagaku, 15, 72-76

Takeshita, R., Yamashita, T. & Itoh, N. (1972) Separation and detection of water-soluble acid dyes on polyamide thin layers. J. Chromat., 73, 173-182

Turner, T.D. & Jones, B.E. (1971) The identification of blue triphenylmethane food dyes by thin-layer chromatography. J. pharm. Pharmacol., 23, 806-807

US Food and Drug Administration (1976) Food and drugs. US Code of Federal Regulations, Title 21, parts 8.501, 9.2, 9.23

US International Trade Commission (1977) Synthetic Organic Chemicals, US Production and Sales, 1975, USITC Publication 804, Washington DC, US Government Printing Office, p. 70

US Tariff Commission (1928) Census of Dyes and of Other Synthetic Organic Chemicals, 1927, Tariff Information Series No. 37, Washington DC, US Government Printing Office, p. 67

WHO (1970) Specifications for the Identity and Purity of Food Additives and their Toxicological Evaluation: Some Food Colours, Emulsifiers, Stabilizers, Anticaking Agents, and Certain Other Substances. Wld Hlth Org. techn. Rep. Ser., No. 445, p. 32

Zuckerman, S. (1964) Colors for foods, drugs, and cosmetics. In: Kirk, R.E. & Othmer, D.F., eds, Encyclopedia of Chemical Technology, 2nd ed., Vol. 5, New York, John Wiley and Sons, pp. 865-866

GUINEA GREEN B

1. Chemical and Physical Data

1.1 Synonyms and trade names

Colour Index No.: 42085

Colour Index Names: Acid Green 3; Food Green 1

Chem. Abstr. Services Reg. No.: 4680-78-8

Chem. Abstr. Name: *N*-Ethyl-*N*-(4-[(4-{ethyl[(3-sulfophenyl)methyl]-amino}phenyl)phenylmethylene]-2,5-cyclohexadien-1-ylidene)-3-sulfobenzenemethanaminium hydroxide inner salt, sodium salt

Acid green; C.I. Acid Green 3, sodium salt; FD and C Green No. 1; FD and C Green Number 1; FDC Green 1; guinea green; monosodium salt of 4-[4-(*N*-ethyl-*para*-sulphobenzylamino)diphenylmethylene]-[1-(*N*-ethyl-*N*-*para*-sulphoniumbenzyl)-$\Delta^{2,5}$-cyclohexadienimine]

Acidal Green G; Acid Green B; Acid Green G; Acid Green 2G; Acid Green L; Acid Green S; Acid Leather Green F; Acid Leather Green 3G; Acilan Green B; A.F. Green No. 1; Amacid Green B; Brilliant Green 3EMBL; Bucacid Guinea Green BA: Calcocid Green G; Fenazo Green L; Guinea Green; Guinea Green BA; Hidacid Emerald Green; Hispacid Green GB; Intracid Green F; Kiton Green F; Kiton Green FC; Leather Green B; Lissamine Green G; Merantine Green G; Naphthalene Green G; Naphthalene Lake Green G; Naphthalene Leather Green G; Neran Brilliant Green G; Pontacyl Green BL; Sulfacid Brilliant Green 1B; Sulpho Green 2B; Vondacid Green L

1.2 Chemical formula and molecular weight

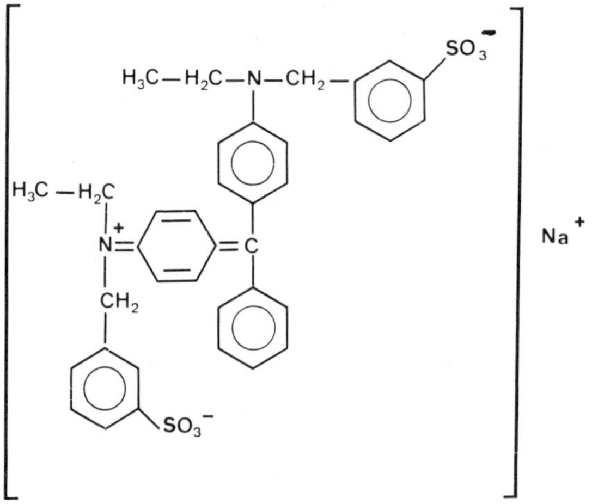

SO_3^-
$H_3C-H_2C-N-CH_2-$
H_3C-H_2C
$N^+=$
$=C$
CH_2
SO_3^-
Na^+

$C_{37}H_{35}N_2O_6S_2.Na$ Mol. wt: 690.8

1.3 Chemical and physical properties of the pure substance

(a) Description: A dull, dark-green powder or a bright, crystalline solid (Windholz, 1976)

(b) Solubility: Soluble in water; slightly soluble in ethanol (Windholz, 1976)

1.4 Technical products and impurities

In Japan, guinea green B has the following specifications: purity, 85% min.; inorganic salt, 4% max.; and water content, 10% max.

2. Production, Use, Occurrence and Analysis

For background information on this section, see preamble, p. 15.

2.1 Production and use

(a) Production

The colouring properties of guinea green B were reported in 1883 by Schultz and Streng. It may be prepared commercially by condensing benzaldehyde with α-(*N*-ethylanilino)-*meta*-toluenesulphonic acid, followed by oxidation and conversion to the sodium salt (The Society of Dyers and Colourists, 1971a).

Guinea green B has been produced commercially in the US for over fifty years (US Tariff Commission, 1919). Four US companies reported production of approximately 93.5 thousand kg and sales of 74.9 thousand kg in 1973 (US International Trade Commission, 1975); sales of 74.5 thousand kg were reported in 1974 (US International Trade Commission, 1976), but these dropped to only 32.2 thousand kg in 1975 (US International Trade Commission, 1977).

Guinea green B was first produced commercially in Japan in 1940. It is believed that there is one producer currently; however, no production has been reported since 1973, when 200 kg were manufactured. There are no Japanese exports or imports of this chemical.

(b) Use

Guinea green B is used to dye wool, silk (The Society of Dyers and Colourists, 1971b) and leather (The Society of Dyers and Colourists, 1971c). It is also used to dye paper, in wood stains, as a biological stain and as an indicator. Its barium and aluminium salts are used as pigments (The Society of Dyers and Colourists, 1971b).

In the past, this colour was used as a food, drug and cosmetic dye to colour gelatine desserts, frozen desserts (ice-creams and sherbets), sweets and confections which did not contain fats and oils, bakery products and cereals, and drug capsules (Zuckerman, 1964). However, its use as colour additive for foods, drugs and cosmetics was forbidden in the US in late 1966 (US Food and Drug Administration, 1976), and its use as a food additive was forbidden in Japan in 1967 (Japanese Union of Food Additives Associations, 1974). It is considered to be unsafe for use in food throughout the world (Codex Alimentarius Commission, 1973).

In western Europe, guinea green B can provisionally be used in cosmetics that do not come into contact with mucous membranes (EEC, 1976); and in Japan, it is used in externally applied cosmetics.

2.2 Occurrence

Guinea green B is not known to occur as a natural product. No data on its occurrence in the environment were available to the Working Group.

2.3 Analysis

Chromatographic and spectrophotometric methods for the identification of food dyes, including guinea green B, have been reviewed (FAO, 1963). Prior to its cancellation for use in foods, drugs and cosmetics, the Association of Official Analytical Chemists published recommended methods for use in batch certifications of guinea green B so that it meet specifications defined by the US Food and Drug Administration (Horwitz, 1965). It has similarly published Official First Actions for dyes, including guinea green B, in foods by selective solvent extraction (Horwitz, 1970) or by column chromatography and spectrophotometric identification (Horwitz, 1975).

A titration method for determining the dye content in a sample of guinea green B has been published (FAO, 1963). Nuclear magnetic resonance spectroscopy has been evaluated as a means of identifying and differentiating food colours (Marmion, 1974).

Paper chromatography has been used to separate and identify dyes, including guinea green B, used to dye paper (Wallace *et al.*, 1967) and foods using spectrodensitometric quantitation (Sasaki & Iwata, 1972).

Dyes used in foods, including guinea green B, have been separated and identified by thin-layer chromatography on polyamide-silica gel and polyamide-diatomaceous earth (Hsu *et al.*, 1971), on polyamide plates (Takeshita *et al.*, 1972) and on cellulose MN after isolation of the dyes from the food sample by adsorption on wool (de Sousa Carvalho & Figueira, 1967). Adsorption on diethylaminoethyl-Sephadex has also been used to isolate dyes (Takeshita & Sakagami, 1969).

3. Biological Data Relevant to the Evaluation of Carcinogenic Risk to Man

3.1 Carcinogenicity and related studies in animals

(a) Oral administration

Mouse: Groups of C_3HeB/FeJ mice were fed diets containing 0% (101 males, 101 females), 1.0% (50 males, 53 females) or 2% (51 males, 49 females) guinea green B for 2 years. The numbers of survivors at 78 weeks were 100, 72 and 68 and those at 104 weeks, 23, 23 and 15 for the 3 levels, respectively. The total tumour incidence in the 3 groups did not differ (P>0.05) from that in the controls. In all 3 groups, neoplasms occurred much more often in males than in females. Primary hepatic tumours were the most common, but no statistically significant difference was seen among the groups (Hansen *et al.*, 1966) [The Working Group noted the inadequacy of the experiment].

Rat: Four groups of 15 male and 15 female Wistar rats, 4-6-weeks old, were fed diets containing 0 (control), 0.03, 0.3 or 3% guinea green B for 75 weeks, at which time the experiment was terminated. The numbers of survivors at this time were 23 (control), 27, 25 and 20 rats, in the four groups, respectively; tissues from 5 male and 5 female survivors in each group were examined histologically. Malignant tumours developed in 1 male (kidney carcinoma) of the median dose level group and in 3 females fed the highest dose (1 liposarcoma and 2 squamous-cell carcinomas of the skin). In addition, 2 females at the 3% level had benign skin lesions (kerato-acanthomas). In the control group, 1 female had a fibrosarcoma (Mannell *et al.*, 1962) [The Working Group noted the inadequate histological examination of tissues in this experiment].

Five groups of 4-week-old Osborne-Mendel rats, 50 per group and evenly divided by sex, were fed diets containing 0 (control), 0.5, 1.0, 2.0 or 5.0% guinea green B for 2 years. The numbers of females still alive at 2 years were 7 controls and 14, 14 and 17 in the treated groups, respectively. There were no significant differences in tumour incidence, except that primary hepatic tumours did not occur in controls or in animals at the 0.5% dose level. Six males (2 at 1% and 4 at 5% level) and 4 females (1 at 2% and 3 at 5% level) developed hepatic-cell adenomas. Three males (2 at 2% and 1 at 5% levels) and 1 female (5% level) had carcinomas of the liver (Hansen *et al.*, 1966).

(b) Subcutaneous and/or intramuscular administration

Rat: Eighteen 4-week-old Osborne-Mendel rats of both sexes received weekly s.c. injections of approximately 15 mg (in 1 ml of an aqueous solution) guinea green B for life. Fourteen animals survived for 18 months and 6 for 2 years. Four rats developed injection-site fibrosarcomas. No injection-site tumours occurred in 64 controls (51 alive at 18 months and 19 at 2 years) injected with saline (Hansen *et al.*, 1966) [P<0.01].

(c) Other experimental systems

Rat: A group of 20 male and 20 female 4-week-old Wistar rats had continual skin/fur contact with guinea green B for 100 weeks: the tray of the cage was sprinkled with 3 g of the dye, and the rats were placed directly on the tray; the tray was cleaned twice weekly, and the same amount of dye was added after each cleaning. Three of 18 females developed skin fibromas, as did 3/14 female controls. One exposed male developed an epidermoid carcinoma, compared with 0/12 male controls. The increased incidence of benign mammary tumours in the test group (7/18) compared with that in controls (1/14) was statistically significant (P<0.05) (Mannell *et al.*, 1964).

3.2 Other relevant biological data

(a) Experimental systems

The oral LD_{50} of guinea green B in rats is greater than 2.0 g/kg bw (Lu & Lavalle, 1964).

Studies in rats involving oral administration showed that only small amounts of the dye (<5%) are absorbed and excreted unchanged in the bile (Hess & Fitzhugh, 1953, 1954, 1955; Minegishi & Yamaha, 1974). Following its i.v. injection in rats, about 75% of the dye was excreted in the bile within 4 hours (Iga *et al.*, 1971; Minegishi & Yamaha, 1974).

Groups of 2 male and 2 female beagle dogs, 6-9-months old, were fed diets containing 0, 1.0 or 2.0% guinea green B (reaching daily amounts of 205-405 mg/kg bw and 520-714 mg/kg bw at the 1 and 2% levels, respectively) for 2 years, at which time the experiment was terminated. Mild leucopenia, weight loss and anorexia were observed in some dogs (Hansen *et al.*, 1966).

No data on the embryotoxicity, teratogenicity or mutagenicity of this colour were available to the Working Group.

(b) Man

No data were available to the Working Group.

3.3 Case reports and epidemiological studies

No data were available to the Working Group.

4. Comments on Data Reported and Evaluation[1]

4.1 Animal data

Guinea green B has been tested in mice and rats by oral administration and in rats by ambient exposure and subcutaneous injection. It was carcinogenic in rats, producing hepatic tumours after its oral administration, local sarcomas after its subcutaneous injection and benign mammary tumours following ambient exposure.

4.2 Human data

No case reports or epidemiological studies were available to the Working Group.

---

[1]See also the section, 'Animal Data in Relation to the Evaluation of Risk to Man' in the introduction to this volume, p. 13.

5. References

Codex Alimentarius Commission (1973) List of Additives Evaluated for Their Safety-In-Use In Food, 1st Series, Rome, Joint FAO/WHO Food Standards Programme, CAC/FAL 1-1973, p. 45

EEC (1976) Législation. Journal Officiel des Communautés Européennes, No. L 262, p. 199

FAO (1963) Specifications for Identity and Purity of Food Additives, Vol. II, Food Colors, Rome, pp. 101-102, 119-121, 128-129

Hansen, W.H., Long, E.L., Davis, K.J., Nelson, A.A. & Fitzhugh, O.G. (1966) Chronic toxicity of three food colourings: guinea green B, light green SF yellowish and fast green FCF in rats, dogs and mice. Fd Cosmet. Toxicol., 4, 389-410

Hess, S.M. & Fitzhugh, O.G. (1953) Metabolism of coal-tar dyes. I. Triphenylmethane dyes (Abstract No. 1090). Fed. Proc., 12, 330-331

Hess, S.M. & Fitzhugh, O.G. (1954) Metabolism of coal-tar colors. II. Bile studies (Abstract No. 1201). Fed. Proc., 13, 365

Hess, S.M. & Fitzhugh, O.G. (1955) Absorption and excretion of certain triphenylmethane colors in rats and dogs. J. Pharmacol. exp. Ther., 114, 38-42

Horwitz, W., ed. (1965) Official Methods of Analysis of the Association of Official Agricultural Chemists, 10th ed., Washington DC, Association of Official Agricultural Chemists, pp. 679-680, 682-683, 685-687, 696, 698-699

Horwitz, W., ed. (1970) Official Methods of Analysis of the Association of Official Analytical Chemists, 11th ed., Washington DC, Association of Official Analytical Chemists, pp. 575-578

Horwitz, W., ed. (1975) Official Methods of Analysis of the Association of Official Analytical Chemists, 12th ed., Washington DC, Association of Official Analytical Chemists, pp. 629-630

Hsu, H.-Y., Chiang, H.-C. & Yang, P.-C. (1971) Polyamine mixed layer chromatography of food dyes. T'ai-Wan Yao Hsueh Tsa Chih, 23, 34-36

Iga, T., Awazu, S. & Nogami, H. (1971) Pharmacokinetic study of biliary excretion. II. Comparison of excretion behaviour in triphenylmethane dyes. Chem. pharm. Bull., 19, 273-281

Japanese Union of Food Additives Associations (1974) The Japanese Standards of Food Additives, 3rd ed., Tokyo, p. 12

Lu, F.C. & Lavalle, A. (1964) The acute toxicity of some synthetic colours used in drugs and foods. Canad. pharm. J., 97, 30

Mannell, W.A., Grice, H.C. & Allmark, M.G. (1962) Chronic toxicity studies on food colours. V. Observations on the toxicity of brilliant blue FCF, guinea green B and benzyl violet 4B in rats. J. pharm. Pharmacol., 14, 378-384

Mannell, W.A., Grice, H.C. & Dupuis, I. (1964) The effect on rats of long term exposure to guinea green B and benzyl violet 4B. Fd Cosmet. Toxicol., 2, 345-347

Marmion, D.M. (1974) Applications of nuclear magnetic resonance spectroscopy to certifiable food colors. J. Ass. off. analyt. Chem., 57, 495-507

Minegishi, K.-I. & Yamaha, T. (1974) Metabolism of triphenylmethane colors. I. Absorption, excretion, and distribution of guinea green B (FD and C green no. 1) in rats. Chem. pharm. Bull., 22, 2042-2047

Sasaki, H. & Iwata, T. (1972) Analytical studies of food dyes. IV. Direct densitometry of paper chromatograms of food and other dyes by transparent methods. Shokuhin Eiseigaku Zasshi, 13, 120-126

The Society of Dyers and Colourists (1971a) Colour Index, 3rd ed., Vol. 4, Yorkshire, UK, p. 4385

The Society of Dyers and Colourists (1971b) Colour Index, 3rd ed., Vol. 1, Yorkshire, UK, p. 1397

The Society of Dyers and Colourists (1971c) Colour Index, 3rd ed., Vol. 2, Yorkshire, UK, p. 2812

de Sousa Carvalho, M.M.S. & Figueira, M.A. (1967) Thin-layer chromatography for the separation and identification of synthetic food dyes. Rev. Port. Farm., 17, 438-447

Takeshita, R. & Sakagami, Y. (1969) Food additives. XII. Detection of some water-soluble acid dyes in foods. Eisei Kagaku, 15, 72-76

Takeshita, R., Yamashita, T. & Itoh, N. (1972) Separation and detection of water-soluble acid dyes on polyamide thin layers. J. Chromat., 73, 173-182

US Food and Drug Administration (1976) Food and drugs. US Code of Federal Regulations, Title 21, part 8.510

US International Trade Commission (1975) Synthetic Organic Chemicals, US Production and Sales, 1973, ITC Publication 728, Washington DC, US Government Printing Office, pp. 54, 65

US International Trade Commission (1976) Synthetic Organic Chemicals, US Production and Sales, 1974, USITC Publication 776, Washington DC, US Government Printing Office, pp. 52, 63

US International Trade Commission (1977) Synthetic Organic Chemicals, US Production and Sales of Dyes, 1975, USITC Publication 804, Washington DC, US Government Printing Office, p. 48

US Tariff Commission (1919) Report on Dyes and Related Coal-Tar Chemicals, 1918, revised ed., Washington DC, US Government Printing Office, p. 41

Wallace, M.R., Milliken, L.T. & Toner, S.D. (1967) Identification of dyes in paper by extraction and chromatographic analysis. Techn. Ass. Pulp Paper Industr., 50, 121A-124A

Windholz, M., ed. (1976) The Merck Index, 9th ed., Rahway, NJ, Merck & Co., p. 595

Zuckerman, S. (1964) Colors for foods, drugs, and cosmetics. In: Kirk, R.E. & Othmer, D.F., eds, Encyclopedia of Chemical Technology, 2nd ed., Vol. 5, New York, John Wiley and Sons, pp. 865-866

LIGHT GREEN SF

1. Chemical and Physical Data

1.1 Synonyms and trade names

Colour Index No.: 42095

Colour Index Names: Acid Green 5; Food Green 2

Chem. Abstr. Services Reg. No.: 5141-20-8

Chem. Abstr. Name: *N*-Ethyl-*N*-{4-[(4-[(3-sulfophenyl)methyl]amino}-phenyl)-(4-sulfophenyl)methylene]-2,5-cyclohexadien-1-ylidene}-3-sulfobenzenemethanaminium hydroxide inner salt, disodium salt

C.I. Acid Green 5, disodium salt; D and C Green No. 4; disodium salt of 4-{[4-(*N*-ethyl-*para*-sulphobenzylamino)phenyl][4-sulphoniumbenzyl]methylene}(1-*N*-ethyl-*N*-*para*-sulphobenzyl)-$\Delta^{2,5}$-cyclohexadienimine; ethyl(4-{*para*-[ethyl(*meta*-sulphobenzyl)amino]-α-(*para*-sulphophenyl)benzylidene}-2,5-cyclohexadien-1-ylidene)(*meta*-sulphobenzyl)ammonium hydroxide inner salt, disodium salt; FD and C Green No. 2; light green SF yellowish

Acidal Light Green SF; Acid Brilliant Green SF; Acid Green; Acid Green A; Acid Green G; Acilan Green SFG; A.F. Green No. 2; Amacid Green G; Fast Acid Green N; Fenazo Green 7G; Leather Green SF; Light Green CF; Light Green G; Light Green 2GN; Light Green Lake; Light Green S; Light Green SFA; Light Green SFD; Light Green Yellowish; Light Green SF Yellowish; Light SF Yellowish; Lissamine Green SF; Lissamine Lake Green SF; Merantine Green SF; Pencil Green SF; Sulfo Green J; Sumitomo Light Green SF Yellowish; Wool Brilliant Green SF

1.2 Chemical formula and molecular weight

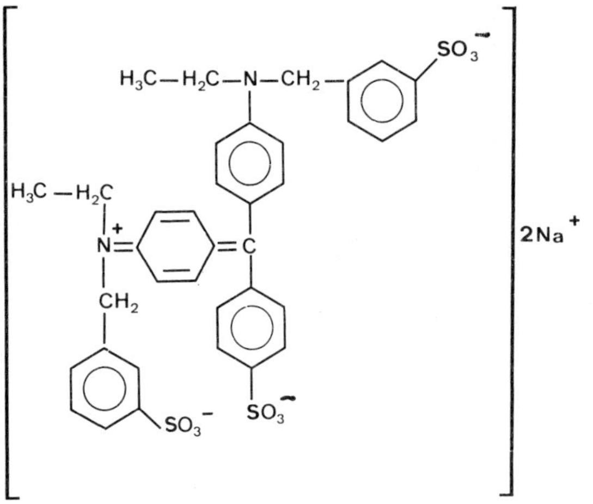

$C_{37}H_{34}N_2O_9S_3 \cdot 2Na$

Mol. wt: 792.9

1.3 Chemical and physical properties of the pure substance

(a) Description: Reddish-brown powder (Windholz, 1976)

(b) Solubility: Soluble in water; almost insoluble in ethanol (The Society of Dyers and Colourists, 1971a)

(c) Surface activity: A 3% aqueous solution showed a depression of surface tension of water of 32% at 37°C (Gangolli *et al.*, 1967).

1.4 Technical products and impurities

No data were available to the Working Group.

2. Production, Use, Occurrence and Analysis

For background information on this section, see preamble, p. 15.

2.1 Production and use

(a) Production

The colouring properties of light green SF were reported in 1879 by Köhler (The Society of Dyers and Colourists, 1971a). Light green SF is prepared commercially by condensing ethylbenzylaniline with benzaldehyde, followed by sulphonation and oxidation, and, finally, conversion to the sodium salt (Witterholt, 1969).

Light green SF has been produced commercially in the US for over fifty years (US Tariff Commission, 1919); in 1975, only one US company reported an undisclosed amount (see preamble, p. 15) (US International Trade Commission, 1977a). US imports of light green SF through principal US customs districts were reported as 350 kg in 1974 (US International Trade Commission, 1976) and 48 kg in 1975 (US International Trade Commission, 1977b).

Light green SF has not been produced commercially in Japan since its use as a food colour was forbidden in 1970.

(b) Use

Light green SF is used to dye wool, nylon, silk (The Society of Dyers and Colourists, 1971b) and leather (The Society of Dyers and Colourists, 1971c). It is also used as a biological stain, and its barium salt is used to colour paper coatings and printing inks (The Society of Dyers and Colourists, 1971b).

In the past, light green SF was primarily used in food and drugs for colouring products such as gelatine desserts, maraschino cherries, ice-creams and sherbets, sweets and confections, bakery products and cereals, and gelatine capsules for pharmaceutical preparations (Zuckerman, 1964); however, it is no longer believed to be used in these applications. Certification of light green SF for use as a colour additive in foods, drugs or cosmetics was cancelled in the US in 1966 (US Food and Drug Administration, 1976) and in Japan in 1970 (Japanese Union of Food Additives Associations, 1974). Light green SF is not included in the list of colourants permitted for use in foods in the European Economic Communities (The Society of Dyers and Colourists, 1975), however, it may provisionally be used in cosmetics which come into contact with mucous membranes (EEC, 1976).

2.2 Occurrence

Light green SF is not known to occur as a natural product. No data on its occurrence in the environment were available to the Working Group.

2.3 Analysis

Prior to its cancellation for use in foods, drugs or cosmetics, the Association of Official Analytical Chemists (AOAC) published recommended methods for use in batch certifications of light green SF so that it meet specifications defined by the US Food and Drug Administration (Horwitz, 1965). A titration method with titanous chloride for determining the purity of a sample of light green SF has been published (FAO, 1963), and an Official First Action has been published by the AOAC for the detection of dyes, including light green SF, in foods by selective solvent extractions (Horwitz, 1970) or by separating the dyes on a column chromatograph followed by spectrophotometric identification (Horwitz, 1975).

Chromatographic and spectrophotometric methods for the identification of food dyes, including light green SF, have been reviewed (FAO, 1963). Food dyes, including light green SF, have been separated by paper chromatography and quantitated by spectrodensitometry (Sasaki & Iwata, 1972).

Thin-layer chromatography was used: to separate and identify food dyes, including light green SF, on polyamide-diatomaceous earth (Hsu *et al.*, 1971), to separate and identify food dye mixtures of light green SF and brilliant blue FCF on microcrystalline cellulose powder (Hayes *et al.*, 1972) and to separate and identify cosmetic dyes, including light green SF, on silica gel G with two developing agents (Pinter *et al.*, 1968) and on silica gel with one solvent system (Pinter & Kramer, 1969). Water-soluble dyes, including light green SF, have been detected in food by thin-layer chromatography using diethylaminoethyl cellulose to isolate the dyes initially (Takeshita & Sakagami, 1969; Takeshita *et al.*, 1972).

3. Biological Data Relevant to the Evaluation of Carcinogenic Risk to Man

3.1 Carcinogenicity and related studies in animals

(a) Oral administration

Mouse: Groups of 50 male and 50 female C_3HeB/FeJ mice were fed diets containing 1% or 2% light green SF for 2 years; 100 males and 100 females served as controls. After 78 weeks, 50 controls and 39 and 22 mice in the two treated groups, respectively, were still alive; 8, 13 and 3 mice were still alive at 104 weeks. There was no difference in the total incidence of tumours among the three groups (Hansen *et al.*, 1966) [The Working Group noted the inadequacy of the experiment].

Rat: Five groups of 4-week-old Osborne-Mendel rats, 50 per group and divided evenly by sex, were fed diets containing 0 (control), 0.5, 1.0, 2.0 or 5.0% light green SF for 2 years. Data on survival rates were not reported. The numbers of animals with malignant tumours and the types of tumours were not different among all groups when compared with controls (Hansen *et al.*, 1966) [The Working Group noted the inadequacy of the experiment].

(b) Subcutaneous and/or intramuscular administration

Rat: Seven rats (strain and sex not specified) received once or twice weekly s.c. injections of 1-3 ml of a 2% or 3% aqueous solution of light green SF. The injections were adapted to the condition of the animal. Towards the end of the first year, 4/7 rats developed fibrosarcomas at the injection site. Three untreated rats served as controls, but no further information on these animals was given (Schiller, 1937).

Thirty Wistar rats (sex not specified) were given 59 repeated s.c. injections of a 2% aqueous solution of light green SF (commercial sample). The first injection consisted of 2 ml, then this was reduced to doses of 1 ml, usually twice weekly; after 1 month (9 injections), the concentration was increased to 3%. At the 33rd week, injections were discontinued (total dose, 1.7 g dye/animal). Of 24 surviving rats, 15 developed fibrosarcomas at the injection site (Harris, 1947) [The Working Group noted that no data on controls were reported].

Ten 6-month-old rats received s.c. injections of 15 mg light green SF (purity, 75%; pH 7.2) in saline twice weekly for 11½ months (total dose, 1.4 g): 8 developed injection-site sarcomas. In addition, 10 rats were treated for a period of 3 months (total dose, 0.4 g), and one fibrosarcoma was found 3½ months after the end of the treatment. In another experiment, of 12 rats treated with the same dose schedule, with treatment interrupted at various times between 2 and 9 months, one developed a fibrosarcoma. In a total of 59 control rats treated with saline solution and observed up to a maximum of 29 months, no injection-site sarcomas were observed (Gross, 1955, 1961).

Two groups of 16 female Fischer rats, approximately 1-month old, were given repeated s.c. injections of 0.5 ml of a 3% or 6% aqueous solution of light green SF (the dye contained 92% of the disodium salt; solution adjusted to pH 7.1). Initially, injections of 6% were given twice or thrice weekly for 7 weeks to one group and injections of 3% twice weekly to the other. Treatment in both groups was then continued twice weekly at the 3% level for a further 39 weeks (total doses, 1.6 and 1.4 g/animal). All of the rats in both groups developed injection-site fibrosarcomas; most of the tumours appeared between 12-13 months after the start of treatment. No injection-site tumours occurred in 10 control rats injected with distilled water. Five female rats received a single s.c. injection of 0.5 ml of a 3% aqueous solution of the dye, and no tumours developed within the 16 months' observation period (Hesselbach & O'Gara, 1960).

Thirty Wistar rats received weekly s.c. injections of approximately 30 mg light green SF weekly in 1 ml of aqueous solution. No animals survived at 18 months; 23 developed injection-site fibrosarcomas, the first of which appeared after 40 weeks of treatment. No injection-site tumours occurred in 10 Wistar and 54 Osborne-Mendel rats (in total, 19 alive at 2 years) injected with saline (Hansen *et al.*, 1966; Nelson & Hagan, 1953).

A group of 63 male and 63 female Ash/CFE rats, weighing about 100 g, were given twice weekly injections of 1 ml of a 2% aqueous solution, buffered to pH 7.0, for different periods (1-112 injections per animal). Ninety-two rats were alive when the first tumours appeared; 29 animals that received more than 28 injections developed injection-site sarcomas after about 47 weeks (Hooson *et al.*, 1973).

(c) Intraperitoneal administration

Eighteen weanling Osborne-Mendel rats were given thrice weekly i.p. injections of 20 mg light green SF as a 2% aqueous solution for 2 years; no tumours were observed (Hansen *et al.*, 1966; Nelson & Davidow, 1957) [The Working Group noted the inadequacy of the experiment].

3.2 Other relevant biological data

(a) Experimental sytems

The s.c. LD_{50} of light green SF in mice is 525 mg/kg bw (Gross, 1961). The oral LD_{50} in rats is more than 2.0 g/kg bw (Lu & Lavalle, 1964).

Administration of 3.0% light green SF for 65 weeks to rats produced testicular tubular atrophy and incomplete spermatogenesis (Allmark *et al.*, 1956).

Studies in rats and dogs involving oral administration of light green SF indicate that only small amounts of the dye (<5%) are absorbed and excreted, mainly in the faeces (Hess & Fitzhugh, 1954, 1955). Following its i.v. injection in rats, about 20% of the dye was excreted in the bile within 4 hours (Iga *et al.*, 1971a).

Following repeated i.v. injections in rats of 2-4 mg/animal light green SF in saline, sodium, potassium and water excretion were increased (Heller & Horáček, 1971). Similar effects were observed in infusion experiments in dogs (Lynch *et al.*, 1973).

Groups of 2 male and 2 female beagle dogs, 6-9-months old, were fed diets containing 0, 1.0 or 2.0% light green SF for 2 years. The amounts of colour consumed were approximately 600 and 485 in males and females fed 2.0% in the diet and 315 and 265 mg/kg per day in males and females fed 1.0%. No toxic effects were detected during this period (Hansen *et al.*, 1966).

Light green SF exhibits high plasma protein binding (Gangolli *et al.*, 1967, 1972; Grossmann & Frey, 1969; Iga *et al.*, 1971b).

No data on the embryotoxicity, teratogenicity or mutagenicity of this colour were available to the Working Group.

(b) Man

No data were available to the Working Group.

3.3 Case reports and epidemiological studies

No data were available to the Working Group.

4. Comments on Data Reported and Evaluation[1]

4.1 Animal data

Light green SF has been tested in mice and rats by oral administration and in rats by subcutaneous and intraperitoneal injection. It was carcinogenic in rats after its subcutaneous injection, producing local sarcomas.

4.2 Human data

No case reports or epidemiological studies were available to the Working Group.

---

[1]See also the section, 'Animal Data in Relation to the Evaluation of Risk to Man' in the introduction to this volume, p. 13.

5. References

Allmark, M.G., Grice, H.C. & Mannell, W.A. (1956) Chronic toxicity studies on food colours. II. Observations on the toxicity of FD & C green no. 2 (light green SF yellowish), FD & C orange no. 2 (orange SS) and FD & C red no. 32 (oil red XO) in rats. J. Pharm. Pharmacol., 8, 417-424

EEC (1976) Législation. Journal Officiel des Communautés Européennes, No. L 262, p. 199

FAO (1963) Specifications for Identity and Purity of Food Additives, Vol. II, Food Colors, Rome, pp. 103-104, 119-121, 128-129

Gangolli, S.D., Grasso, P. & Golberg, L. (1967) Physical factors determining the early local tissue reactions produced by food colourings and other compounds injected subcutaneously. Fd Cosmet. Toxicol., 5, 601-621

Gangolli, S.D., Grasso, P., Golberg, L. & Hooson, J. (1972) Protein binding by food colourings in relation to the production of subcutaneous sarcoma. Fd Cosmet. Toxicol., 10, 449-462

Gross, E. (1955) Über den Triphenylmethanfarbstoff Lichtgrün SF als Sarkomerreger bei der Ratte. Naunyn-Schmiedeberg's Arch. exp. Path. Pharmak., 225, 175-179

Gross, E. (1961) Über die Erzeugung von Sarkomen durch die besonders gereinigten Triphenylmethanfarbstoffe Lichtgrün SF und Patentblau AE bei der wiederholten subcutanen Injektion an der Ratte. Z. Krebsforsch., 64, 287-304

Grossmann, D.F. & Frey, J. (1969) Renal clearance of lissamine green in the rat. In: Peters, G., ed., Proceedings of a Symposium on Progress in Nephrology, 5th, 1967, Berlin, Springer, pp. 391-394

Hansen, W.H., Long, E.L., Davis, K.J., Nelson, A.A. & Fitzhugh, O.G. (1966) Chronic toxicity of three food colourings: guinea green B, light green SF yellowish and fast green FCF in rats, dogs and mice. Fd Cosmet. Toxicol., 4, 389-410

Harris, P.N. (1947) Production of sarcoma in rats with light green SF. Cancer Res., 7, 35-36

Hayes, W.P., Nyaku, N.Y. & Burns, D.T. (1972) Separation and identification of food dyes. III. Improved resolution of selected dye pairs. J. Chromat., 71, 585-587

Heller, J. & Horáček, V. (1971) The influence of lissamine green on tubular reabsorption of electrolytes and water in rats. Pflügers Arch., 323, 27-33

Hess, S.M. & Fitzhugh, O.G. (1954) Metabolism of coal-tar colors. II. Bile studies (Abstract No. 1201). Fed. Proc., 13, 365

Hess, S.M. & Fitzhugh, O.G. (1955) Absorption and excretion of certain triphenylmethane colors in rats and dogs. J. Pharmacol. exp. Ther., 114, 38-42

Hesselbach, M.L. & O'Gara, R.W. (1960) Fast green- and light green-induced tumors: induction, morphology, and effect on host. J. nat. Cancer Inst., 24, 769-793

Hooson, J., Grasso, P. & Gangolli, S.D. (1973) Injection site tumours and preceding pathological changes in rats treated subcutaneously with surfactants and carcinogens. Brit. J. Cancer, 27, 230-244

Horwitz, W., ed. (1965) Official Methods of Analysis of the Association of Official Agricultural Chemists, 10th ed., Washington DC, Association of Official Agricultural Chemists, pp. 679-680, 682-683, 685-687, 696, 698-699

Horwitz, W., ed. (1970) Official Methods of Analysis of the Association of Official Analytical Chemists, 11th ed., Washington DC, Association of Official Analytical Chemists, pp. 576-578

Horwitz, W., ed. (1975) Official Methods of Analysis of the Association of Official Analytical Chemists, 12th ed., Washington DC, Association of Official Analytical Chemists, pp. 629-630

Hsu, H.-Y., Chiang, H.-C. & Yang, P.-C. (1971) Polyamine mixed layer chromatography of food dyes. T'ai-Wan Yao Hsueh Tsa Chih, 23, 34-36

Iga, T., Awazu, S. & Nogami, H. (1971a) Pharmacokinetic study of biliary excretion. II. Comparison of excretion behavior in triphenylmethane dyes. Chem. pharm. Bull., 19, 273-281

Iga, T., Awazu, S., Hanano, M. & Nogami, H. (1971b) Pharmacokinetic studies of biliary excretion. IV. The relationship between the biliary excretion behavior and the elimination from plasma of azo dyes and triphenylmethane dyes in rat. Chem. pharm. Bull., 19, 2609-2616

Japanese Union of Food Additives Associations (1974) The Japanese Standards of Food Additives, 3rd ed., pp. 11-12

Lu, F.C. & Lavalle, A. (1964) The acute toxicity of some synthetic colours used in drugs and foods. Canad. pharm. J., 97, 30

Lynch, R.E., Schneider, E.G., Strandhoy, J.W., Willis, L.R. & Knox, F.G. (1973) Effect of lissamine green dye on renal sodium reabsorption in the dog. J. appl. Physiol., 35, 169-171

Nelson, A.A. & Davidow, B. (1957) Injection site fibrosarcoma production in rats by food colors (Abstract No. 1571). Fed. Proc., 16, 367

Nelson, A.A. & Hagan, E.C. (1953) Production of fibrosarcomas in rats at site of subcutaneous injection of various food dyes (Abstract No. 1307). Fed. Proc., 12, 397-398

Pinter, I. & Kramer, M. (1969) Simultaneous, thin-layer chromatographic detection of several dyes in cosmetics. Parfuem. Kosmet., 50, 129-134

Pinter, I., Kramer, M. & Kleeberg, J. (1968) Thin-layer-chromatographic method for detecting various cosmetic dyes in mixtures. Elelmiszervizsgalati Kozlem., 14, 169-175

Sasaki, H. & Iwata, T. (1972) Analytical studies of food dyes. IV. Direct densitometry of paper chromatograms of food and other dyes by transparent methods. Shokuhin Eiseigaku Zasshi, 13, 120-126

Schiller, W. (1937) Rat sarcoma produced by the injection of the dye, light green FS. Amer. J. Cancer, 31, 486-490

The Society of Dyers and Colourists (1971a) Colour Index, 3rd ed., Vol. 4, Yorkshire, UK, p. 4385

The Society of Dyers and Colourists (1971b) Colour Index, 3rd ed., Vol. 1, Yorkshire, UK, p. 1398

The Society of Dyers and Colourists (1971c) Colour Index, 3rd ed., Vol. 2, Yorkshire, UK, p. 2812

The Society of Dyers and Colourists (1975) Colour Index, revised 3rd ed., Vol. 6, Yorkshire, UK, pp. 6235-6236

Takeshita, R. & Sakagami, Y. (1969) Food additives. XII. Detection of some water-soluble acid dyes in foods. Eisei Kagaku, 15, 72-76

Takeshita, R., Yamashita, T. & Itoh, N. (1972) Separation and detection of water-soluble acid dyes on polyamide thin layers. J. Chromat., 73, 173-182

US Food and Drug Administration (1976) Cancellation of certificates. US Code of Federal Regulations, Title 21, part 8.510

US International Trade Commission (1976) Imports of Benzenoid Chemicals and Products, 1974, USITC Publication 762, Washington DC, US Government Printing Office, p. 45

US International Trade Commission (1977a) Synthetic Organic Chemicals, US Production and Sales, 1975, USITC Publication 804, Washington DC, US Government Printing Office, p. 57

US International Trade Commission (1977b) Imports of Benzenoid Chemicals and Products, 1975, USITC Publication 806, Washington DC, US Government Printing Office, p. 43

US Tariff Commission (1919) Report on Dyes and Related Coal-Tar Chemicals, 1918, revised ed., Washington DC, US Government Printing Office, p. 41

Windholz, M., ed. (1976) The Merck Index, 9th ed., Rahway, NJ, Merck & Co., p. 717

Witterholt, V.G. (1969) Triphenylmethane and related dyes. In: Kirk, R.E. & Othmer, D.F., eds, Encyclopedia of Chemical Technology, 2nd ed., Vol. 20, New York, John Wiley and Sons, p. 707

Zuckerman, S. (1964) Colors for foods, drugs, and cosmetics. In: Kirk, R.E. & Othmer, D.F., eds, Encyclopedia of Chemical Technology, 2nd ed., Vol. 5, New York, John Wiley and Sons, pp. 865-866

RHODAMINE B

1. Chemical and Physical Data

1.1 Synonyms and trade names

Colour Index No.: 45170

Colour Index Names: Basic Violet 10; Food Red 15

Chem. Abstr. Services Reg. No.: 81-88-9

Chem. Abstr. Name: *N*-[9-(2-Carboxyphenyl)-6-(diethylamino)-3*H*-xanthen-3-ylidene]-*N*-ethylethanaminium chloride

[9-(*ortho*-Carboxyphenyl)-6-(diethylamino)-3*H*-xanthen-3-ylidene]diethylammonium chloride; D and C Red No. 19; 3-ethochloride of 9-*ortho*-carboxyphenyl-6-diethylamino-3-ethylimino-3-isoxanthene; FD and C Red No. 19; rhodamine; rhodamine B chloride; tetraethylrhodamine

Acid Brilliant Pink B; ADC Rhodamine B; Aizen Rhodamine BH; Aizen Rhodamine BHC; Akiriku Rhodamine B; Brilliant Pink B; Calcozine Red BX; Calcozine Rhodamine BL; Calcozine Rhodamine BX; Calcozine Rhodamine BXP; Cerise Toner X1127; Certiqual Rhodamine; Cogilor Red 321.10; Cosmetic Brilliant Pink Bluish D conc; Diabasic Rhodamine B; Edicol Supra Rose B; Edicol Supra Rose BS; Elcozine Rhodamine B; Eriosin Rhodamine B; Geranium Lake N; Hexacol Rhodamine B Extra; Ikada Rhodamine B; Iragen Red L-U; Mitsui Rhodamine BX; 11411 Red; Red No. 213; Rheonine B; Rhodamine BA; Rhodamine BA Export; Rhodamine B Extra; Rhodamine B Extra M 310; Rhodamine B Extra S; Rhodamine BF; Rhodamine B500 Hydrochloride; Rhodamine BL; Rhodamine BN; Rhodamine BS; Rhodamine BX; Rhodamine BXL; Rhodamine BXP; Rhodamine FB; Rhodamine Lake Red B; Rhodamine O; Sicilian Cerise Toner A-7127; Symulex Magenta F; Symulex Pink F; Symulex Rhodamine B Toner F; Takaoka Rhodamine B

1.2 Chemical formula and molecular weight

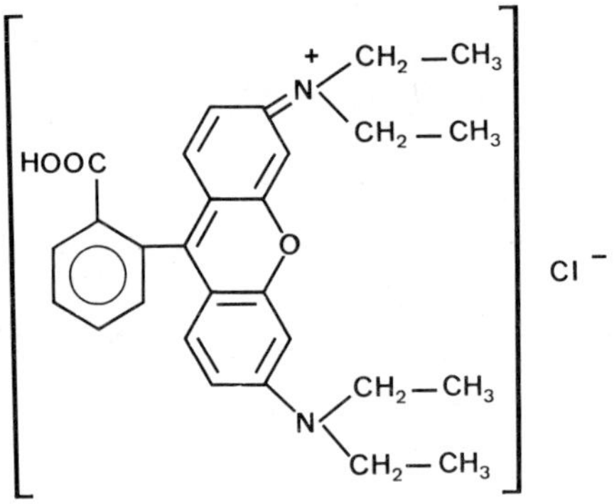

$C_{28}H_{31}N_2O_3.Cl$ Mol. wt: 479.0

1.3 Chemical and physical properties of the pure substance

From Weast (1976), unless otherwise specified

(a) Description: Green crystals or reddish-violet powder

(b) Melting-point: 165°C (basic anhydrous form)

(c) Spectroscopy data: λ_{max} 546 nm (E^1_1 = 1639) (basic form); nuclear magnetic resonance (Horobin & Murgatroyd, 1969) and mass spectral data (Davie *et al.*, 1974) have been published.

(d) Solubility: Soluble in water, ethanol and benzene

1.4 Technical products and impurities

Rhodamine B, certified for use as a colour additive in drugs and cosmetics in the US by the US Food and Drug Administration, contains at least 92.0% pure dye (as determined by titanium trichloride titration) and not more than: 5.0% volatile matter (at 135°C), 2.0% sodium chloride and sodium sulphate, 1.0% mixed oxides, 1.0% water-insoluble matter, 0.5% ether extracts (from acid solution), 0.2% diethyl-3-aminophenol, 30 mg/kg heavy metals other than lead and arsenic (by precipitation as sulphides), 20 mg/kg lead and 2 mg/kg arsenic (as As_2O_3) (US Food and Drug Administration, 1976).

2. Production, Use, Occurrence and Analysis

For background information on this section, see preamble, p. 15.

2.1 Production and use

(a) Production

The colouring properties of rhodamine B were reported in 1887 by Céresole and in 1888 by Homolka and Boedeker (The Society of Dyers and Colourists, 1971a). It is produced commercially by condensing *N,N*-diethyl-3-aminophenol with phthalic anhydride, followed by treatment with dilute hydrochloric acid, or by reacting 3',6'-dichlorofluoran with diethylamine under pressure (Cesark, 1970).

Rhodamine B has been produced commercially in the US for over fifty years (US Tariff Commission, 1919). In 1974, seven companies reported its production; four of them produced 9500 kg of a grade suitable for use in drugs and cosmetics, and three companies produced an undisclosed amount (see preamble, p. 15) of general purpose rhodamine B (US International Trade Commission, 1976a). In 1973, these same three companies reported production of 143 thousand kg of general purpose rhodamine B (US International Trade Commission, 1975). In 1975, five US companies reported production of 5500 kg of the drug and cosmetic grade, and three other companies reported an undisclosed amount (see preamble, p. 15) of general purpose rhodamine B (US International Trade Commission, 1977a). US imports through principal US customs districts amounted to about 60 thousand kg per year during 1972-1974 (US International Trade Commission, 1976b; US Tariff Commission, 1973, 1974) and 27.7 thousand kg in 1975 (US International Trade Commission, 1977b).

Rhodamine B is produced commercially in Japan by one company. Japanese production amounted to 44 thousand kg in 1972 and to 18 thousand kg in 1975 (Japan Dyestuff Industry Association, 1976). Japan exported 14 thousand kg in 1972 and 11 thousand kg in 1974; there are no Japanese imports of this chemical (Japan Tariff Association, 1976).

An estimated 500 thousand kg rhodamine B are produced annually in western Europe.

(b) Use

Rhodamine B can be used to dye silk, cotton, wool, bast fibres, nylon, acetate fibres, paper, spirit inks and lacquers, soap, wood stains, feathers, leather and distempers on china clay (The Society of Dyers and Colourists, 1971b). In the US, it has been used as a drug and cosmetic colour in aqueous drug solutions, tablets, capsules, toothpaste, soap, hair-waving fluids, bath salts, lipsticks and rouges (Zuckerman, 1964). This colour has also been used as a tracing agent in water pollution studies, as a dye for waxes and antifreeze (Cesark, 1970) and as an analytical reagent for antimony, bismuth, cobalt, niobium, gold, manganese, mercury, molybdenum, tantalum, thallium and tungsten (Windholz, 1976).

The free base of rhodamine B, C.I. Solvent Red 49 (9-*ortho*-carboxyphenyl-6-diethylamino-3-diethylamino-3-isoxanthene), is used as a dye in various non-textile applications (e.g., in inks, solvents, fats, etc.) (The Society of Dyers and Colourists, 1971c). In 1975, four US companies reported production of 29 thousand kg C.I. Solvent Red 49 (US International Trade Commission, 1977a).

A mixture of the phosphomolybdic acid and phosphotungstic acid salts of rhodamine B is marketed as C.I. Pigment Violet 1, a commercially significant pigment used mainly in printing inks (The Society of Dyers and Colourists, 1971c). In 1975, nine US companies produced over 54.5 thousand kg of the various forms of this pigment (US International Trade Commission, 1977a).

In Japan and Switzerland, rhodamine B is used as a dye.

In the US, this colour is provisionally listed for use in drugs and cosmetics, subject to certification, by the US Food and Drug Administration. Certified rhodamine B must meet the product specifications as promulgated by law (see section 1.4, p. 222). Its use in drug products for internal use and in mouthwashes, dentrifrices and proprietary products is limited to a maximum tolerance level of 0.75 mg in the amount of product reasonably expected to be ingested in one day. In lipsticks, the maximum amount of all colours used, including rhodamine B, is limited to a maximum of 6% pure dye by weight of each lipstick. It may be used without tolerance levels in other externally applied cosmetics and drugs (US Food and Drug Administration, 1976).

In western Europe, it may be used in cosmetics, including those which may be in contact with mucous membranes (EEC, 1976).

2.2 Occurrence

Rhodamine B is not known to occur as a natural product. No data on its occurrence in the environment were available to the Working Group.

2.3 Analysis

The Association of Official Analytical Chemists has published methods recommended for use in batch certifications of rhodamine B so that it meet specifications as defined by the US Food and Drug Administration (Horwitz, 1975). Methods used for determining the amount of basic dyes, including rhodamine B, in a sample have been reviewed (Burgess *et al.*, 1973).

Rhodamine B has been determined in amounts as low as 0.1 ng by chemiluminescence (Bowman & Alexander, 1966). Fluorescent dyes, including rhodamine B, used as hydrological tracer substances have been determined fluorimetrically with a detection limit of 0.004 μg/l (Behrens, 1971), and in sea water with a detection limit of 0.05 μg/l (Bernhard *et al.*, 1972).

Paper chromatography has been used to separate and identify basic dyes, including rhodamine B, using two solvent systems (Fogg *et al.*, 1973). In another study, 160 solvents were tested (Tajiri, 1969). Two-dimensional paper chromatography using two solvent systems has also been used (Dobrecky & De Carnevale Bonino, 1971).

Thin-layer chromatography has been used to separate and identify rhodamine B as follows: (1) in the presence of other water-soluble fluorescent compounds on silica gel and unmodified cellulose layers with two solvent systems (Gill, 1967), (2) in the presence of other low-polarity dyes using a zigzag development path on silica gel plates (Jimeno de Ossó, 1971), and (3) in the presence of other red food dyes on a polyamide-silica gel mixed layer using five solvent systems, with a detection limit of approximately 2 μg (Chiang, 1969). Thin-layer chromatography has also been used to separate rhodamine B and other dyes as part of a method of analysis in which the separated dyes have been: (1) identified by their nuclear magnetic resonance or visible spectra (Horobin & Murgatroyd, 1969), (2) identified by their absorption or fluorescence spectra and determined fluorimetrically (Ruedt, 1969), and (3) identified by colour under ultra-violet or ordinary light and determined with a spectrodensitometer (Penner, 1968).

3. Biological Data Relevant to the Evaluation of Carcinogenic Risk to Man

3.1 Carcinogenicity and related studies in animals

(a) Oral administration

Mouse: Fifteen male and 15 female stock mice were given 0.05% rhodamine B in the drinking-water for 52 weeks; the weekly intake per mouse was 17 mg, and the total dose was 884 mg/animal. Mice were allowed to survive as long as possible. Two of the male mice that died between 50 and 69 weeks had intestinal tumours, one of which was malignant; 3 mice had polyposis of the stomach, and 7 had lymphomas (Bonser *et al.*, 1956) [The Working Group noted that no contemporary controls were used].

Rat: Twenty-one rats of both sexes of a mixed Saitama strain, weighing approximately 150 g, were fed a rice diet that initially contained 0.13% rhodamine B, increased to 0.2% after 7 months. Eighteen rats were alive at 300 days, and one survived for 661 days. None of the rats developed tumours (Umeda, 1956) [The Working Group noted the insufficient duration of the experiment].

(b) Subcutaneous and/or intramuscular administration

Mouse: Twenty-five mice of both sexes of a mixed Saitama strain, weighing approximately 20 g, were given twice weekly s.c. injections of 0.1 ml of a 0.5% aqueous solution of rhodamine B. The concentration of the solution was reduced to 0.25% after 1.5 weeks and increased to the original concentration after 4 months; six mice were alive at 150 days, and 1 survived for 260 days. In a second experiment, 120 male and 180 female mice were given 0.1 ml of a 0.5% solution 2 or 3 times a week for

4 weeks; the dose was increased to 0.7% for one month and then reduced to the original concentration. After 6 months of treatment, the concentration of the solution was gradually decreased. Subsequently, injections were discontinued for about 2 months. Thirty-seven mice were alive at 200 days, and 10 survived more than 280 days. None of the mice in either experiment developed tumours at the site of injection (Umeda, 1956) [The Working Group noted the high mortality rate among treated animals].

Rat: Albino rats of a mixed Saitama strain received s.c. injections, 2-3 times a week, of 1 ml of an aqueous solution of rhodamine B. In the first experiment, with 20 rats, the concentration was gradually increased from 0.5% to 0.75% and then to 1.5% over one year, after which the concentration was gradually decreased. Six of the 10 rats that survived 420 days developed fibrosarcomas at the site of injection between 420 and 630 days. In a second experiment, 17 male and 13 female Wistar-descendant rats were injected 2-3 times a week with a 0.5% solution for 2 months; the concentration was then increased to 0.75% and after 5 months to 1.0%. Two of the 11 rats alive at 436 days had fibrosarcomas at the site of injection (Umeda, 1956) [The Working Group noted that no data on controls were reported].

3.2 Other relevant biological data

(a) Experimental systems

The i.v. LD_{50} of rhodamine B in propylene glycol in rats is 89.5 mg/kg bw (Webb *et al.*, 1961).

It was reported in an abstract that groups of 5 male and 5 female weanling albino rats ingested diets containing 0, 0.1, 0.25, 0.5, 1.0 or 2.0% rhodamine B for 18 weeks. Growth was retarded in all animals except those given the lowest dose; all of the rats given the highest dose died within 6 weeks with evidence of liver damage (Hansen *et al.*, 1958).

Rhodamine was metabolized similarly in dogs, rats and rabbits, as evidenced by chromatographic analysis of their urine and faecal extracts. The compound was extensively absorbed from the gastrointestinal tract: of 1% given in the diet, only 3-5% was recovered unchanged in the urine and faeces. Rhodamine B is de-ethylated enzymatically; two of three observed metabolites were identified as *N,N'*-diethyl-3,6-diaminofluoran and 3,6-diaminofluoran (Webb & Hansen, 1961).

Rhodamine B exhibits high plasma protein binding (Gangolli *et al.*, 1972).

No data on the embryotoxicity, teratogenicity or mutagenicity of this colour were available to the Working Group.

(b) Man

No data were available to the Working Group.

3.3 Case reports and epidemiological studies

No data were available to the Working Group.

4. Comments on Data Reported and Evaluation[1]

4.1 Animal data

Rhodamine B has been tested in mice and rats by subcutaneous injection and, in inadequate studies, by oral administration. It was carcinogenic in rats when injected subcutaneously, producing local sarcomas.

4.2 Human data

No case reports or epidemiological studies were available to the Working Group.

---

[1]See also the section, 'Animal Data in Relation to the Evaluation of Risk to Man' in the introduction to this volume, p. 13.

5. References

Behrens, H. (1971) Quantitative detection of fluorescent dyes applied as hydrological tracer substances. Geol. Bavarica, 64, 120-131

Bernhard, M., Cagnetti, P. & Zattera, A. (1972) Preliminary results with rhodamine B as a pollutant tracer in shallow sea water. G. Fis. Sanit. Prot. Radiaz., 16, 71-80

Bonser, G.M., Clayson, D.B. & Jull, J.W. (1956) The induction of tumours of the subcutaneous tissues, liver and intestine in the mouse by certain dye-stuffs and their intermediates. Brit. J. Cancer, 10, 653-667

Bowman, R.L. & Alexander, N. (1966) Ozone-induced chemiluminescence of organic compounds. Science, 154, 1454-1456

Burgess, C., Fogg, A.G. & Burns, D.T. (1973) An evaluation of methods for the purification and quantitative determination of basic dyestuffs used in spectrophotometry. Lab. Pract., 22, 472-475

Cesark, F.F. (1970) Xanthene dyes. In: Kirk, R.E. & Othmer, D.F., eds, Encyclopedia of Chemical Technology, 2nd ed., Vol. 22, New York, John Wiley and Sons, p. 434, 436

Chiang, H.-C. (1969) Polyamide-silica gel thin-layer chromatography of red food dyes. J. Chromat., 40, 189-190

Davie, E., Morris, J.H. & Smith, W.E. (1974) Electron-impact and thermal degradation of rhodamine F5G chloride and related compounds. Org. Mass Spectr., 9, 763-773

Dobrecky, J. & De Carnevale Bonino, R.C.D. (1971) Separation, identification, and determination of 6 food dyes considered hazardous for addition to foods and drugs. Rev. Asoc. Bioquím. Argent., 36, 143-145

EEC (1976) Législation. Journal Officiel des Communautés Européennes, No. L 262, p. 189

Fogg, A.G., Willcox, A. & Burns, D.T. (1973) Separation and identification of certain xanthene and other red basic dyes. J. Chromat., 77, 237-238

Gangolli, S.D., Grasso, P., Golberg, L. & Hooson, J. (1972) Protein binding by food colourings in relation to the production of subcutaneous sarcoma. Fd Cosmet. Toxicol., 10, 449-462

Gill, J.E. (1967) Partition and ion-exchange thin-layer chromatography of water-soluble fluorescent compounds. J. Chromat., 26, 315-319

Hansen, W.H., Fitzhugh, O.G. & Williams, M.W. (1958) Subacute oral toxicity of nine D and C coal-tar colors (Abstract No. 7789). J. Pharmacol. exp. Ther., 122, 29A

Horobin, R.W. & Murgatroyd, L.B. (1969) The identification and purification of pyronin and rhodamine dyes. Stain Technol., 44, 297-302

Horwitz, W., ed. (1975) Official Methods of Analysis of the Association of Official Analytical Chemists, 12th ed., Washington DC, Association of Official Analytical Chemists, pp. 631-632, 634, 638

Japan Dyestuff Industry Association (1976) Statistics of Synthetic Dyes, 1972-1975, Tokyo

Japan Tariff Association (1976) Japan Export and Import, Commodity By Country, 1972-1975, Tokyo

Jimeno de Ossó, F. (1971) Very long Z-shaped thin-layer chromatography and its application to the separation of dyes. J. Chromat., 58, 294-296

Penner, M.H. (1968) Thin-layer chromatography of certified coal tar color additives. J. pharm. Sci., 57, 2132-2135

Ruedt, U. (1969) Fluorimetric determination of xanthene coloring materials in lipsticks. Fette, Seifen, Anstrichm., 71, 982-985

The Society of Dyers and Colourists (1971a) Colour Index, 3rd ed., Vol. 4, Yorkshire, UK, p. 4420

The Society of Dyers and Colourists (1971b) Colour Index, 3rd ed., Vol. 1, Yorkshire, UK, p. 1651

The Society of Dyers and Colourists (1971c) Colour Index, 3rd ed., Vol. 3, Yorkshire, UK, pp. 3328-3329, 3334-3335, 3596-3597

Tajiri, H. (1969) Paper chromatography of dyes. Senshoku Kogyo, 17, 513-519

Umeda, M. (1956) Experimental study of xanthene dyes as carcinogenic agents. Gann, 47, 51-78

US Food and Drug Administration (1976) Food and drugs. US Code of Federal Regulations, Title 21, parts 8.501, 8.502, 8.503, 9.164

US International Trade Commission (1975) Synthetic Organic Chemicals, US Production and Sales, 1973, ITC Publication 728, Washington DC, US Government Printing Office, pp. 56, 68

US International Trade Commission (1976a) Synthetic Organic Chemicals, US Production and Sales, 1974, USITC Publication 776, Washington DC, US Government Printing Office, pp. 57, 66, 76

US International Trade Commission (1976b) Imports of Benzenoid Chemicals and Products, 1974, USITC Publication 762, Washington DC, US Government Printing Office, p. 51

US International Trade Commission (1977a) Synthetic Organic Chemicals, US Production and Sales, 1975, USITC Publication 804, Washington DC, US Government Printing Office, pp. 52-53, 61, 71, 73, 80, 84

US International Trade Commission (1977b) Imports of Benzenoid Chemicals and Products, 1975, USITC Publication 806, Washington DC, US Government Printing Office, p. 51

US Tariff Commission (1919) Report on Dyes and Related Coal-Tar Chemicals, 1918, revised ed., Washington DC, US Government Printing Office, p. 42

US Tariff Commission (1973) Imports of Benzenoid Chemicals and Products, 1972, TC Publication 601, Washington DC, US Government Printing Office, p. 49

US Tariff Commission (1974) Imports of Benzenoid Chemicals and Products, 1973, TC Publication 688, Washington DC, US Government Printing Office, p. 47

Weast, R.C., ed. (1976) CRC Handbook of Chemistry and Physics, 57th ed., Cleveland, Ohio, Chemical Rubber Co., p. C-492

Webb, J.M. & Hansen, W.H. (1961) Studies of the metabolism of rhodamine B. Toxicol. appl. Pharmacol., 3, 86-95

Webb, J.M., Hansen, W.H., Desmond, A. & Fitzhugh, O.G. (1961) Biochemical and toxicologic studies of rhodamine B and 3,6-diaminofluoran. Toxicol. appl. Pharmacol., 3, 696-706

Windholz, M., ed. (1976) The Merck Index, 9th ed., Rahway, NJ, Merck & Co., p. 1061

Zuckerman, S. (1964) Colors for foods, drugs, and cosmetics. In: Kirk, R.E. & Othmer, D.F., eds, Encyclopedia of Chemical Technology, 2nd ed., Vol. 5, New York, John Wiley and Sons, pp. 865, 870, 872

RHODAMINE 6G

1. Chemical and Physical Data

1.1 Synonyms and trade names

Colour Index No.: 45160

Colour Index Name: Basic Red 1

Chem. Abstr. Services Reg. No.: 989-38-8

Chem. Abstr. Name: 2-[6-(Ethylamino)-3-(ethylimino)-2,7-dimethyl-3*H*-xanthen-9-yl]benzoic acid ethyl ester, monohydrochloride

Basic rhodamine yellow; C.I. Basic Red 1, monohydrochloride; *ortho*-[6-(ethylamino)-3-(ethylimino)-2,7-dimethyl-3*H*-xanthen-9-yl]benzoic acid ethyl ester, monohydrochloride

Basic Rhodaminic Yellow; Calcozine Red 6G; Calcozine Rhodamine 6GX; Elcozine Rhodamine 6GDN; Eljon Pink Toner; Fanal Pink GFK; Fanal Red 25532; Heliostable Brilliant Pink B extra; Mitsui Rhodamine 6GCP; Nyco Liquid Red GF; Rhodamine 69DN Extra; Rhodamine F4G; Rhodamine F5G; Rhodamine F5G chloride; Rhodamine 6GB; Rhodamine 6GBN; Rhodamine 6GCP; Rhodamine 4GD; Rhodamine 6GD; Rhodamine GDN; Rhodamine 5GDN; Rhodamine 6GDN; Rhodamine 6GDN Extra; Rhodamine 6GEx ethyl ester; Rhodamine 6G Extra; Rhodamine 6G Extra Base; Rhodamine 4GH; Rhodamine 6GH; Rhodamine 5GL; Rhodamine 6G lake; Rhodamine 6GX; Rhodamine 6JH; Rhodamine 7JH; Rhodamine Lake Red 6G; Rhodamine Y 20-7425; Rhodamine Zh; Rhodamine 6ZH; Rhodamine 6Zh-DN; Silosuper Pink B

1.2 Chemical formula and molecular weight

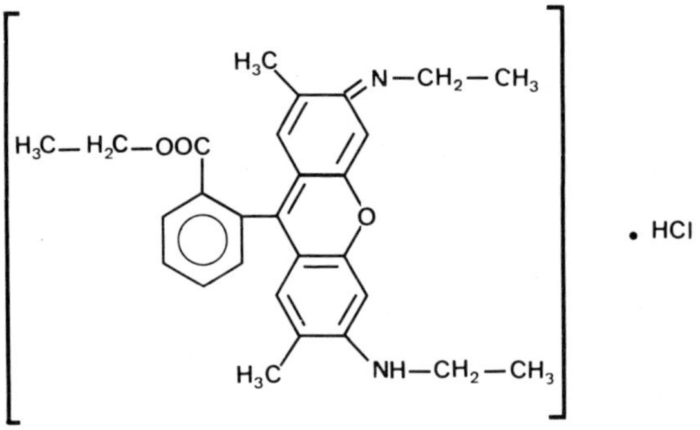

$C_{28}H_{30}N_2O_3.HCl$ Mol. wt: 479.0

1.3 Chemical and physical properties of the pure substance

(a) Spectroscopy data: λ_{max} 535 nm; nuclear magnetic resonance (Horobin & Murgatroyd, 1969) and mass spectral data (Davie *et al.*, 1974) have been published.

(b) Solubility: Soluble in water and ethanol (The Society of Dyers and Colourists, 1971a)

(c) Stability: Volatilizes at <200°C (Davie *et al.*, 1974)

1.4 Technical products and impurities

No data were available to the Working Group.

2. Production, Use, Occurrence and Analysis

For background information on this section, see preamble, p. 15.

2.1 Production and use

(a) Production

The colouring properties of rhodamine 6G were reported in 1892 by Bernthsen and by Schmid and Rey (The Society of Dyers and Colourists, 1971a). Rhodamine 6G is prepared commercially by condensing 3-ethylamino-4-methylphenol with phthalic anhydride, followed by esterification with ethyl chloride under pressure (Cesark, 1970).

Rhodamine 6G has been produced commercially in the US for over fifty years (US Tariff Commission, 1922). Only two US companies reported an undisclosed amount (see preamble, p. 15) in 1975 (US International Trade Commission, 1977a). US imports through principal US customs districts were reported as 69 thousand kg in 1964 (US International Trade Commission, 1976a) and 10.4 thousand kg in 1975 (US International Trade Commission, 1976b).

Rhodamine 6G is produced commercially in Japan by one company, and production amounted to 15 thousand kg in 1974 and 2 thousand kg in 1975. There are no Japanese exports or imports.

No data on its production in Europe were available to the Working Group.

(b) Use

Rhodamine 6G can be used to dye silk, cotton, wool, bast fibres and paper (The Society of Dyers and Colourists, 1971b). It has also been used as a tracing agent in water pollution studies, in leather dyeing (Cesark, 1970) and as an adsorption indicator, especially in very acid solutions (Matsuyama, 1966).

A mixture of the phosphomolybdic and phosphotungstic acid salts of rhodamine 6G is marketed as C.I. Pigment Red 81, an important pigment used primarily in printing inks (The Society of Dyers and Colourists, 1971c). In 1975, 12 US companies reported production of more than 250 thousand kg of this pigment (US International Trade Commission, 1976b).

The copper ferrocyanide complex of rhodamine 6G, C.I. Pigment Red 169, is used in printing inks (The Society of Dyers and Colourists, 1971c); however, it does not appear to be produced commercially in the US. Imports of C.I. Pigment Red 169 through principal US customs districts amounted to 1124 kg in 1974 (US International Trade Commission, 1976a) and 25 kg in 1975 (US International Trade Commission, 1977b).

In western Europe, rhodamine 6G may provisionally be used in cosmetics which may come into contact with mucous membranes (EEC, 1976).

In Japan, rhodamine 6G is used as a dye.

2.2 Occurrence

Rhodamine 6G is not known to occur as a natural product. No data on its occurrence in the environment were available to the Working Group.

2.3 Analysis

A method for detecting rhodamine 6G at very low levels (<10^{-12} to 10^{-6} M) has been described in which aqueous or ethanolic solutions of the dye are excited by a pulsed nitrogen laser; the resulting fluorescence is measured (Bradley & Zare, 1976).

Paper chromatography has been used to separate and identify basic dyes, including rhodamine 6G, using two or more solvent systems (Fogg *et al.*, 1973; Tajiri, 1969).

Dyes used in lipsticks, including rhodamine 6G, have been separated by thin-layer chromatography and identified by their absorption and fluorescence spectra or determined fluorimetrically (Ruedt, 1969). Thin-layer chromatography on unactivated, precoated silica, using one solvent system, has been used to separate biological stains, including rhodamine 6G. The separated dyes were then identified by their nuclear magnetic resonance or visible spectra (Horobin & Murgatroyd, 1969).

Gel permeation chromatography on Sephadex LH-20 resin has been used to separate and identify biological stains, including rhodamine 6G (Horobin, 1971).

3. Biological Data Relevant to the Evaluation of Carcinogenic Risk to Man

3.1 Carcinogenicity and related studies in animals

(a) Oral administration

Mouse: Of 40 mice of both sexes of a mixed Saitama strain, weighing about 20 g, fed a rice diet containing 0.05, then 0.01, then 0.02% rhodamine 6G, 12 survived for 100 days or more and 1 for 223 days (total dose, 90.6 mg). None of the mice developed tumours (Umeda, 1956) [The Working Group noted the high mortality and short duration of the experiment].

(b) Subcutaneous and/or intramuscular administration

Rat: Sixteen rats (sex unspecified) of a mixed Saitama strain, weighing about 200 g, received repeated s.c. injections of 1 ml of a 0.02% aqueous solution of rhodamine 6G two to three times per week. After 4 months, the injections were discontinued for one month and then resumed, until a total of 100 injections had been given. Four of 7 rats that survived for 487 days developed fibrosarcomas, 2 of which were transplanted and grew in other rats. One rat that did not have a s.c. sarcoma had a spindle-cell sarcoma of the liver (Umeda, 1956) [The Working Group noted that no data on controls were reported].

3.2 Other relevant biological data

(a) Experimental systems

Rhodamine 6G exhibits high plasma protein binding (Gangolli *et al.*, 1972).

No adequate data on the embryotoxicity, teratogenicity or mutagenicity of this colour were available to the Working Group.

(b) Man

No data were available to the Working Group.

3.3 Case reports and epidemiological studies

No data were available to the Working Group.

4. Comments on Data Reported and Evaluation[1]

4.1 Animal data

Rhodamine 6G is carcinogenic in rats after its subcutaneous injection: it produces sarcomas. It has been inadequately tested in mice by oral administration.

4.2 Human data

No case reports or epidemiological studies were available to the Working Group.

---

[1]See also the section, 'Animal Data in Relation to the Evaluation of Risk to Man' in the introduction to this volume, p. 13.

5. References

Bradley, A.B. & Zare, R.N. (1976) Laser fluorimetry: sub-part-per-trillion detection of solutes. J. Amer. chem. Soc., 98, 620-621

Cesark, F.F. (1970) Xanthene dyes. In: Kirk, R.E. & Othmer, D.F., eds, Encyclopedia of Chemical Technology, 2nd ed., Vol. 22, New York, John Wiley and Sons, pp. 434, 436

Davie, E., Morris, J.H. & Smith, W.E. (1974) Electron-impact and thermal degradation of rhodamine F5G chloride and related compounds. Org. Mass Spectr., 9, 763-773

EEC (1976) Législation. Journal Officiel des Communautés Européennes, No. L 262, p. 197

Fogg, A.G., Willcox, A. & Burns, D.T. (1973) Separation and identification of certain xanthene and other red basic dyes. J. Chromat., 77, 237-238

Gangolli, S.D., Grasso, P. & Golberg, L. & Hooson, J. (1972) Protein binding by food colourings in relation to the production of subcutaneous sarcoma. Fd Cosmet. Toxicol., 10, 449-462

Horobin, R.W. (1971) Analysis and purification of biological stains by gel filtration. Stain Technol., 46, 297-304

Horobin, R.W. & Murgatroyd, L.B. (1969) The identification and purification of pyronin and rhodamine dyes. Stain Technol., 44, 297-302

Matsuyama, G. (1966) Indicators. In: Kirk, R.E. & Othmer, D.F., eds, Encyclopedia of Chemical Technology, 2nd ed., Vol. 11, New York, John Wiley and Sons, p. 558

Ruedt, U. (1969) Fluorimetric determination of xanthene coloring materials in lipsticks. Fette, Seifen, Anstrichm., 71, 982-985

The Society of Dyers and Colourists (1971a) Colour Index, 3rd ed., Vol. 4, Yorkshire, UK, p. 4420

The Society of Dyers and Colourists (1971b) Colour Index, 3rd ed., Vol. 1, Yorkshire, UK, p. 1633

The Society of Dyers and Colourists (1971c) Colour Index, 3rd ed., Vol. 3, Yorkshire, UK, pp. 3314-3315, 3328-3329

Tajiri, H. (1969) Paper chromatography of dyes. Senshoku Kogyo, 17, 513-519

Umeda, M. (1956) Experimental study of xanthene dyes as carcinogenic agents. Gann, 47, 51-78

US International Trade Commission (1976a) Imports of Benzenoid Chemicals and Products, 1974, USITC Publication 762, Washington DC, US Government Printing Office, pp. 50, 74

US International Trade Commission (1976b) Synthetic Organic Chemicals, US Production and Sales of Organic Pigments, 1975, Preliminary, Washington DC, US Government Printing Office, pp. 2, 6, 8

US International Trade Commission (1977a) Synthetic Organic Chemicals, US Production and Sales, 1975, USITC Publication 804, Washington DC, US Government Printing Office, p. 60

US International Trade Commission (1977b) Imports of Benzenoid Chemicals and Products, 1975, USITC Publication 806, Washington DC, US Government Printing Office, pp. 50, 52, 75

US Tariff Commission (1922) Census of Dyes and Other Synthetic Organic Chemicals, 1921, Tariff Information Series No. 26, Washington DC, US Government Printing Office, p. 66

MISCELLANEOUS INDUSTRIAL CHEMICALS

5-AMINOACENAPHTHENE

1. Chemical and Physical Data

1.1 Synonyms and trade names

Chem. Abstr. Reg. Serial No.: 4657-93-6

Chem. Abstr. Name: 1,2-Dihydro-5-acenaphthylenamine

5-Acenaphthenamine

1.2 Chemical formula and molecular weight

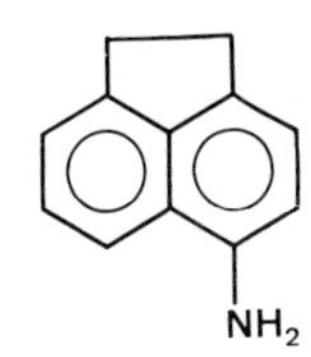

$C_{12}H_{11}N$ Mol. wt: 169.2

1.3 Chemical and physical properties of the pure substance

From Weast (1976), unless otherwise specified

(a) Description: Colourless needles

(b) Melting-point: 108°C

(c) Spectroscopy data: λ_{max} 321 nm (E_1^1 = 429), 240 nm (E_1^1 = 1355), 216 nm (E_1^1 = 2686) in methanol; infra-red and nuclear magnetic spectra have been tabulated (Grasselli, 1973).

(d) Solubility: Soluble in ethanol

(e) Stability: See 'General Remarks on the Substances Considered', p. 27.

1.4 Technical products and impurities

No data were available to the Working Group.

2. Production, Use, Occurrence and Analysis

For background information on this section, see preamble, p. 15.

2.1 Production and use

(a) Production

5-Aminoacenaphthene was prepared by Quincke in 1888 by reducing 5-nitroacenaphthene with tin and hydrochloric acid or with alcoholic ammonium sulphide (Prager *et al.*, 1929). It may also be prepared by reducing 5-nitroacenaphthene with a palladium catalyst and hydrazine hydrate in 95% ethanol (Cava *et al.*, 1965).

No evidence was found that 5-aminoacenaphthene has ever been produced commercially in the US or Japan.

(b) Use

No evidence was found that 5-aminoacenaphthene has ever had any commercial applications in the US or Japan.

It has been investigated experimentally for use as a disinfectant (Rotmistrov *et al.*, 1960), as an agent to control powdery mildew in cucumbers (Ohtsuka *et al.*, 1973), as a diazo component of insoluble azo dyes (Tochilkin & Glushkova, 1962) and in the synthesis of polymers suitable for use in electrophotographic films (Saito & Tsuchiya, 1974).

2.2 Occurrence

5-Aminoacenaphthene is not known to occur as a natural product. No data on its occurrence in the environment were available to the Working Group.

2.3 Analysis

Although no specific studies on this compound were available to the Working Group, the accompanying monographs describe a number of methods generally applicable to the analysis of aromatic amines.

3. Biological Data Relevant to the Evaluation of Carcinogenic Risk to Man

3.1 Carcinogenicity and related studies in animals

(a) Intraperitoneal administration

Mouse: In an abstract, it was reported that a group of 20 female dd mice received i.p. injections of 5-aminoacenaphthene in arachis oil at a dose of 6 mg/kg bw twice weekly for a period of 18 months; a group of 20 female controls were given arachis oil alone. In 11 mice that survived treatment with 5-aminoacenaphthene, 3 myeloid leukaemias and 1 lymphosarcoma were found. No tumours were detected in the controls (Takemura, 1970) [The Working Group noted the inadequacy of the study].

(b) Other experimental systems

Bladder implantation: In the same abstract, it was reported that a group of 90 female dd mice received single bladder implantations of a paraffin wax pellet mixed with 5-aminoacenaphthene, and 123 mice received implantations of paraffin wax pellets alone. After 40 weeks, carcinomas of the bladder epithelium were observed in 12/49 survivors that had received 5-aminoacenaphthene, compared with 6/82 controls ($P<0.05$) (Takemura, 1970).

3.2 Other relevant biological data

No data were available to the Working Group.

3.3 Case reports and epidemiological studies

No data were available to the Working Group.

4. Comments on Data Reported and Evaluation

4.1 Animal data

Data reported in an abstract indicate that 5-aminoacenaphthene increased the incidence of bladder carcinomas in mice following implantation of paraffin wax pellets containing this compound into the bladder and that it produced leukaemias following its intraperitoneal injection.

No evaluation of the carcinogenicity of 5-aminoacenaphthene can be made on the basis of the available data.

4.2 Human data

No case reports or epidemiological studies were available to the Working Group.

5. References

Cava, M.P., Merkel, K.E. & Schlessinger, R.H. (1965) Pleiadene systems. II. On the mechanism of acepleiadylene formation - a vinylogous elimination in the acenaphthene series. Tetrahedron Lett., 21, 3059-3064

Grasselli, J.G., ed. (1973) CRC Atlas of Spectral Data and Physical Constants for Organic Compounds, Cleveland, Ohio, Chemical Rubber Co., p. B-91

Ohtsuka, T., Takahashi, N., Satake, K. & Funayama, S. (1973) Acenaphthenes for controlling powdery mildew. Japanese Patent 73 00,051, January 5 (to Kureha Chemical Industry Co., Ltd)

Prager, B., Jacobson, P., Schmidt, P. & Stern, D., eds (1929) Beilsteins Handbuch der Organischen Chemie, 4th ed., Vol. 12, Syst. No. 1734, Berlin, Springer-Verlag, p. 1322

Rotmistrov, M.N., Stetsenko, A.V., Kulik, G.V., Vasilevskaya, I.A., Baĭsheva, V.G. & Gamaleya, N.F. (1960) Disinfectant properties of some acenaphthene derivatives. Mikrobiologiya, 29, 756-761

Saito, H. & Tsuchiya, Y. (1974) Organic photoconductive material for electrophotography. Japanese Patent 74 01,585, January 14 (to Canon K. K.)

Takemura, N. (1970) Carcinogenic action of the intermediates of fluorescent whitening agents, 5-nitroacenaphthene and 5-aminoacenaphthene (Abstract No. 123). In: Cumley, R.W., ed., Tenth International Cancer Congress, Houston, Texas, Houston, Medical Arts Publ. Co., p. 79

Tochilkin, A.I. & Glushkova, G.A. (1962) Amine derivatives of acenaphthene as diazo components for ice colors. In: Proceedings of All-Union Scientific and Technical Conference on the Problems of Synthesis and Use of Organic Dyes, 1961, pp. 3-9

Weast, R.C., ed. (1976) CRC Handbook of Chemistry and Physics, 57th ed., Cleveland, Ohio, Chemical Rubber Co., p. C-81

para-AMINOBENZOIC ACID

1. Chemical and Physical Data

1.1 Synonyms and trade names

Chem. Abstr. Services Reg. No.: 150-13-0

Chem. Abstr. Name: 4-Aminobenzoic acid

1-Amino-4-carboxybenzene; anticanitic vitamin; anti-chromotrichia factor; bacterial vitamin H'; 4-carboxyaniline; *para*-carboxyaniline; *para*-carboxyphenylamine; chromotrichia factor; PAB; PABA; trichochromogenic factor; vitamin B_x; vitamin H'

Amben; Paraminol; Paranate

1.2 Chemical formula and molecular weight

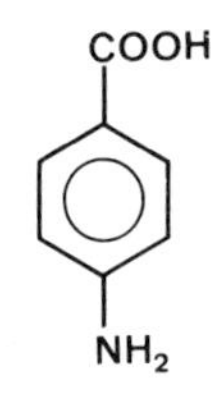

$C_7H_7NO_2$ Mol. wt: 137.1

1.3 Chemical and physical properties of the pure substance

(a) Description: Colourless crystals (Duncker, 1964)

(b) Melting-point: 188-189°C (Weast, 1976)

(c) Spectroscopy data: λ_{max} 289 nm (E_1^1 = 1328), 219 nm (E_1^1 = 624) in methanol; infra-red and nuclear magnetic resonance spectra have been tabulated (Grasselli, 1973).

(d) Solubility: Soluble in water (6 g/l at 25°, 11 g/l at 100°); very soluble in ethanol (125 g/l) (Duncker, 1964); and soluble in ether (20 g/l) (Weast, 1976)

(e) Stability: See 'General Remarks on the Substances Considered', p. 27.

1.4 Technical products and impurities

para-Aminobenzoic acid is available in the US as a technical and as a US Pharmacopeia (USP) grade. The latter contains 98.5-101.5% pure substance and a maximum of 2 mg/kg heavy metals and 0.2% moisture. Solutions are available that contain 5% *para*-aminobenzoic acid in 65-75% ethanol (US Pharmacopeial Convention, Inc., 1975). Tablets and solutions of the sodium salt are also available (Greengard, 1970).

In western Europe, technical grade *para*-aminobenzoic acid contains a minimum of 98.5% pure substance; ash, 0.2% max.; loss on drying, 0.2% max.; and a melting-point of 185°C min.

In Japan, *para*-aminobenzoic acid has a minimum purity of 98% and contains 4-nitrobenzoic acid and aminobenzoic acid isomers as purities.

2. Production, Use, Occurrence and Analysis

For background information on this section, see preamble, p. 15.

2.1 Production and use

(a) Production

para-Aminobenzoic acid was prepared by Fischer in 1863 by reducing 4-nitrobenzoic acid with ammonium sulphide (Prager *et al.*, 1931). It is manufactured commercially by the catalytic hydrogenation of 4-nitrobenzoic acid using a platinum or palladium catalyst or by reduction with tin or iron and hydrochloric acid (Duncker, 1964).

para-Aminobenzoic acid has been produced commercially in the US for over fifty years (US Tariff Commission, 1919). Three US companies reported undisclosed amounts (see preamble, p. 15) in 1975 (US International Trade Commission, 1977a). US imports through principal US customs districts amounted to 84.4 thousand kg in 1972 (US Tariff Commission, 1973), 81.2 thousand kg in 1973 (US Tariff Commission, 1974), 55.7 thousand kg in 1974 (US International Trade Commission, 1976) and 53.4 thousand kg in 1975 (US International Trade Commission, 1977b).

It is estimated that five companies in western Europe produce a total of 100 thousand to 1 million kg *para*-aminobenzoic acid per year. Annual production in the Federal Republic of Germany, Italy, Scandinavia and the UK is in the range of 10-100 thousand kg. Benelux, the Federal Republic of Germany, France, Spain and the UK import less than 10 thousand kg annually; Switzerland is believed to import 10-100 thousand kg; and the Federal Republic of Germany exports less than 10 thousand kg. Annual production in eastern Europe is in the range of 10-100 thousand kg, with imports and exports of less than 10 thousand kg.

para-Aminobenzoic acid has been produced commercially in Japan since 1961. In 1975, two Japanese manufacturers produced an estimated 30 thousand kg, and about 6 thousand kg were imported from Sweden and the US (Japan Tariff Association, 1976).

(b) Use

para-Aminobenzoic acid is used at concentrations of 0.5-5% (Harvey, 1975) as a topical sunscreen agent (US Pharmacopeial Convention, 1975). It has also been used in conjunction with salicylates, to prolong their effect, in the treatment of rheumatic fever (Harvey, 1975). Total US sales of *para*-aminobenzoic acid for use in human medicine are estimated to be less than 25 thousand kg annually.

para-Aminobenzoic acid is also used as a sulphonamide antagonist in laboratories, and in veterinary medicine for the treatment of eczema of the nose in dogs (Windholz, 1976).

It can be used as an intermediate for the synthesis of pharmaceuticals and dyes and for the manufacture of folic acid, a member of the vitamin B complex, and of several *para*-aminobenzoic acid derivatives commonly used as local anaesthetics, e.g., novocaine and benzocaine (Duncker, 1964). *para*-Aminobenzoic acid is also used for the production of two dyes, C.I. Direct Orange 102 and C.I. Vat Orange 13, believed to have commercial world significance (The Society of Dyers and Colourists, 1971a). Only C.I. Direct Orange 102 is produced commercially in the US, and in 1975, five US companies sold over 150 thousand kg (US International Trade

Commission, 1977a); it is used to dye cellulose fibres (The Society of Dyers and Colourists, 1971b).

In the UK, 10-100 thousand kg *para*-aminobenzoic acid are used annually for the manufacture of dyestuffs. In Switzerland it is used as a dyestuff intermediate.

In Japan, *para*-aminobenzoic acid is primarily used as an intermediate in the manufacture of pharmaceuticals, and minor amounts are used as a feed additive.

2.2 Occurrence

para-Aminobenzoic acid occurs naturally, especially in bakers' yeast (5-6 mg/kg) and brewers' yeast (10-100 mg/kg) (Duncker, 1964) and as a metabolite in the growth of a variety of microorganisms (Greengard, 1970). It is also present in small amounts in cereal, eggs, milk, meat and human blood, spinal fluid, urine and sweat. No further data on its occurrence in the environment were available to the Working Group.

2.3 Analysis

A review of chromatographic methods for determining *para*-aminobenzoic acid in pharmaceutical preparations has been published (Galecka, 1973), and the Association of Official Analytical Chemists has published an Official Final Action for the colorimetric detection and spectrophotometric determination of *para*-aminobenzoic acid in animal feeds (Horwitz, 1975).

Primary aromatic amines, including *para*-aminobenzoic acid, have been detected in solution or on thin-layer chromatography plates by determining the fluorescence spectra of their fluorescamine derivatives; the method has a limit of detection of 3 μg/l when in solution and 0.2 μg when on chromatographic plates (de Silva & Strojny, 1975). Visualization with ultra-violet light after reaction with glutaconic aldehyde in pyridine has a detection limit of 0.1-0.3 μg (Ohkuma & Sakai, 1975).

Thin-layer chromatography on weak and strong cation and anion exchangers (Lepri *et al.*, 1972, 1974), on rice starch (Canić & Perišić-Janjić, 1974) and on silica gel impregnated with cadmium salts using three solvent systems

(Yasuda, 1972) has been used for the separation of *para*-aminobenzoic acid in mixtures of aromatic amines. Silica gel thin-layers have also been used to separate *para*-aminobenzoic acid in pharmaceutical preparations, with a detection limit of 1-5 μg (Thompson & Johnson, 1974), and with subsequent elution and colorimetric determination (Bićan-Fišter & Dražin, 1973). Tryptophan metabolites, including *para*-aminobenzoic acid, in urine have been separated and detected by two-dimensional thin-layer chromatography (Haworth & Walmsley, 1972).

High-pressure liquid chromatography has been used for determining *para*-aminobenzoic acid and its metabolites in urine and serum, with a limit of detection of 5 ng (Brown *et al.*, 1974).

Electrophoresis has been used to separate and identify aromatic compounds, including *para*-aminobenzoic acid, in mixtures of organic intermediates (Akimova *et al.*, 1972), in blood and urine (Radyuk *et al.*, 1971) and in aromatic amine sample mixtures (Lepri *et al.*, 1972). Paper chromatography has similarly been used, to detect amounts greater than 1 μg by diazotization and coupling with 8-hydroxyquinoline-5-sulphonic acid (Balica & Stefan, 1973), and to detect amounts greater than 100 μg by its complex with iron III (Firmin & Gray, 1974).

Aromatic amines, including *para*-aminobenzoic acid, may be determined colorimetrically by their reaction with peroxydisulphate (Gupta & Srivastava, 1971). Aniline derivatives, including *para*-aminobenzoic acid, have been determined in blood and urine by titration with *N*-bromosuccinimide. The relative error was less than 2% for 0.1-10 mg samples (Barakat *et al.*, 1973).

An ultra-violet spectrophotometric method for its assay in tablets has been described (Paul, 1973); organic compounds, including *para*-benozic acid, may be identified by measuring their ultra-violet spectra in acid, basic and alcoholic media (Siek, 1974). A method for detecting aromatic compounds, including *para*-aminobenzoic acid, in which the phosphorescent spectral characteristics of the compound are measured at room temperature has a detection limited of 0.1 ng (Wellons *et al.*, 1974).

para-Aminobenzoic acid has been determined indirectly by atomic absorption spectrophotometry, measuring its ternary complex with copper and bathophenanthroline (Kidani *et al.*, 1975).

3. Biological Data Relevant to the Evaluation of Carcinogenic Risk to Man

3.1 Carcinogenicity and related studies in animals[1]

Skin application

Mouse: Groups of 50 7-week old female Swiss mice received 0.02 ml of a 1, 5 or 10% solution of *para*-aminobenzoic acid in acetone applied to the dorsal skin twice weekly for up to 110 weeks. A further group of 50 mice were painted with acetone alone, and 150 mice were left untreated. At 80 weeks, 23, 29 and 28 mice in the treated groups, 25 acetone-treated controls and 82 untreated mice were still alive. The percentages of tumour-bearing mice were 44% in acetone-treated controls, 42% in untreated controls, and 36, 44 and 40% in the treated groups. Two acetone-treated control mice, 3 untreated mice and 1 treated mouse had skin tumours. The incidences of lymphomas, lung adenomas and liver haemangiomas were similar in *para*-aminobenzoic acid-treated and control mice (Stenbäck & Shubik, 1974).

3.2 Other relevant biological data

(a) Experimental systems

The oral LD_{50} of *para*-aminobenzoic acid in dogs is 1-3 g/kg bw, in mice 2.4-3.2 g/kg bw, and in rats greater than 6 g/kg bw (Scott & Robbins, 1942). It was reported in an abstract that after oral administration of 10 g/kg bw *para*-aminobenzoic acid, 90% of immature and 45% of adult rats died (Robin *et al.*, 1947). The i.v. LD_{50} of the sodium salt in mice is 4.4-4.8 and in rats 2.5-3.0 g/kg bw (Scott & Robbins, 1942). The oral LD_{50} in rabbits is 1.7-1.9 g/kg bw (Cronheim, 1951).

---

[1]The Working Group was aware of carcinogenicity tests in progress on *para*-aminobenzoic acid involving skin application to rabbits (IARC, 1976).

Acute necrosis of the liver was found in dogs given 2-3 g/kg bw *para*-aminobenzoic acid orally; acute gastroenteritis with haemorrhage was found in dogs given lethal doses (Scott & Robbins, 1942). It was reported in an abstract that rats maintained on a synthetic diet containing 4% *para*-aminobenzoic acid for 2 months exhibited moderate leucocytosis; of 40-day old rats maintained on the same diet containing 6% *para*-aminobenzoic acid, 63% died within a month, but no significant changes were observed in erythrocyte, leucocyte or platelet counts (Robin *et al.*, 1947).

Six female rats were fed diets containing 1% or 2% *para*-aminobenzoic acid from 23 to 60 days of age, at which time they were mated with males of proven fertility. No effects on reproduction or growth were observed (Ershoff, 1946).

The metabolism of *para*-aminobenzoic acid was reviewed by Williams (1959). Very little is excreted as such. Less than 10% of an i.p. dose of 18 or 105 mg/kg bw is excreted within 24 hours in the bile of rats (Abou-el-Makarem *et al.*, 1967). The two major metabolic pathways in a variety of species (guinea-pigs, rabbits and rats, but not dogs) are acetylation of the amino group and conjugation of the carboxy group, either with glycine or with glucuronic acid. Acetylation occurs in liver, heart, lung, blood and kidneys of rats (Koivusalo & Luukkainen, 1959; Koivusalo *et al.*, 1958; Terp, 1951) and in the mucous membranes of the gastrointestinal tract of cattle (Kusen & Moravsky, 1967). *N*-Acetyl-transferase activity in the presence of *para*-aminobenzoic acid is similar in liver and lung tissue of rabbits (Gram *et al.*, 1974). Acetylation not only of *para*-aminobenzoic acid (30-40%) but also of *para*-aminohippuric acid (70%) takes place in the kidney of rabbits (Wan *et al.*, 1972); acetylation of *para*-aminohippurate also occurs in the kidney of guinea-pigs (Setchell & Blanch, 1961).

Acetylation is dose-dependent. In rats given up to 5 mg/kg bw, 75% of the metabolites were acetylated; with higher doses, the extent of acetylation decreased down to 40% (Riggs & Hegsted, 1951). An inverse relationship exists between acetylation and glycine conjugation: when acetylation decreases, glycine conjugation increases (Koivusalo *et al.*,

1958); such a decrease is seen in pantothenic acid-deficient rats (Riggs & Hegsted, 1948). Male rats excreted a larger amount of acetylated conjugates in the urine than females (Luukkainen, 1958).

para-Aminobenzoic acid can decrease thyroid activity and protect against liver damage associated with choline and cystine deficiencies (Handler & Follis, 1948). It is a protective agent in ozone toxicity (Goldstein *et al.*, 1972) and antagonizes the bacteriostatic effect of sulphonamides (Strauss *et al.*, 1941).

A 44-fold molar excess of *para*-aminobenzoic acid (8100 mg/kg of diet) over *N*-2-acetylaminofluorene (AAF) (300 mg/kg of diet) decreased the toxicity of AAF to male Fisher rats but had only a weak effect in preventing its carcinogenic effect (Yamamoto *et al.*, 1970).

para-Aminobenzoic acid (1000 μg/ml) did not mutate *Escherichia coli* from streptomycin sensitivity to resistance or from histidine, methionine and tryptophan requirements to prototrophy (no liver activation was used in these experiments) (Greer, 1958). The ethyl ester of *para*-aminobenzoic acid did not induce mutations in *Salmonella typhimurium* TA 98 or TA 100 in the presence or absence of a liver post-mitochondrial supernatant fraction from rats pretreated with polychlorinated biphenyl (Kanechlor 500) or phenobarbital, nor in *E. coli* WP2 in the absence of a rat liver preparation (Sugimura *et al.*, 1976).

(b) Man

Of an oral dose of 1 g *para*-aminobenzoic acid, 82% was excreted in the urine of 3 male volunteers within 4 hours; *para*-aminohippuric acid and acetyl-*para*-aminohippuric acid were the principal metabolites. Concurrent administration of sodium benzoate totally abolished the excretion of these glycine conjugates (Koivusalo *et al.*, 1959).

After oral doses of 500 mg *para*-aminobenzoic acid, 88% of the metabolites are acetylated; this fraction decreases with higher doses (Gershberg & Kuhl, 1950); glycine conjugation is low in the newborn (Vest & Salzberg, 1965). *para*-Aminobenzoic acid is acetylated by homogenates of human placenta (Juchau *et al.*, 1968).

3.3 Case reports and epidemiological studies

No data were available to the Working Group.

4. Comments on Data Reported and Evaluation

4.1 Animal data

In the only study available, *para*-aminobenzoic acid was tested in mice by skin application; no carcinogenic effects were observed. The available data are insufficient for an evaluation of the carcinogenicity of this compound to be made.

4.2 Human data

No case reports or epidemiological studies were available to the Working Group.

5. References

Abou-el-Makarem, M.M., Millburn, P., Smith, R.L. & Williams, R.T. (1967) Biliary excretion of foreign compounds: benzene and its derivatives in the rat. Biochem. J., 105, 1269-1274

Akimova, T.G., Dedkova, V.P. & Savvin, S.B. (1972) Electrophoretic and chromatographic methods of analyzing organic reagents. IV. Identification of aromatic amines and naphthol sulfonic acids used in organic reagent synthesis. Zh. analit. Khim., 27, 1396-1400

Balica, G. & Stefan, I. (1973) New reactant used in chromatographic identification of novocaine, anesthesin, and *p*-aminobenzoic acid. Rev. Chim. (Bucharest), 24, 390-391

Barakat, M.Z., Fayzalla, A.S. & El-Aassar, S. (1973) Microdetermination of aniline salts and certain derivatives. Microchem. J., 18, 308-323

Bičan-Fišter, T. & Dražin, V. (1973) Quantitative analysis of water-soluble vitamins in multicomponent pharmaceutical forms. J. Chromat., 77, 389-395

Brown, N.D., Lofberg, R.T. & Gibson, T.P. (1974) High-performance liquid chromatographic method for the determination of *p*-aminobenzoic acid and some of its metabolites. J. Chromat., 99, 635-641

Canić, V.D. & Perišić-Janjić, N.U. (1974) Separation of aliphatic and aromatic organic acids on starch thin-layers. Chromatographic behaviour and chemical structure. Z. analyt. Chem., 270, 16-19

Cronheim, G. (1951) Acute toxicity of salicyclic, *p*-aminobenzoic and *p*-aminosalicylic acid alone and in combination (Abstract). Fed. Proc., 10, 289-290

Duncker, C. (1964) Benzoic acid. In: Kirk, R.E. & Othmer, D.F., eds, Encyclopedia of Chemical Technology, 2nd ed., Vol. 3, New York, John Wiley and Sons, p. 435-436

Ershoff, B.H. (1946) Effects of massive doses of *p*-aminobenzoic acid and inositol on reproduction in the rat. Proc. Soc. exp. Biol. (Wash.), 63, 479-480

Firmin, J.L. & Gray, D.O. (1974) A versatile and sensitive method for the detection of organic acids and organic phosphates on paper chromatograms. J. Chromat., 94, 294-297

Galecka, H. (1973) Water-soluble vitamins. *p*-Aminobenzoic acid. In: Borkowski, B., ed., Chromatogra. Cienkowarstwowa Anal. Farm., Warsaw, Panstw. Zakl. Wydawn. Lek., pp. 269-271

Gershberg, H. & Kuhl, W.J., Jr (1950) Acetylation studies in human subjects with metabolic disorders. J. clin. Invest., 29, 1625-1632

Goldstein, B.D., Levine, M.R., Cuzzi-Spada, R., Cardenas, R., Buckley, R.D. & Balchum, O.J. (1972) *p*-Aminobenzoic acid as a protective agent in ozone toxicity. Arch. environm. Hlth, 24, 243-247

Gram, T.E., Litterst, C.L. & Mimnaugh, E.G. (1974) Enzymatic conjugation of foreign chemical compounds by rabbit lung and liver. Drug. Metab. Disp., 2, 254-258

Grasselli, J.G., ed. (1973) CRC Atlas of Spectral Data and Physical Constants for Organic Compounds, Cleveland, Ohio, Chemical Rubber Co., p. B-280

Greengard, P. (1970) Water-soluble vitamins. I. The vitamin B complex. In: Goodman, L.S. & Gilman, A., eds, The Pharmacological Basis of Therapeutics, 4th ed., London, Macmillan, pp. 1662-1663

Greer, S.B. (1958) Growth inhibitors and their antagonists as mutagens and antimutagens in *Escherichia coli*. J. gen. Microbiol., 18, 543-564

Gupta, R.C. & Srivastava, S.P. (1971) Oxidation of aromatic amines by peroxodisulphate ion. II. Identification of aromatic amines on the basis of absorption maxima of coloured oxidation products. Z. analyt. Chem., 257, 275-277

Handler, P. & Follis, R.H., Jr (1948) The role of thyroid activity in the pathogenesis of hepatic lesions due to choline and cystine deficiency. J. Nutr., 35, 669-687

Harvey, S.C. (1975) Topical drugs. In: Osol, A. *et al*., eds, Remington's Pharmaceutical Sciences, 15th ed., Easton, Pennsylvania, Mack Publishing Co., pp. 725-726

Haworth, C. & Walmsley, T.A. (1972) A study of the chromatographic behaviour of tryptophan metabolites and related compounds by chromatography on thin layers of silica gel. I. Qualitative separation. J. Chromat., 66, 311-319

Horwitz, W., ed. (1975) Official Methods of Analysis of the Association of Official Analytical Chemists, 12th ed., Washington DC, Association of Official Analytical Chemists, p. 778

IARC (1976) Information Bulletin on the Survey of Chemicals Being Tested for Carcinogenicity, No. 6, Lyon, p. 229

Japan Tariff Association (1976) Japan Exports and Imports, Commodity By Country, Tokyo

Juchau, M.R., Niswander, K.R. & Yaffe, S.J. (1968) Drug metabolizing systems in homogenates of human immature placentas. Amer. J. Obstet. Gynecol., 100, 348-356

Kidani, Y., Saotome, T., Inagaki, K. & Koike, H. (1975) Quantitative microdetermination of drugs by atomic absorption spectrometry. Indirect determination of *p*-aminobenzoic acid by atomic absorption spectrometry. Bunseki Kagaku, 24, 463-466

Koivusalo, M. & Luukkainen, T. (1959) Acetylation of *p*-aminobenzoic acid in various animal tissues *in vitro*. Acta physiol. scand., 45, 278-282

Koivusalo, M., Luukkainen, T., Miettinen, T. & Pispa, J. (1958) Effect of benzoic acid on the metabolic conjugations of *p*-aminobenzoic acid in the rat. Acta chem. scand., 12, 1919-1922

Koivusalo, M., Luukkainen, T., Miettinen, T. & Pispa, J. (1959) Urinary metabolites of *p*-aminobenzoic acid in man. Ann. Med. exp. Fenn., 37, 93-99

Kusen, S.I. & Moravsky, B.P. (1967) The acetylating ability of digestive tract's mucous membrane of cattle. Dopov. Akad. Nauk Ukr. SSR, Ser. B. 29, 945-947

Lepri, L., Desideri, P.G. & Coas, V. (1972) Chromatographic and electrophoretic behavior of purines and pyrimidines on layers of weak and strong cation exchangers. J. Chromat., 64, 271-284

Lepri, L., Desideri, P.G. & Coas, V. (1974) Chromatographic behavior of aromatic amino acids on thin layers of anion and cation exchangers. J. Chromat., 88, 331-339

Luukkainen, T. (1958) Studies in the acetylation of *p*-aminobenzoic acid and its regulation by the sex glands in the rat organism. Acta physiol. scand., 45, Suppl. 154, 1-100

Ohkuma, S. & Sakai, I. (1975) Detection of aromatic primary amines by a photochemical reaction with pyridine. Bunseki Kagaku, 24, 385-387

Paul, W.L. (1973) UV spectrophotometric analysis of aminobenzoic acid tablets. J. pharm. Sci., 62, 660-662

Prager, B., Jacobson, P., Schmidt, P. & Stern, D., eds (1931) Beilsteins Handbuch der Organischen Chemie, 4th ed., Vol. 14, Syst. No. 1905, Berlin, Springer-Verlag, p. 418

Radyuk, M.I., Rudakov, A.F. & Pesakhovich, L.V. (1971) Detection and determination of novocaine in biological material. Nauch. Tr., Irkutsk. Gos. Med. Inst., 113, 108-110

Riggs, T.R. & Hegsted, D.M. (1948) The effect of pantothenic acid deficiency on acetylation in rat. J. biol. Chem., 172, 539-545

Riggs, T.R. & Hegsted, D.M. (1951) Size of dose and acetylation of aromatic amines in the rat. J. biol. Chem., 193, 669-673

Robin, E.D., Tabor, C.W. & Smith, P.K. (1947) Studies on the toxicity of *para*-aminobenzoic acid in rats (Abstract). Fed. Proc., 6, 366

Scott, C.C. & Robbins, E.B. (1942) Toxicity of *p*-aminobenzoic acid. Proc. Soc. exp. Biol. (Wash.), 49, 184-186

Setchell, B.P. & Blanch, E. (1961) Conjugation of *p*-aminohippurate by the kidney and effective renal plasma-flow. Nature (Lond.), 189, 230-231

Siek, T.J. (1974) Identification of drugs and other toxic compounds from their ultraviolet spectra. I. Ultraviolet absorption properties of sixteen structural groups. J. Forensic Sci., 19, 193-214

de Silva, J.A.F. & Strojny, N. (1975) Spectrofluorometric determination of pharmaceuticals containing aromatic or aliphatic primary amino groups as their fluorescamine (fluram) derivatives. Analyt. Chem., 47, 714-718

The Society of Dyers and Colourists (1971a) Colour Index, 3rd ed., Vol. 4, Yorkshire, UK, p. 4745

The Society of Dyers and Colourists (1971b) Colour Index, 3rd ed., Vol. 2, Yorkshire, UK, p. 2096

Stenbäck, F. & Shubik, P. (1974) Lack of toxicity and carcinogenicity of some commonly used cutaneous agents. Toxicol. appl. Pharmacol., 30, 7-13

Strauss, E., Lowell, F.C. & Finland, M. (1941) Observations on the inhibition of sulfonamide action by *para*-aminobenzoic acid. J. clin. Invest., 20, 189-197

Sugimura, T., Sato, S., Nagao, M., Yahagi, T., Matsushima, T., Seino, Y., Takeuchi, M. & Kawachi, T. (1976) Overlapping of carcinogens and mutagens. In: Magee, P.N., Takayama, S., Sugimura, T. & Matsushima, T., eds, Fundamentals in Cancer Prevention, Baltimore, University Park Press, pp. 191-215

Terp, P. (1951) Studies on elimination of procaine. IV. Hydrolysis and acylation in animals and perfused organs. Acta pharmacol. toxicol., 7, 381-394

Thompson, R.D. & Johnson, G.L. (1974) Thin-layer chromatography of various analgesics, antipyretics and anti-inflammatory agents. J. Chromat., 88, 361-363

US International Trade Commission (1976) Imports of Benzenoid Chemicals and Products, 1974, USITC Publication 762, Washington DC, US Government Printing Office, p. 8

US International Trade Commission (1977a) Synthetic Organic Chemicals, US Production and Sales, 1975, USITC Publication 804, Washington DC, US Government Printing Office, pp. 23, 50, 63, 98

US International Trade Commission (1977b) Imports of Benzenoid Chemicals and Products, 1975, USITC Publication 806, Washington DC, US Government Printing Office, p. 9

US Pharmacopeial Convention, Inc. (1975) The US Pharmacopeia, 19th rev., Rockville, Maryland, pp. 23-24

US Tariff Commission (1919) Report on Dyes and Related Coal-Tar Chemicals, 1918, revised ed., Washington DC, US Government Printing Office, p. 26

US Tariff Commission (1973) Imports of Benzenoid Chemicals and Products, 1972, TC Publication 601, Washington DC, US Government Printing Office, pp. 8, 82

US Tariff Commission (1974) Imports of Benzenoid Chemicals and Products, 1973, TC Publication 688, Washington DC, US Government Printing Office, pp. 9, 78

Vest, M.F. & Salzberg, R. (1965) Conjugation reactions in the newborn infant: the metabolism of *para*-aminobenzoic acid. Arch. Dis. Child., 40, 97-105

Wan, S.H., von Lehman, B. & Riegelman, S. (1972) Renal contribution to overall metabolism of drugs. III. Metabolism of *p*-aminobenzoic acid. J. pharm. Sci., 61, 1288-1292

Weast, R.C., ed. (1976) CRC Handbook of Chemistry and Physics, 57th ed., Cleveland, Ohio, Chemical Rubber Co., p. C-183

Wellons, S.L., Paynter, R.A. & Winefordner, J.D. (1974) Room temperature phosphorimetry of biologically-important compounds adsorbed on filter paper. Spectrochim. acta, 30A, 2133-2140

Williams, R.T. (1959) Detoxification Mechanisms, 2nd ed., London, Chapman & Hall

Windholz, M., ed. (1976) The Merck Index, 9th ed., Rahway, NJ, Merck & Co., p. 57

Yamamoto, R.S., Frankel, H.H. & Weisburger, J.H. (1970) Effects of isomers of acetotoluidide and aminobenzoic acid on the toxicity and carcinogenicity of *N*-2-fluorenylacetamide. Toxicol. appl. Pharmacol., 17, 98-106

Yasuda, K. (1972) Thin-layer chromatography of aromatic amines on cadmium acetate impregnated silica gel thin layers. J. Chromat., 72, 413-420

ANTHRANILIC ACID

1. Chemical and Physical Data

1.1 Synonyms and trade names

Chem. Abstr. Services Reg. No.: 118-92-3

Chem. Abstr. Name: 2-Aminobenzoic acid

ortho-Aminobenzoic acid; 1-amino-2-carboxybenzene; 2-carboxyaniline; *ortho*-carboxyaniline; vitamin L_1

1.2 Chemical formula and molecular weight

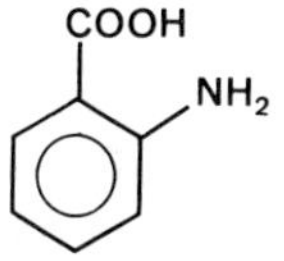

COOH
NH2

$C_7H_7NO_2$ Mol. wt: 137.1

1.3 Chemical and physical properties of the pure substance

From Weast (1976), unless otherwise specified

(a) Description: White to pale-yellow, crystalline powder (Windholz, 1976)

(b) Boiling-point: Sublimes

(c) Melting-point: 146-147°C

(d) Spectroscopy data: λ_{max} 335 nm (E_1^1 = 320), 247 nm (E_1^1 = 491), 217 nm (E_1^1 = 2072) in methanol; infra-red, nuclear magnetic resonance and mass spectral data have been tabulated (Grasselli, 1973).

(e) Solubility: Soluble in water, ethanol and ether; slightly soluble in benzene

(f) Stability: See 'General Remarks on the Substances Considered', p. 27.

1.4 Technical products and impurities

Anthranilic acid is available in the US as a 95-98% pure technical grade and as a 99% pure laboratory grade (Hawley, 1971).

In Japan, it is available with a minimum purity of 98% and a maximum of 0.5% moisture and contains phthalic acid as an impurity.

2. Production, Use, Occurrence and Analysis

For background information on this section, see preamble, p. 15.

2.1 Production and use

(a) Production

Anthranilic acid was prepared by Fischer in 1896 by heating 2-nitrosobenzoic acid with aqueous ammonium sulphide at 100°C (Prager *et al.*, 1931). It is produced commercially by the Hofmann reaction, in which phthalimide is reacted with sodium hypochlorite and caustic soda (Duncker, 1964).

Anthranilic acid has been produced commercially in the US for over fifty years (US Tariff Commission, 1919). In 1975, only one US company reported an undisclosed amount (see preamble, p. 15) (US International Trade Commission, 1977a). US imports through principal US customs districts were reported as 102 thousand kg in 1972 (US Tariff Commission, 1973), 7 thousand kg in 1973 (US Tariff Commission, 1974), 105 thousand kg in 1975 (US International Trade Commission, 1976) and 5 thousand kg in 1975 (US International Trade Commission, 1977b).

In the UK, 10-100 thousand kg anthranilic acid are produced annually. In Switzerland, it is believed that 100-100 thousand kg are imported per year.

In Japan, five companies produced an estimated 57 thousand kg anthranilic acid in 1973, 85 thousand kg in 1974 and 44 thousand kg in 1975. There have been no Japanese imports or exports of this chemical.

(b) Use

Anthranilic acid can be used for the production of fourteen dyes and four pigments which are believed to have commercial world significance (The Society of Dyers and Colourists, 1971). Only the following are produced commercially in the US (US International Trade Commission, 1977a): C.I. Acid Yellow 121, C.I. Direct Brown 112, Mordant Yellow 8, Mordant Red 9, Mordant Brown 40, Vat Violet 13, Vat Blue 1 and Pigment Red 60.

Anthranilic acid has also been used as a raw material in the manufacture of saccharin (Beck, 1969) and as a mold inhibitor in soya sauce; its esters have found applications in perfumes, synthetic flavours, bird and insect repellents, and as fungicides for syrups (Duncker, 1964). The cadmium salt has been reported to be useful in veterinary medicine as an ascaricide in swine (Windholz, 1976).

In the UK, 10-100 thousand kg anthranilic acid are used annually for dyestuff manufacture. In Switzerland, anthranilic acid is used as a dyestuff intermediate.

In Japan, anthranilic acid is used as an intermediate in the manufacture of dyes (70%), pharmaceuticals and perfumes (20%) and pigments (10%).

2.2 Occurrence

Anthranilic acid occurs as a metabolite of tryptophan, and it is an intermediate in the biosynthesis of this amino acid in some microorganisms (Thompson, 1972). No data on its occurrence in the environment were available to the Working Group.

2.3 Analysis

Aniline derivatives, including anthranilic acid, have been determined in blood and urine by direct titration with *N*-bromosuccinimide. The error was less than 2% for 0.1-10 mg samples (Barakat *et al.*, 1973).

Aromatic amines, including anthranilic acid, have been determined colorimetrically with peroxydisulphate (Gupta & Srivastava, 1971).

Anthranilic acid has been determined indirectly by the atomic absorption spectrophotometric measurement of its iron complex (Kidani *et al.*, 1972) or of its ternary complex with cobalt and bathophenanthroline (Kidani *et al.*, 1975).

Electrophoresis has been used to separate aromatic compounds, including anthranilic acid, with detection of nanogram amounts by fluorescence (Wachter *et al.*, 1972) or diazotization and coupling with chromotropic acid (Akimova *et al.*, 1972).

Thin-layer chromatography (TLC) has been used to separate aromatic amines, including anthranilic acid, on rice starch (Canić & Perišić-Janjić, 1974), on cellulose and polystyrene-based ion exchangers (Lepri *et al.*, 1974) and on silica gel impregnated with cadmium salts (Yasuda, 1972). TLC has also been used to separate and identify tryptophan metabolites, including anthranilic acid, in urine on ECTEOLA-cellulose with ultra-violet light visualization (Humbel & Marsault, 1973), in urine using a two dimensional system (Haworth & Walmsley, 1972) and in faeces on silica gel with an estimated accuracy of ±10% in the μg/kg range (Anderson, 1975). TLC with visualization with ultra-violet light after reaction with glutaconic aldehyde in pyridine has also been used. The detection limit is 0.1-0.3 μg (Ohkuma & Sakai, 1975).

High-pressure liquid chromatography has been used to separate and determine benzenecarboxylic acids, including anthranilic acid (Aurenge, 1973). It has been determined in urine by gas chromatography after conversion to its trifluoroacetyl derivative (Hirano *et al.*, 1972).

3. Biological Data Relevant to the Evaluation of Carcinogenic Risk to Man

3.1 Carcinogenicity and related studies in animals[1]

Other experimental systems

Bladder implantation: Cholesterol pellets containing 20% anthranilic acid were implanted into the bladders of mice. Carcinomas of the bladder were observed in 10/75 survivors within 12-15 months. Of controls that received implants of cholesterol alone, 4/54 mice developed bladder carcinomas (>0.05) (Bryan, 1969).

3.2 Other relevant biological data

(a) Experimental systems

Of 12 rats fed a synthetic diet containing 0.2% anthranilic acid (daily intake, 20-30 mg/animal) for 33-374 days, bladder lesions were observed in 4/11 animals examined. These lesions included moderate and extensive epithelial metaplasia as well as pronounced epithelial proliferation with a tendency to keratotic papillomas with large cysts (Ekman & Strömbeck, 1949).

The main metabolites of anthranilic acid in rats, rabbits and dogs are the glucuronic acid and glycine conjugates (Charconnet-Harding *et al.*, 1953; Knoefel *et al.*, 1959). No acetylation of anthranilic acid was observed in rabbits (Bray *et al.*, 1948); 1.2-2.5% anthranilamide was found in rats injected intraperitoneally with anthranilic acid (Sutamihardja *et al.*, 1972). In rabbits, there is suggestive evidence that it is hydroxylated in the 3- and 5-positions (Bray *et al.*, 1948; Kotake & Shirai, 1953); hydroxylation in the 3-position was observed in rat liver homogenates (Wiss & Hellmann, 1953). The instability of 3-hydroxyanthranilic acid makes the determination of small amounts difficult (Pipkin *et al.*, 1967).

---

[1]The Working Group was aware of carcinogenicity tests in progress on anthranilic acid involving oral administration in mice and rats and s.c. injection in mice (IARC, 1976).

Dimerization of the hydroxylated derivatives leads to phenoxazinone (Butenandt *et al.*, 1957), a chromophore which is also present in carcinogenic fungal pigments like the actinomycins.

Anthraniloylanthranilic acid (Hagihara & Okazaki, 1956) and 2-methyl-1,2,3,4-tetrahydro-4-quinazolone (Ishiguro *et al.*, 1974) have also been identified as metabolites of anthranilic acid.

Less than 10% of a dose (18 or 101 mg/kg bw) of anthranilic acid given by i.p. injection was excreted in the bile of rats (Abou-el-Makarem *et al.*, 1967).

Anthranilic acid and 3-hydroxyanthranilic acid are metabolites of tryptophan and occur normally in the urine with other tryptophan metabolites.

A dose of 380 mg/kg bw anthranilic acid given by s.c. injection on day 9 of pregnancy to a group of 10 female Sprague-Dawley rats produced no teratogenic effects (Koshakji & Schulert, 1973).

When anthranilic acid was added at a level of 20 mM (2.75 g) to 5 ml culture medium, the incidence of melanocytic tumours in second instar larvae of the *tu, bw ; st su-tu* strain of *Drosophila melanogaster* was increased (Burnet & Sang, 1968). Similar effects were observed by Kanehisa (1956) and Plaine & Glass (1955).

No data on the mutagenicity of anthranilic acid were available to the Working Group.

3-Hydroxyanthranilic acid has been reported to induce chromatid breaks in cultured human embryo cells at a concentration of 30 mg/l. One chromosome translocation was observed in 338 metaphase cells examined; none were seen in 300 control metaphase cells (Kuznezova, 1969).

(b) Man

Anthranilic acid is a metabolite of tryptophan.

A significant percentage of patients with non-occupational 'spontaneous' bladder cancer excrete higher amounts of kynurenine, kynurenic acid, acetyl-kynurenine, 3-hydroxykynurenine and 3-hydroxyanthranilic acid than normal control individuals, particularly after a loading dose of tryptophan (Boyland & Williams, 1955; Brown *et al.*, 1955). This effect was not seen in patients with bladder cancers that could be traced to industrial exposure to aromatic amines (Brown *et al.*, 1972).

Levels of tryptophan metabolites are also increased in patients with other diseases, including scleroderma, porphyria and mental disease. Furthermore, they are higher in women who take oestrogen analogue oral contraceptives, in patients with bilharziasis and in habitual smokers (see Arcos & Argus, 1974).

3.3 Case reports and epidemiological studies

No data were available to the Working Group.

4. Comments on Data Reported and Evaluation

4.1 Animal data

Anthranilic acid has been tested in mice by bladder implantation. The available data do not allow an evaluation of the carcinogenicity of this compound to be made.

4.2 Human data

No case reports or epidemiological studies were available to the Working Group.

5. References

Abou-el-Makarem, M.M., Millburn, P., Smith, R.L. & Williams, R.T. (1967) Biliary excretion of foreign compounds. Benzene and its derivatives in the rat. Biochem. J., 105, 1269-1274

Akimova, T.G., Dedkova, V.P. & Savvin, S.B. (1972) Electrophoretic and chromatographic methods of analyzing organic reagents. IV. Identification of aromatic amines and naphthol sulfonic acids used in organic reagent synthesis. Zh. analit. Khim., 27, 1396-1400

Anderson, G.M. (1975) Quantitation of tryptophan metabolites in rat feces by thin-layer chromatography. J. Chromat., 105, 323-328

Arcos, J.C. & Argus, M.F. (1974) Chemical Induction of Cancer, Vol. IIB, New York, Academic Press, pp. 65-93

Aurenge, J. (1973) Séparation et dosage par chromatographie liquide sous haute pression des acides benzènepolycarboxyliques. Utilisation de la programmation d'éluant. J. Chromat., 84, 285-298

Barakat, M.Z., Fayzalla, A.S. & El-Aassar, S. (1973) Microdetermination of aniline salts and certain derivatives. Microchem. J., 18, 308-323

Beck, K.M. (1969) Sweeteners, nonnutritive. In: Kirk, R.E. & Othmer, D.F., eds, Encyclopedia of Chemical Technology, 2nd ed., Vol. 19, New York, John Wiley and Sons, p. 598

Boyland, E. & Williams, D.C. (1955) The estimation of tryptophan metabolites in the urine of patients with cancer of the bladder. Biochem. J., 60,

Bray, H.G., Lake, H.J., Neale, F.C., Thorpe, W.V. & Wood, P.B. (1948) The fate of certain organic acids and amides in the rabbit. IV. The aminobenzoic acids and their amides. Biochem. J., 42, 434-443

Brown, R.R., Price, J.M. & Wear, J.B. (1955) The metabolism of tryptophan in bladder tumor patients. Proc. Amer. Ass. Cancer Res., 2, 7

Brown, R.R., Friedell, G.H. & Leklem, J.E. (1972) Tryptophan metabolism in patients with bladder cancer of occupational etiology. Amer. industr. Hyg. Ass. J., 33, 217-222

Bryan, G.T. (1969) Role of tryptophan metabolites in urinary bladder cancer. Amer. industr. Hyg. Ass. J., 30, 27-34

Burnet, B. & Sang, J.H. (1968) Physiological genetics of melanotic tumors in *Drosophila melanogaster*. V. Amino acid metabolism and tumor formation in the *tu bw; st su-tu* strain. Genetics, 59, 211-235

Butenandt, A., Biekert, E. & Neubert, G. (1957) Über 3-Hydroxy-5-acetyl-phenoxazon-(2) und 1.6-Diacetyl-triphendioxazin ein neuer Weg zur Darstellung von Phenoxazonen. Liebigs Ann. Chem., 602, 72-80

Canić, V.D. & Perišić-Janjić, N.U. (1974) Separation of aliphatic and aromatic organic acids on starch thin-layers. Chromatographic behaviour and chemical structure. Z. analyt. Chem., 270, 16-19

Charconnet-Harding, F., Dalgliesh, C.E. & Neuberger, A. (1953) The relation between riboflavin and tryptophan metabolism, studied in the rat. Biochem. J., 53, 513-521

Duncker, C. (1964) Benzoic acid. In: Kirk, R.E. & Othmer, D.F., eds, Encyclopedia of Chemical Technology, 2nd ed., Vol. 3, New York, John Wiley and Sons, pp.434-435

Ekman, B. & Strömbeck, J.P. (1949) The effect of some splitproducts of 2,3'-azotoluene on the urinary bladder in the rat and their excretion on various diets. Acta path. microbiol. scand., 26, 447-471

Grasselli, J.G., ed. (1973) CRC Atlas of Spectral Data and Physical Constants for Organic Compounds, Cleveland, Ohio, Chemical Rubber Co., p. B-278

Gupta, R.C. & Srivastava, S.P. (1971) Oxidation of aromatic amines by peroxodisulphate ion. II. Identification of aromatic amines on the basis of absorption maxima of coloured oxidation products. Z. analyt. Chem., 257, 275-277

Hagihara, F. & Okazaki, Y. (1956) Isolation of anthraniloylanthranilic acid as the metabolic products of anthranilic acid. J. pharm. Soc. Japan, 76, 362-364

Hawley, G.G., ed. (1971) The Condensed Chemical Dictionary, 8th ed., New York, Van Nostrand-Reinhold, p. 66

Haworth, C. & Walmsley, T.A. (1972) A study of the chromatographic behaviour of tryptophan metabolites and related compounds by chromatography on thin layers of silica gel. J. Chromat., 66, 311-319

Hirano, K., Mori, K., Tsuboi, N., Kawai, S. & Ohno, T. (1972) Gas chromatography of urinary anthranilic acid and 3-hydroxyanthranilic acid by solvent extraction method. Chem. pharm. Bull., 20, 1412-1416

Humbel, R. & Marsault, C. (1973) Thin-layer chromatography of kynurenin metabolites. J. Chromat., 79, 347-348

IARC (1976) Information Bulletin on the Survey of Chemicals Being Tested for Carcinogenicity, No. 6, Lyon, pp. 113, 161

Ishiguro, I., Naito, J., Shinohara, R. & Ishikura, A. (1974) Isolation and identification of the 2-methyl-1,2,3,4-tetrahydro-4-quinazolone as a new metabolite in the urine of anthranilic acid injected rats. Yakugaku Zasshi, 94, 1232-1239

Kanehisa, T. (1956) Relation between the formation of melanotic tumors and tryptophane metabolism involving eye-colour in *Drosophila*. Ann. Zool. Jap., 29, 97-100

Kidani, Y., Inagaki, K. & Koike, H. (1972) Atomic absorption spectrophotometric determination of anthranilic acid. Nagoya Shiritsu Daigaku Yakugakubu Kenkyu Nempo, 20, 21-25

Kidani, Y., Osugi, N., Inagaki, K. & Koike, H. (1975) Indirect determination of anthranilic acid by atomic absorption spectrometry. Bunseki Kagaku, 24, 218-221

Knoefel, P.K., Huang, K.C. & Despopoulos, A. (1959) Conjugation and excretion of the amino and acetamido benzoic acids. Amer. J. Physiol., 196, 1224-1230

Koshakji, R.P. & Schulert, A.R. (1973) Biochemical mechanisms of salicylate teratology in the rat. Biochem. Pharmacol., 22, 407-416

Kotake, Y. & Shirai, Y. (1953) Über die Entstehung der 5-Oxy-anthranilsäure aus Anthranilsäure im tierischen Organismus. Hoppe Seyler's Z. physiol. Chem., 295, 160-163

Kuznezova, L.E. (1969) Mutagenic effect of 3-hydroxykynurenine and 3-hydroxyanthranilic acid. Nature (Lond.), 222, 484-485

Lepri, L., Desideri, P.G. & Coas, V. (1974) Chromatographic behaviour of aromatic amino acids on thin layers of anion and cation exchangers. J. Chromat., 88, 331-339

Ohkuma, S. & Sakai, I. (1975) Detection of aromatic primary amines by a photochemical reaction with pyridine. Bunseki Kagaku, 24, 385-387

Pipkin, G.E., Nishimura, R., Banowsky, L. & Schlegel, J.U. (1967) Stabilization of urinary 3-hydroxyanthranilic acid by oral administration of L-ascorbic acid. Proc. Soc. exp. Biol. (Wash.), 126, 702-704

Plaine, H.L. & Glass, B. (1955) Influence of tryptophan and related compounds upon the action of a specific gene and the induction of melanotic tumours in *Drosophila melanogaster*. J. Genet., 53, 244-261

Prager, B., Jacobson, P., Schmidt, P. & Stern, D., eds (1931) Beilsteins Handbuch der Organischen Chemie, 4th ed., Vol. 14, Syst. No. 1885-1889, Berlin, Springer-Verlag, p. 310

The Society of Dyers and Colourists (1971) Colour Index, 3rd ed., Vol. 4, Yorkshire, UK, p. 4707

Sutamihardja, T.M., Ishikura, A., Naito, J. & Ishiguro, I. (1972) Studies on the new metabolic pathway of anthranilic acid in the rat. I. Isolation and urinary excretion of anthranilamide as a new metabolite of anthranilic acid. Chem. pharm. Bull., 20, 2694-2700

Thompson, A.R. (1972) Organic nitrogen compounds. In: Chemical Technology: An Encyclopedic Treatment, Vol. 4, Chapter 15, New York, Barnes & Noble, pp. 521-522

US International Trade Commission (1976) Imports of Benzenoid Chemicals and Products, 1974, USITC Publication 762, Washington DC, US Government Printing Office, p. 11

US International Trade Commission (1977a) Synthetic Organic Chemicals, US Production and Sales, 1975, USITC Publication 804, Washington DC, US Government Printing Office, pp. 26, 55, 65, 71-72, 75, 85

US International Trade Commission (1977b) Imports of Benzenoid Chemicals and Products, 1975, USITC Publication 806, Washington DC, US Government Printing Office, p. 11

US Tariff Commission (1919) Report on Dyes and Related Coal-Tar Chemicals, 1918, revised ed., Washington DC, US Government Printing Office, p. 26

US Tariff Commission (1973) Imports of Benzenoid Chemicals and Products, 1972, TC Publication 601, Washington DC, US Government Printing Office, p. 11

US Tariff Commission (1974) Imports of Benzenoid Chemicals and Products, 1973, TC Publication 688, Washington DC, US Government Printing Office, p. 11

Wachter, H., Grassmayr, K., Guetter, W., Hausen, A. & Sallaberger, G. (1972) Identification of fluorescent aromatic acids by spectral fluorometry *in situ* after electrophoresis. Mikrochim. acta, 6, 861-875

Weast, R.C., ed. (1976) CRC Handbook of Chemistry and Physics, 57th ed., Cleveland, Ohio, Chemical Rubber Co., p. C-182

Windholz, M., ed. (1976) The Merck Index, 9th ed., Rahway, NJ, Merck & Co., pp. 57, 1292

Wiss, O. & Hellmann, H. (1953) Über die Einführung phenolischer Hydroxylgruppen beim oxydativen Tryptophan-Stoffwechsel. Z. Naturforsch., 8B, 70-76

Yasuda, K. (1972) Thin-layer chromatography of aromatic amines on cadmium acetate impregnated silica gel thin layers. J. Chromat., 72, 413-420

para-CHLORO-*ortho*-TOLUIDINE (HYDROCHLORIDE)

1. Chemical and Physical Data

para-Chloro-*ortho*-toluidine

1.1 Synonyms and trade names

Colour Index No.: 37085

Colour Index Name: Azoic Diazo component 11, base

Chem. Abstr. Services Reg. No.: 95-69-2

Chem. Abstr. Name: 4-Chloro-2-methylbenzeneamine

2-Amino-5-chlorotoluene; asymmetric *meta*-chloro-*ortho*-toluidine; 5-chloro-2-aminotoluene; 4-chloro-2-methylaniline; 4-chloro-6-methylaniline; 4-chloro-2-toluidine; 4-chloro-*ortho*-toluidine; 2-methyl-4-chloroaniline

Amarthol Fast Red TR Base; Azoene Fast Red TR Base; Azogene Fast Red TR; Brentamine Fast Red TR Base; Daito Red Base TR; Deval Red K; Deval Red TR; Diazo Fast Red TRA; Fast Red Base TR; Fast Red 5CT Base; Fast Red TR; Fast Red TR11; Fast Red TR Base; Fast Red TRO Base; Kako Red TR Base; Kambamine Red TR; Mitsui Red TR Base; Red Base Ciba IX; Red Base Irga IX; Red Base NTR; Red TR Base; Sanyo Fast Red TR Base; Tulabase Fast Red TR

1.2 Chemical formula and molecular weight

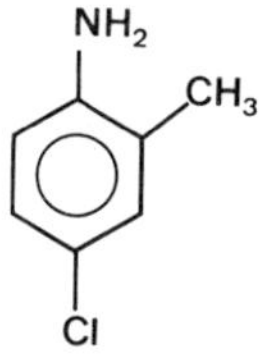

NH_2

CH_3

Cl

C_7H_8NCl Mol. wt: 141.6

1.3 Chemical and physical properties of the pure substance

From Grasselli (1973), unless otherwise specified

(a) Description: Crystalline, fused, grayish-white solid (commercial product) (Pfister Chemical Inc., 1975)

(b) Boiling-point: 241°C

(c) Melting-point: 29-30°C

(d) Refractive index: n_D^{20} 1.5848 (Aldrich Chemical Company, 1976)

(e) Spectroscopy data: λ_{max} 296 nm (E_1^1 = 180), 242 nm (E_1^1 = 939) in methanol; infra-red and nuclear magnetic resonance spectra have been tabulated.

(f) Solubility: Sparingly soluble in water, soluble in ethanol and dilute acids (Hawley, 1971)

(g) Stability: See 'General Remarks on the Substances Considered' p. 27.

1.4 Technical products and impurities

para-Chloro-*ortho*-toluidine is available in the US as a technical grade with a minimum purity of 99% (Pfister Chemical Inc., 1975).

In Japan, *para*-chloro-*ortho*-toluidine is available with a minimum purity of 99.0% and contains dichloro-*ortho*-toluidine and chlorotoluidine isomers as impurities.

para-Chloro-*ortho*-toluidine hydrochloride

1.1 Synonyms and trade names

Colour Index No.: 37085

Colour Index Name: Azoic diazo component 11

Chem. Abstr. Services Reg. No.: 3165-93-3

Chem. Abstr. Name: 4-Chloro-2-methylbenzenamine hydrochloride

2-Amino-5-chlorotoluene hydrochloride; asymmetric *meta*-chloro-*ortho*-toluidine hydrochloride; 5-chloro-2-aminotoluene hydrochloride;

4-chloro-2-methylaniline hydrochloride; 4-chloro-6-methylaniline hydrochloride; 4-chloro-2-toluidine hydrochloride; 4-chloro-*ortho*-toluidine hydrochloride; 2-methyl-4-chloroaniline hydrochloride

Amarthol Fast Ref TR Salt; Azanil Red Salt TRD; Azoene Fast Red TR Salt; Brentamine Fast Red TR Salt; Daito Red Salt TR; Devol Red TA Salt; Diazo Fast Red TR; Diazo Fast Red TRA; Fast Red 5CT Salt; Fast Red Salt TR; Fast Red Salt TRA; Fast Red Salt TRR; Fast Red TR Salt; Hindasol Red TR Salt; Natasol Fast Red TR Salt; Ofna-Perl Salt RRA; Red Salt Ciba IX; Red Salt Irga IX; Red TRS Salt; Sanyo Fast Red Salt TR

1.2 Empirical formula and molecular weight

$C_7H_8NCl.HCl$ Mol. wt: 178.1

1.3 Chemical and physical properties of the pure substance

No data were available to the Working Group.

1.4 Technical products and impurities

No data were available to the Working Group.

2. Production, Use, Occurrence and Analysis

For background information on this section, see preamble, p. 15.

2.1 Production and use

(a) Production

para-Chloro-*ortho*-toluidine was prepared by Beilstein & Kuhlberg in 1870 by reducing 2-nitrotoluene with tin and hydrochloric acid (Prager *et al.*, 1929). It has also been prepared by the ring chlorination of *ortho*-toluidine (Schimelpfenig, 1975) and by the chlorination of *ortho*-acetotoluidide and saponification to the free amine (The Society of Dyers and Colourists, 1971a).

Commercial production of *para*-chloro-*ortho*-toluidine in the US was first reported in 1939 (US Tariff Commission, 1940), and that of *para*-chloro-*ortho*-toluidine hydrochloride in 1941-1943 (US Tariff Commission,

1945). In 1975, only one US company reported an undisclosed amount (see preamble, p. 15) of *para*-chloro-*ortho*-toluidine, and three US companies produced undisclosed amounts of *para*-chloro-*ortho*-toluidine hydrochloride (US International Trade Commission, 1977a). US imports of *para*-chloro-*ortho*-toluidine through principal US customs districts amounted to 86.3 thousand kg in 1973 (two thousand kg of the free base and hydrochloride combined were also imported that year) (US Tariff Commission, 1974), 73.3 thousand kg in 1974 (US International Trade Commission, 1976) and 2.4 thousand kg in 1975 (US International Trade Commission, 1977b).

In Switzerland, *para*-chloro-*ortho*-toluidine and its salts were produced from 1956 until the end of 1976. During 1976, between 100-200 thousand kg were produced.

Neither *para*-chloro-*ortho*-toluidine nor *para*-chloro-*ortho*-toluidine hydrochloride has ever been produced commercially in Japan. Japanese imports of *para*-chloro-*ortho*-toluidine hydrochloride from the Federal Republic of Germany for the years 1973-1975 were 14 thousand kg, 3 thousand kg and 4 thousand kg per year, respectively.

(b) Use

para-Chloro-*ortho*-toluidine and its hydrochloride are used to produce azo dyes for cotton, silk, acetate and nylon (The Society of Dyers and Colourists, 1971b).

para-Chloro-*ortho*-toluidine can also be used as an intermediate for the production of two pigments and one dye that are believed to be of commercial world significance (The Society of Dyers and Colourists, 1971a): C.I. Pigment Red 7, C.I. Pigment Yellow 49 and C.I. Azoic Coupling Component 8. Only C.I. Pigment Red 7 and C.I. Azoic Coupling Component 8 are produced commercially in the US (US International Trade Commission, 1977a).

para-Chloro-*ortho*-toluidine is also believed to be used in the production of chlordimeform (also known as chlorphenamidine [*N*'-(4-chloro-2-methylphenyl)-*N*,*N*-dimethylformamidine]) (Sittig, 1977), an insecticide and acaricide used on cotton, deciduous fruit and nut trees and vegetables, which was used in the US to the extent of about 600 thousand kg in 1974. All of the product used to date in the US was imported from Switzerland.

In Switzerland, *para*-chloro-*ortho*-toluidine and its salts were used as dyestuff intermediates. *para*-Chloro-*ortho*-toluidine was also used as an intermediate in the production of a cotton insecticide. *para*-Chloro-*ortho*-toluidine and its hydrochloride are not permitted for use in cosmetics in the European Economic Communities (EEC, 1976).

In Japan, *para*-chloro-*ortho*-toluidine is used in the production of azo dyes.

2.2 Occurrence

para-Chloro-*ortho*-toluidine and its hydrochloride are not known to occur as natural products. However, *para*-chloro-*ortho*-toluidine can be formed from chlordimeform, an insecticide and acaricide, by enzymes present in the leaves of apple seedlings (Sen Gupta & Knowles, 1969) as well as those of cotton plants (Bull, 1973).

No further data on its occurrence in the environment were available to the Working Group.

2.3 Analysis

Isomeric chloroanilines and their derivatives, including *para*-chloro-*ortho*-toluidine, have been detected colorimetrically in their mixtures by coupling with diazotized sulphanilic acid (Légrádi, 1967).

Paper chromatography using six different developing solvent systems and fifteen detecting reagents has been used to separate and identify 240 compounds of biochemical interest, including *para*-chloro-*ortho*-toluidine (Reio, 1970). Chloromethylaniline isomers, including *para*-chloro-*ortho*-toluidine, have been separated by gas chromatography (Townley *et al.*, 1970).

Methods of analysis for *para*-chloro-*ortho*-toluidine as a metabolite of the systemic acaricide, chlordimeform, in plants and soils have been described in which the amine is determined either by thin-layer chromatography and colorimetry or by gas chromatography (Kossmann *et al.*, 1971). In another study of the pesticide chlordimeform, *para*-chloro-*ortho*-toluidine was determined by ultra-violet reflectance after separation by thin-layer chromatography on silica gel (Kynast, 1970). Thin-layer chromatography was also used to separate and identify primary aromatic amines, including *para*-chloro-*ortho*-toluidine, on silica gel buffered with sodium acetate (Bassl *et al.*, 1967).

3. Biological Data Relevant to the Evaluation of Carcinogenic Risk to Man

3.1 Carcinogenicity and related studies[1]

Oral administration

Mouse: In an abstract, it was reported that in lifetime feeding studies in CD mice of both sexes, *para*-chloro-*ortho*-toluidine hydrochloride induced 40 haemangiosarcomas in various organs in 82 mice; no such tumours occurred in controls (Homburger *et al.*, 1972) [No further details were available to the Working Group].

3.2 Other relevant biological data

para-Chloro-*ortho*-toluidine inhibited RNA synthesis in HeLa cells (Murakami & Fukami, 1974).

No data on the metabolism, embryotoxicity, teratogenicity or mutagenicity of *para*-chloro-*ortho*-toluidine were available to the Working Group.

3.3 Case reports and epidemiological studies

Uebelin & Pletscher (1954) studied the occurrence of bladder tumours in workers at a factory in Switzerland producing dyestuff intermediates (for details, see monograph on *ortho*-toluidine). The authors distinguished a group of 35 men who prepared *para*-chloro-*ortho*-toluidine from *ortho*-toluidine during the years 1924-1953; no bladder tumours were found among these men [Insufficient details on person-years at risk or follow-up were provided to evaluate the significance of this observation].

---

[1]The Working Group was aware of carcinogenicity tests in progress on *para*-chloro-*ortho*-toluidine hydrochloride involving oral administration in mice and rats (IARC, 1976).

4. Comments on Data Reported and Evaluation

4.1 Animal data

The only data on *para*-chloro-*ortho*-toluidine hydrochloride that were available to the Working Group were reported in an abstract. Although the results of this experiment indicate a carcinogenic effect, the inadequacy of reporting does not allow an evaluation of the carcinogenicity of *para*-chloro-*ortho*-toluidine or its hydrochloride to be made.

4.2 Human data

Insufficient information is available to permit the Working Group to evaluate the carcinogenicity of *para*-chloro-*ortho*-toluidine or its hydrochloride.

5. References

Aldrich Chemical Company, Inc. (1976) The 1977-1978 Aldrich Catalog/Handbook of Organic and Biochemicals (Catalog No. 18), Milwaukee, Wisconsin, p. 179

Bassl, A., Heckemann, H.J. & Baumann, E. (1967) Thin-layer chromatography of primary aromatic amines. I. J. prakt. Chem., 36, 265-273

Bull, D.L. (1973) Metabolism of chlordimeform in cotton plants. Environm. Entomol., 2, 869-871

EEC (1976) Législation. Journal Officiel des Communautés Européennes, No. L 262, p. 175

Grasselli, J.G., ed. (1973) CRC Atlas of Spectral Data and Physical Constants for Organic Compounds, Cleveland, Ohio, Chemical Rubber Co., p. B-944

Hawley, G.G., ed. (1971) The Condensed Chemical Dictionary, 8th ed., New York, Van Nostrand-Reinhold p. 43

Homburger, F., Friedell, G.H., Weisburger, E.K. & Weisburger, J.H. (1972) Carcinogenicity of simple aromatic amine derivatives in mice and rats (Abstract No. 14). Toxicol. appl. Pharmacol., 22, 280-281

IARC (1976) Information Bulletin on the Survey of Chemicals Being Tested for Carcinogenicity, No. 6, Lyon, p. 181

Kossmann, K., Geissbühler, H. & Boyd, V.F. (1971) Specific determination of chlorphenamidine [*N*'-(4-chloro-*o*-tolyl)-*N*,*N*-dimethylformamidine] in plants and soil material by colorimetry and thin-layer and electron capture gas chromatography. J. agric. Fd Chem., 19, 360-364

Kynast, G. (1970) Specific and quantitative determination of impurities in pesticides on thin-layer chromatograms by reflectance measurements. Fresenius' Z. analyt. Chem., 251, 161-166

Légrádi, L. (1967) Nachweis isomerer Chloraniline und irher Derivate. Mikrochim. acta, 4, 676-684

Murakami, M. & Fukami, J.-I. (1974) Effects of chlorphenamidine and its metabolites on HeLa cells. Bull. environm. Contam. Toxicol., 11, 184-188

Pfister Chemical Inc. (1975) Technical Data (Red TR Base) 5-Chloro-2-Amino Toluene, Ridgefield, NJ

Prager, B., Jacobson, P., Schmidt, P. & Stern, D., eds (1929) Beilsteins Handbuch der Organischen Chemie, 4th ed., Vol. 12, Syst. No. 1680-1681, Berlin, Springer-Verlag, p. 835

Reio, L. (1970) Third supplement for the paper chromatographic separation and identification of phenol derivatives and related compounds of biochemical interest using a 'reference system'. J. Chromat., 47, 60-85

Schimelpfenig, C.W. (1975) Ring chlorination of *o*-toluidine. US Patent 3,890,388, June 17 (to E.I. du Pont de Nemours and Co.)

Sen Gupta, A.K. & Knowles, C.O. (1969) Metabolism of *N*'-(4-chloro-*o*-tolyl)-*N*,*N*-dimethylformamidine by apple seedlings. J. agric. Fd Chem., 17, 595-600

Sittig, M. (1977) Pesticide Process Encyclopedia, Park Ridge, NJ, Noyes Data Corporation, pp. 100-101

The Society of Dyers and Colourists (1971a) Colour Index, 3rd ed., Vol. 4, Yorkshire, UK, pp. 4025, 4037, 4042, 4348, 4357, 4857

The Society of Dyers and Colourists (1971b) Colour Index, 3rd ed., Vol. 1, Yorkshire, UK, pp. 1569, 1585

Townley, E., Perez, I. & Kabasakalian, P. (1970) Gas-liquid chromatographic separation of the ten geometric isomers of chloromethylaniline using Ucon oil as stationary phase on a textured glass bead support. Analyt. Chem., 42, 947-948

Uebelin, F. & Pletscher, A. (1954) Ätiologie und Prophylaxe gewerblicher Tumoren in der Farbstoffindustrie. Schw. med. Wschr., 84, 917-928

US International Trade Commission (1976) Imports of Benzenoid Chemicals and Products, 1974, USITC Publication 762, Washington DC, US Government Printing Office, p. 48

US International Trade Commission (1977a) Synthetic Organic Chemicals, US Production and Sales, 1975, USITC Publication 804, Washington DC, US Government Printing Office, pp. 29, 59, 82

US International Trade Commission (1977b) Imports of Benzenoid Chemicals and Products, 1975, USITC Publication 806, Washington DC, US Government Printing Office, p. 47

US Tariff Commission (1940) Synthetic Organic Chemicals, US Production and Sales, 1939, Report No. 140, Second Series, Washington DC, US Government Printing Office, p. 9

US Tariff Commission (1945) Synthetic Organic Chemicals, US Production and Sales, 1941-43, Report No. 153, Second Series, Washington DC, US Government Printing Office, p. 73

US Tariff Commission (1974) Imports of Benzenoid Chemicals and Products, 1973, TC Publication 688, Washington DC, US Government Printing Office, pp. 13,44

CINNAMYL ANTHRANILATE

1. Chemical and Physical Data

1.1 Synonyms and trade names

Chem. Abstr. Services Reg. No.: 87-29-6

Chem. Abstr. Name: 2-Amino-benzoic acid, 3-phenyl-2-propenyl ester

Anthranilic acid, cinnamyl ester; cinnamyl alcohol, anthranilate; cinnamyl 2-aminobenzoate; cinnamyl *ortho*-aminobenzoate; 3-phenyl-2-propenyl anthranilate; 3-phenyl-2-propen-1-yl anthranilate

1.2 Chemical formula and molecular weight

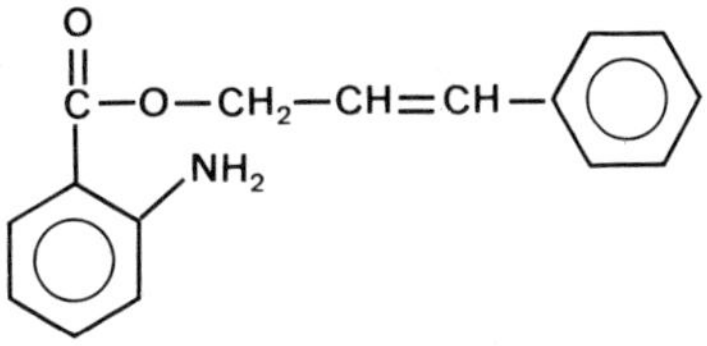

$C_{16}H_{15}NO_2$ Mol. wt: 253.3

1.3 Chemical and physical properties of the pure substance

From Furia & Bellanca (1971), unless otherwise specified

(a) Description: Brownish powder

(b) Boiling-point: 332°C

(c) Melting-point: >60°C (commercial product)

(d) Solubility: Insoluble in water; soluble in ethanol (50 g/l in 95% ethanol), diethyl ether and chloroform (Committee on Specifications, 1972)

(e) Stability: See 'General Remarks on the Substances Considered', p. 27.

(f) Reactivity: Reacts with various aldehydes, forming highly coloured compounds (Bedoukian, 1967)

1.4 Technical products and impurities

Cinnamyl anthranilate is available in the US with a minimum purity of 96% (Givaudan-Delawanna, Inc., 1961).

2. Production, Use, Occurrence and Analysis

For background information on this section, see preamble, p. 15.

2.1 Production and use

(a) Production

Cinnamyl anthranilate can be prepared by reacting isatoic anhydride (*N*-carboxyanthranilic anhydride) with cinnamic alcohol (Opdyke, 1975) or by esterifying anthranilic acid with cinnamyl alcohol (Furia & Bellanca, 1971).

Commercial production of cinnamyl anthranilate in the US was first reported in 1939 (US Tariff Commission, 1940); production figures have not been reported in recent years, but three US companies reported total sales of 454 kg in 1975 (US International Trade Commission, 1977).

No evidence was found that cinnamyl anthranilate has ever been produced commercially or imported in Japan.

No data on its production in Europe were available to the Working Group.

(b) Use

Cinnamyl anthranilate has been used as a flavour and fragrance chemical in the US since 1940 (Opdyke, 1975). It is used as an ingredient in non-alcoholic beverages (6.8 mg/kg), ice-cream and ices (1.7 mg/kg), sweets (4.3 mg/kg), baked goods (5.3 mg/kg), gelatines and puddings (28 mg/kg) and in chewing gum (46-730 mg/kg) (Furia & Bellanca, 1971). It has been used as a fragrance ingredient in various cosmetic products, including soaps (at an average concentration of 0.01%), detergents (0.001%) and creams and lotions (0.005%) (Opdyke, 1975). As a perfume ingredient, cinnamyl anthranilate is used in orange blossom, neroli, cologne and other floral blends (Givaudan-Delawanna, Inc., 1961) at an average concentration of 0.08% (Opdyke, 1975).

The US Food and Drug Administration has classified cinnamyl anthranilate as a GRAS (generally recognized as safe) substance which may be safely used as a flavouring substance in food if used in the minimum quantity required to produce its intended effect (US Food and Drug Administration, 1976). In 1974, The Council of Europe listed cinnamyl anthranilate as an artificial flavouring substance that may be used in food without hazard to public health at levels of 25 mg/kg (Opdyke, 1975).

2.2 Occurrence

Cinnamyl anthranilate is not known to occur as a natural product. No data on its occurrence in the environment were available to the Working Group.

2.3 Analysis

The Committee on Specifications (1972) has published a purity test for cinnamyl anthranilate in which a sample is hydrolysed with alkali, the excess alkali is titrated, and the per cent of ester in the original sample is then calculated.

Although no other specific studies were available to the Working Group, the accompanying monographs describe a number of methods generally applicable to aromatic amines.

3. Biological Data Relevant to the Evaluation of Carcinogenic Risk to Man

3.1 Carcinogenicity and related studies in animals[1]

Intraperitoneal administration

Mouse: Fifteen male and 15 female A/He mice were given i.p. injections of 0.1 ml cinnamyl anthranilate in tricaprylin thrice weekly for 8 weeks (total dose, 12 g/kg bw). At 24 weeks after the start of treatment, 13 males and 15 females were still alive, and 11 males and 10 females had

[1]The Working Group was aware of carcinogenicity tests in progress on cinnamyl anthranilate involving oral administration in mice and rats (IARC, 1976).

lung tumours (2.7 and 2.1 lung tumours/mouse). Of 25 female controls injected with tricaprylin alone, 22 were still alive at 24 weeks, and 10 had lung tumours (0.59 lung tumours/mouse) ($P<0.05$). These results were confirmed in a second series of experiments using the same total dose (12 g/kg bw) (Stoner *et al.*, 1973).

3.2 Other relevant biological data

(a) Experimental systems

The oral LD_{50} in rats and the dermal LD_{50} in rabbits of cinnamyl anthranilate was greater than 5 g/kg bw (Opdyke, 1975). All of 5 mice survived six i.p. injections of 0.5 g/kg bw given over a 2-week period (Stoner *et al.*, 1973).

No data on the metabolism, embryotoxicity, teratogenicity or mutagenicity of cinnamyl anthranilate were available to the Working Group.

(b) Man

No data were available to the Working Group.

3.3 Case reports and epidemiological studies

No data were available to the Working Group.

4. Comments on Data Reported and Evaluation

4.1 Animal data

In the only available study, cinnamyl anthranilate increased the incidence of lung tumours in mice following its intraperitoneal injection. These data are insufficient for an evaluation of the carcinogenicity of this compound to be made.

4.2 Human data

No case reports or epidemiological studies were available to the Working Group.

Bedoukian, P.Z. (1967) Perfumery and Flavoring Synthetics, 2nd revised ed., Amsterdam, Elsevier, pp. 41-47

Committee on Specifications (1972) Food Chemicals Codex, 2nd ed., Committee on Food Protection, National Research Council, Washington DC, National Academy of Sciences, pp. 199, 896, 899, 954-957

Furia, T.E. & Bellanca, N., eds, (1971) CRC Fenaroli's Handbook of Flavor Ingredients, Cleveland, Ohio, Chemical Rubber Co., p. 332

Givaudan-Delawanna, Inc. (1961) The Givaudan Index, 2nd ed., New York, pp. 100, 360

IARC (1976) Information Bulletin on the Survey of Chemicals Being Tested for Carcinogenicity, No. 6, Lyon, p. 283

Opdyke, D.L.J. (1975) Special issue II. Monographs on fragrance raw materials. Fd Cosmet. Toxicol., 13, Suppl., p. 751

Stoner, G.D., Shimkin, M.B., Kniazeff, A.J., Weisburger, J.H., Weisburger, E.K. & Gori, G.B. (1973) Test for carcinogenicity of food additives and chemotherapeutic agents by the pulmonary tumor response in strain A mice. Cancer Res., 33, 3069-3085

US Food and Drug Administration (1976) Synthetic flavoring substances and adjuvants. US Code of Federal Regulations, Title 21, part 121.1164

US International Trade Commission (1977) Synthetic Organic Chemicals, US Production and Sales, 1975, USITC Publication 804, Washington DC, US Government Printing Office, pp. 109, 111

US Tariff Commission (1940) Synthetic Organic Chemicals, US Production and Sales, 1939, Report No. 140, Second Series, Washington DC, US Government Printing Office, pp. 39, 59

N,N'-DIACETYLBENZIDINE

1. Chemical and Physical Data

1.1 Synonyms and trade names

Chem. Abstr. Services Reg. No.: 613-35-4

Chem. Abstr. Name: *N,N'*-(1,1'-Biphenyl)-4,4'-diylbisacetamide

4',4'''-Biacetanilide; *N,N'*-4,4'-biphenylylenebisacetamide; 4,4'-diacetamidobiphenyl; diacetylbenzidine; 4,4'-diacetylbenzidine

1.2 Chemical formula and molecular weight

$CH_3-CO-NH-C_6H_4-C_6H_4-NH-CO-CH_3$

$C_{16}H_{16}N_2O_2$ Mol. wt: 268.3

1.3 Chemical and physical properties of the pure substance

(a) Description: White needles (Carlin & Swakon, 1955)

(b) Melting-point: 327-330°C (Carlin & Swakon, 1955)

(c) Spectroscopy data: λ_{max} 293 nm (E_1^1 = 1355) in 96% ethanol (Zheltov *et al.*, 1975)

(d) Solubility: Soluble in ethanol (Laham *et al.*, 1970) and ethyl acetate (Sciarini & Meigs, 1961)

(e) Stability: See 'General Remarks on the Substances Considered', p. 27.

(f) Polarographic data: The oxidative polarography of *N,N'*-diacetylbenzidine has been studied (Beilis, 1970).

1.4 Technical products and impurities

No data were available to the Working Group.

2. Production, Use, Occurrence and Analysis

For background information on this section, see preamble, p. 15.

2.1 Production and use

(a) Production

N,N-Diacetylbenzidine was prepared by Strakosch in 1872 by reacting benzidine with acetic acid (Prager *et al.*, 1930). It can also be prepared by treating benzidine with acetic anhydride (Carlin & Swakon, 1955).

No evidence was found that *N,N'*-diacetylbenzidine has ever been produced commercially in the US or Japan.

(b) Use

N,N'-Diacetylbenzidine can be used to prepare 3,3'-dichlorobenzidine, which was formerly used principally as an intermediate in the production of pigments [for more detailed information, see monograph on 3,3'-dichlorobenzidine (IARC, 1974], and 3,3'-dinitrobenzidine, which was formerly used in the preparation of C.I. Sulphur Brown 13 (Lurie, 1964); however, it is not known whether it is now used in the commercial manufacture of these compounds.

N,N'-Diacetylbenzidine has been investigated in Japan for use as an intermediate in the manufacture of diaminoazobenzidine, a raw material for azo dyes; however, the production of benzidine and its derivatives has been prohibited in Japan since 1973.

2.2 Occurrence

N,N'-Diacetylbenzidine is not known to occur as a natural product. No data on its occurrence in the environment were available to the Working Group.

2.3 Analysis

Paper chromatography using 10 solvent systems and five spray reagents has been used to separate and identify benzidine, 4,4'-dinitrobiphenyl, and their most important metabolites, including *N,N'*-diacetylbenzidine (Laham *et al.*, 1970).

3. Biological Data Relevant to the Evaluation of Carcinogenic Risk to Man

3.1 Carcinogenicity and related studies in animals

(a) Oral administration

Rat: A group of 10 male and 10 female Holtzman rats (190-210 g) received a diet containing 429 mg/kg of diet (1.6 mM/kg of diet) *N,N'*-diacetylbenzidine for 8 months, during which time 8 males and 8 females died from severe glomerulonephritis. The experiment was terminated at 10 months. Two males developed squamous-cell carcinomas of the ear duct after 5 and 7 months. The authors reported that in another experiment such tumours were found after 4-7 months in 4/20 male rats fed *N,N'*-diacetylbenzidine. No ear-duct tumours occurred in 45 untreated controls that survived up to 8 months (Miller *et al.*, 1956).

(b) Subcutaneous and/or intramuscular administration

Rat: Of 17 male and 23 female rats of the Rappolov nursery (110-140 g) given weekly s.c. injections each of 15 mg *N,N'*-diacetylbenzidine for 9 months (total dose, 870 mg/animal), 13 (10 males and 3 females) were still alive at the appearance of the first tumour; 2 males had hepatomas, and 3 males and 1 female had carcinomas of both the liver and Zymbal gland. Of 25 controls injected with sunflower oil, one developed a sarcoma connected with the walls of a parasitic cyst located in the liver (Pliss, 1962, 1963).

Of 6 female Wistar rats (100-120 g) injected once subcutaneously with 100 mg *N,N'*-diacetylbenzidine in aqueous suspension, 3 developed tumours of the mammary gland and one a skin tumour (average survival time, 12 months). No tumours were seen in 30 controls (average survival time, 9 months); necropsies were performed on 23 animals (Bremner & Tange, 1966).

(c) Intraperitoneal administration

Rat: Among 18 female rats (100-120 g) given one i.p. injection of 100 mg/animal *N,N'*-diacetylbenzidine, 12 tumours of the external auditory canal, 6 mammary tumours (including adenocarcinomas) and 2 skin carcinomas occurred in 11 animals (average survival time, 8.5 months). No tumours were seen in 30 control rats which had an average survival time of 9 months; necropsies were performed on 23 animals (Bremner & Tange, 1966).

3.2 Other relevant biological data

(a) Experimental systems

Mice given diets containing 4 g/kg of diet *N,N'*-diacetylbenzidine had kidney lesions but no damage to the liver (Yanagida, 1969). Lipaemia and glomerular lesions with many fat-filled spaces developed within 2-4.5 months in Buffalo rats fed diets containing 2.5 g/kg of diet *N,N'*-diacetylbenzidine (Dunn *et al.*, 1956).

In 6 rats given single i.p. doses of 200 mg *N,N'*-diacetylbenzidine in aqueous suspension, the average survival time was 5 months; one rat had liver necrosis, and 3 had severe glomerulonephrosis (Bremner & Tange, 1966). Chronic glomerulonephritis was seen in 110 female and 50 male Sprague-Dawley rats fed 430 mg/kg of diet *N,N'*-diacetylbenzidine. The lesions developed more rapidly in female than in male rats (Harman, 1971; Harman *et al.*, 1952).

N,N'-Diacetylbenzidine was a metabolite present in the urine of male Sprague-Dawley rats following 4 i.p. injections of 50 mg 4,4'-dinitrobiphenyl and of male guinea-pigs following 4 i.p. injections of 35 mg (Laham, 1972).

No data on the embryotoxicity, teratogenicity or mutagenicity of *N,N'*-diacetylbenzidine were available to the Working Group.

(b) Man

No data were available to the Working Group.

3.3 Case reports and epidemiological studies

No data were available to the Working Group.

4. Comments on Data Reported and Evaluation[1]

4.1 Animal data

N,N'-Diacetylbenzidine is carcinogenic in rats after its oral, subcutaneous or intraperitoneal administration; it produced carcinomas of the liver, ear duct and mammary gland.

4.2 Human data

No case reports or epidemiological studies were available to the Working Group.

---

[1]See also the section, 'Animal Data in Relation to the Evaluation of Risk to Man' in the introduction to this volume, p. 13.

5. References

Beilis, Y.I. (1970) Investigation of aromatic amines by means of oxidative polarography. III. Aniline derivatives with substituents in the amino group. Zh. obshch. Khim., 40, 1182-1185

Bremner, D.A. & Tange, J.D. (1966) Renal and neoplastic lesions after injection of *N,N'*-diacetylbenzidine. Arch. Path., 81, 146-151

Carlin, R.B. & Swakon, E.A. (1955) Anomalous Ullmann reactions. The unsymmetrical coupling of 2,6-dibromo-4-nitroiodobenzene. J. Amer. chem. Soc., 77, 966-973

Dunn, T.B., Morris, H.P. & Wagner, B.P. (1956) Lipemia and glomerular lesions in rats fed diets containing *N,N'*-diacetyl- and 4,4-4',4'-tetramethylbenzidine. Proc. Soc. exp. Biol. (Wash.), 91, 105-107

Harman, J.W. (1971) Chronic glomerulonephritis and the nephrotic syndrome induced in rats with *N,N'*-diacetylbenzidine. J. Path., 104, 119-128

Harman, J.W., Miller, E.C. & Miller, J.C. (1952) Chronic glomerulonephritis and nephrotic syndrome induced in rats with *N,N'*-diacetylbenzidine (Abstract). Amer. J. Path., 28, 529-530

IARC (1974) IARC Monographs on the Evaluation of Carcinogenic Risk of Chemicals to Man, 4, Some Aromatic Amines, Hydrocarbons and Related Substances, *N*-Nitroso Compounds and Miscellaneous Alkylating Agents, Lyon, pp. 49-55

Laham, S. (1972) Biochemical studies on environmental carcinogens: evidence of a new metabolic pathway in the rat. Ann. occup. Hyg., 15, 203-208

Laham, S., Farant, J.-P. & Potvin, M. (1970) Biochemical determination of urinary bladder carcinogens in human urine. Occup. Hlth Rev., 21, 14-23

Lurie, A.P. (1964) Benzidine and related diaminobiphenyls. In: Kirk, R.E. & Othmer, D.F., eds, Encyclopedia of Chemical Technology, 2nd ed., Vol. 3, New York, John Wiley and Sons, p. 416

Miller, E.C., Sandin, R.B., Miller, J.A. & Rusch, H.P. (1956) The carcinogenicity of compounds related to 2-acetylaminofluorene. III. Aminobiphenyl and benzidine derivatives. Cancer Res., 16, 525-534

Pliss, G.B. (1962) The characteristics of the cancerogenic activity of *N,N'*-diacetylbenzidine. Vop. Onkol., 8, 11-15

Pliss, G.B. (1963) On some regular relationships between carcinogenicity of aminodiphenyl derivatives and the structure of substance. Acta un. int. cancr., 19, 499-501

Prager, B., Jacobson, P., Schmidt, P. & Stern, D., eds (1930) Beilsteins Handbuch der Organischen Chemie, 4th ed., Vol. 13, Syst. No. 1786, Berlin, Springer, p. 227

Sciarini, L.J. & Meigs, J.W. (1961) The biotransformation of benzidine. II. Studies in mouse and man. Arch. environm. Hlth, 2, 423-428

Yanagida, K. (1969) Pathological study on various lesions of mice ingesting *N,N'*-2,7-fluorenylenebisacetamide: especially on inducing process of the general edema. J. Wakayama med. Soc., 20, 17-30

Zheltov, A.Y., Rodionov, V.Y. & Stepanov, B.I. (1975) Investigations in the field of aromatic disulfides. VI. Electronic spectra of 3,8- and 2,9-disubstituted dibenzo[*c,e*]-*o*-dithiins. Zh. org. Khim., 11, 1304-1311

4,4'-DIAMINODIPHENYL ETHER

1. Chemical and Physical Data

1.1 Synonyms and trade names

Chem. Abstr. Services Reg. No.: 101-80-4

Chem. Abstr. Name: 4,4'-Oxybisbenzenamine

Bis(4-aminophenyl)ether; bis(*para*-aminophenyl)ether; 4,4'-diaminobiphenyl ether; 4,4'-diaminobiphenyloxide; diaminodiphenyl ether; 4,4'-diamino diphenyl ether; 4,4'-diaminodiphenylether; *para,para*'-diaminodiphenyl ether; 4,4'-diaminodiphenyl oxide; 4,4'-diaminodiphenyloxide; 4,4'-diaminophenyl ether; oxybis(4-aminobenzene); *para,para*'-oxybis(aniline); oxydianiline; 4,4'-oxydianiline; *para,para*'-oxydianiline; oxydi-*para*-phenylenediamine

1.2 Chemical formula and molecular weight

H_2N–⟨benzene ring⟩–O–⟨benzene ring⟩–NH_2

$C_{12}H_{12}N_2O$ Mol. wt: 200.2

1.3 Chemical and physical properties of the pure substance

From Dean (1973), unless otherwise specified

(a) Description: Colourless crystals

(b) Boiling-point: >300°C (Kazinik *et al.*, 1971a)

(c) Melting-point: 186-187°C

(d) Solubility: Insoluble in water, ethanol, benzene and carbon tetrachloride; soluble in acetone

(e) Stability: See 'General Remarks on the Substances Considered', p. 27.

1.4 Technical products and impurities

In Japan, 4,4'-diaminodiphenyl ether is available as a 99.9% pure commercial grade (melting-point, 190°C), which contains inorganic salts as impurities.

2. Production, Use, Occurrence and Analysis

For background information on this section, see preamble, p. 15.

2.1 Production and use

(a) Production

4,4'-Diaminodiphenyl ether was prepared by Haeussermann & Teichmann in 1896 by treating 4,4'-dinitrodiphenyl ether with tin and hydrochloric acid in the presence of ethanol (Prager *et al.*, 1930). Other methods are variations of the hydrogenation of 4,4'-dinitrodiphenyl ether, using various catalyst systems (Jamieson *et al.*, 1973; Spiegler, 1965).

The commercial production of 4,4'-diaminodiphenyl ether in the US was first reported in 1959 (US Tariff Commission, 1960). Only one US company reported an undisclosed amount (see preamble, p. 15) in 1975 (US International Trade Commission, 1977). US imports through principal US customs districts were reported as 90 kg in 1974 (US International Trade Commission, 1976).

4,4'-Diaminodiphenyl ether has been produced commercially in Japan since 1968; in 1975, one manufacturer produced an estimated 15 thousand kg. There are no Japanese exports or imports.

No data on its production in Europe were available to the Working Group.

(b) Use

4,4'-Diaminodiphenyl ether is used in the US and Japan as an intermediate in the manufacture of high-temperature-resistant, straight polyimide and poly(esterimide) resins (Anon., 1973; Seymour, 1968) capable of withstanding temperatures of up to 480°C for short periods or 260°C for prolonged periods of time. Straight polyimide resins are used in the following applications: as a film (for insulation tape for various electric wires and cables), in molding and machined parts (piston rings, seals, gears, bearings, valve seats and grinding wheels), in adhesives (for bonded-structures in aerospace equipment), in laminating (various laminates, printed circuits and honeycomb structural parts) and in coated fabrics and paper. Poly(esterimide) resins are primarily used as magnet wire enamels for electric motors, generators and switch equipment.

Polyimide and poly(esterimide) resins based on 4,4'-diaminodiphenyl ether do not appear to be produced in western Europe.

2.2 Occurrence

4,4'-Diaminodiphenyl ether is not known to occur as a natural product. No data on its occurrence in the environment were available to the Working Group.

2.3 Analysis

A method for determining the purity of 4,4'-diaminodiphenyl ether by thin-layer chromatography has been described (Eulenhoefer & Bauer, 1969). Thin-layer chromatography using four different solvent systems has been used to separate and identify a group of aromatic diamines, including 4,4'-diaminodiphenyl ether, and *N*-benzamides (products of the thermal degradation of polyamides *in vacuo*) (Krasnov *et al.*, 1970).

A colorimetric method for detecting as little as 0.2 mg 4,4'-diamino-diphenyl ether involves reaction with formaldehyde in formic acid azomethine (Shemyakin & Zelenina, 1968).

Paper chromatography has been used to separate and identify a group of aromatic 4,4'-diamines, including 4,4'-diaminodiphenyl ether (Gasparic & Snobl, 1971).

Gas chromatography with flame ionization detection has been used to separate and identify one group of high-boiling aromatic amines, including 4,4'-diaminodiphenyl ether, with an error of not more than 10-12% (Kazinik *et al.*, 1971a) and another group with an error of not more than 5.8% (Kazinik *et al.*, 1971b).

3. Biological Data Relevant to the Evaluation of Carcinogenic Risk to Man

3.1 Carcinogenicity and related studies in animals[1]

(a) Oral administration

Mouse: A group of 16 male and 24 female CC57W mice received 5 mg 4,4'-diaminodiphenyl ether/animal in 0.2 ml sunflower oil on 5 days per week (total dose, 440 mg/animal); animals were observed up to 472 days. Among 14 mice surviving at the appearance of the first tumour (212 days), 8 developed 10 tumours (6 lymphomas, 2 haemangiomas of the ovaries and 2 lung adenomas). The incidence of lymphomas was 6.8% in untreated mice of that strain (Dzhioev, 1975) [The Working Group noted that no contemporary controls were used].

Rat: A group of 15 male and 33 female white, colony-bred rats received 25 mg/animal 4,4'-diaminodiphenyl ether in 0.5 ml sunflower oil 5 times per week (total dose, 4.12 g/animal); animals were observed for up to 826 days. Of 16 rats surviving at the appearance of the first tumour (540 days), 7 developed 9 tumours (1 kidney carcinoma, 3 reticulum-cell sarcomas, 1 liver fibrosarcoma, 1 neurogenic sarcoma, 2 seminomas and 1 mammary gland fibroadenoma) (Dzhioev, 1975) [The Working Group noted that no controls were used].

4,4'-Diaminodiphenyl ether was administered by stomach tube to 20 female 40-day old Sprague-Dawley rats; 10 equal doses of 40 mg/rat, which was the maximum tolerated dose, were given at 3-day intervals. The experiment was terminated at 9 months. No mammary tumours but 1 squamous metaplasia of the uterus were observed; 11 animals were autopsied. Among 132 controls given sunflower oil that were autopsied, 3 mammary carcinomas were observed (Griswold *et al.*, 1968) [The Working Group noted the short duration of the study].

---

[1]The Working Group was aware of carcinogenicity tests in progress on 4,4'-diaminodiphenyl ether involving subcutaneous injection in rats (IARC, 1976).

(b) Subcutaneous and/or intramuscular administration

Mouse: A group of 15 male and 18 female CC57W mice received s.c. injections of 5 mg/animal 4,4'-diaminodiphenyl ether in 0.2 ml sunflower oil weekly (total dose, 175 mg/animal); animals were observed for up to 316 days. Of 9 mice alive at the appearance of the first tumour (271 days), 3 developed 3 tumours (2 lymphomas and 1 lung adenoma). The incidence of lymphomas in untreated mice of this strain was 6.8% (Dzhioev, 1975) [The Working Group noted that no concurrent controls were used].

Rat: A group of 30 male and 32 female white, colony-bred rats received s.c. injections of 25 mg/animal 4,4'-diaminodiphenyl ether in 0.5 ml sunflower oil weekly (total dose, 2 g/animal); animals were observed for up to 949 days. Among 39 rats alive at the appearance of the first tumour (529 days), 7 developed 7 tumours (2 lymphomas, 1 reticulum-cell sarcoma, 1 liver fibrosarcoma, 1 carcinoma of the kidney and 2 mammary gland fibroadenomas) (Dzhioev, 1975) [The Working Group noted that no controls were used].

3.2 Other relevant biological data

(a) Experimental systems

In mice and rats given oral or s.c. injections of 5 and 25 mg 4,4'-diaminodiphenyl ether weekly or 5 times per week, sclerotic and necrotic damage to the kidneys was noted. Hyperplasia of the bile ducts also occurred (Dzhioev, 1975).

The compound inhibited the growth of spontaneous mammary tumours and transplanted tumours in mice (Boyland, 1946).

No data on the metabolism, embryotoxicity, teratogenicity or mutagenicity of 4,4'-diaminodiphenyl ether were available to the Working Group.

(b) Man

No data were available to the Working Group.

3.3 Case reports and epidemiological studies

No data were available to the Working Group.

4. Comments on Data Reported and Evaluation[1]

4.1 Animal data

4,4'-Diaminodiphenyl ether has been tested in rats and mice by oral and subcutaneous administration, producing tumours at various sites. The results suggest carcinogenic activity, but in the absence of contemporary controls, an evaluation of the carcinogenicity of this compound could not be made.

4.2 Human data

No case reports or epidemiological studies were available to the Working Group.

---

[1]Subsequent to the finalization of this evaluation by the Working Group in June 1977, the Secretariat became aware of a paper by Steinhoff (1977), in which a group of 20 male and 20 female Wistar rats received weekly s.c. injections of 100-300 mg/kg bw 4,4'-diaminodiphenyl ether in saline, up to a total dose of 14.4 g/kg bw, during 670 days. A group of 50 control rats received s.c. injections of saline alone for 907 days. Twenty-two (55%) treated rats died with malignant tumours, compared with 13 (26%) in controls. Benign tumours occurred in 22 treated rats and in 10 controls. Ten (25%) treated rats had malignant tumours of the liver (3 multiple), and 12 had benign liver tumours; no such tumours were found in controls.

5. References

Anon. (1973) New polyimides contain fluorine. Chem. Eng. News, July 30, p. 16

Boyland, E. (1946) Experiments on the chemotherapy of cancer. VI. The effect of aromatic bases. Biochem. J., 40, 55-58

Dean, J.A., ed. (1973) Lange's Handbook of Chemistry, 11th ed., New York, McGraw-Hill Book Co., pp. 7.148-7.149

Dzhioev, F.K. (1975) On carcinogenic activity of 4,4'-diaminodiphenyl ether. Vop. Onkol., 21, 69-73

Eulenhoefer, H.G. & Bauer, M. (1969) Thin-layer chromatographic examination of the purity of bis(*p*-aminophenyl)ether. Fresenius' Z. analyt. Chem., 247, 55-56

Gasparic, J. & Snobl, D. (1971) Identification of organic compounds. LXXIII. Paper chromatography of aromatic *p,p*'-diamines. Sb. Ved. Pr., Vys. Sk. Chemickotechnol., Pardubice, No. 25, 33-40

Griswold, D.P., Jr, Casey, A.E., Weisburger, E.K. & Weisburger, J.H. (1968) The carcinogenicity of multiple intragastric doses of aromatic and heterocyclic nitro or amino derivatives in young female Sprague-Dawley rats. Cancer Res., 28, 924-933

IARC (1976) Information Bulletin on the Survey of Chemicals Being Tested for Carcinogenicity, No. 6, Lyon, p. 41

Jamieson, N.C., McKenzie, L. & Newman, E.R. (1973) 4-Amino-4'-nitro- and 4,4'-diaminodiphenyl ether. Ger. Offen. 2,236,777, February 8, (to Hall, Howard and Co.)

Kazinik, E.M., Gudkova, G.A. & Shcheglova, T.A. (1971a) Gas-liquid chromatographic analysis of some high-boiling aromatic amines. Zh. analit. Khim., 26, 154-157

Kazinik, E.M., Gudkova, G.A., Mesh, L.Y., Shcheglova, T.A. & Ivanov, A.V. (1971b) Gas-chromatographic analysis of high-boiling diamines - diphenyl, diphenyl sulfide, and diphenyl sulfone derivatives. Zh. analit. Khim., 26, 1920-1923

Krasnov, E.P., Logunova, V.I. & Shumakova, V.D. (1970) Thin-layer chromatography of aromatic diamines and *N*-benzamides. In: Pakshver, A.B., ed, Volokna Sin. Polim., Moscow, Khimiya, pp. 292-294

Prager, B., Jacobson, P., Schmidt, P. & Stern, D., eds (1930) Beilsteins Handbuch der Organischen Chemie, 4th ed., Vol. 13, Syst. No. 1844-1846, Berlin, Springer-Verlag, p. 441

Seymour, R.B. (1968) Plastics technology. In: Kirk, R.E. & Othmer, D.F., eds, Encyclopedia of Chemical Technology, 2nd ed., Vol. 15, New York, John Wiley and Sons, p. 796

Shemyakin, F.M. & Zelenina, E.N. (1968) Determination of small concentrations of 4,4'-diaminodiphenyl ether and selective determination of aldehydes. Opred. Mikroprimesei, 2, 106-109

Spiegler, L. (1965) Bis(nitrophenyl) and bis(aminophenyl)ethers. US Patent 3,192,263, June 29 (to E.I. du Pont de Nemours & Co.)

Steinhoff, D. (1977) Cancerogene Wirkung von 4,4'-Diamino-diphenyläther bei Ratten. Naturwissenschaften, 64, 394

US International Trade Commission (1976) Imports of Benzenoid Chemicals and Products, 1974, USITC Publication 762, Washington DC, US Government Printing Office, p. 16

US International Trade Commission (1977) Synthetic Organic Chemicals, US Production and Sales, 1975, USITC Publication 804, Washington DC, US Government Printing Office, p. 39

US Tariff Commission (1960) Synthetic Organic Chemicals, US Production and Sales, 1959, Report No. 206, Second Series, Washington DC, US Government Printing Office, p. 79

3,3'-DICHLORO-4,4'-DIAMINODIPHENYL ETHER

1. Chemical and Physical Data

1.1 Synonyms and trade names

Chem. Abstr. Reg. Serial No.: 28434-86-8

Chem. Abstr. Name: 4,4'-Oxybis(2-chloro-benzenamine)

Bis(4-amino-3-chlorophenyl)ether; 4,4'-oxybis(2-chloroaniline)

1.2 Chemical formula and molecular weight

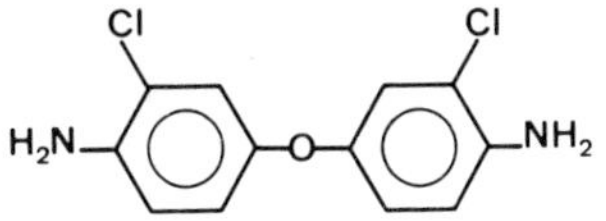

$C_{12}H_{10}ON_2Cl_2$ Mol. wt: 269.1

1.3 Chemical and physical properties of the pure substance

(a) Melting-point: 128-129°C (Farbenfabriken Bayer A.-G., 1970)

(b) Stability: See 'General Remarks on the Substances Considered', p. 27.

1.4 Technical products and impurities

No data were available to the Working Group.

2. Production, Use, Occurrence and Analysis

For background information on this section, see preamble, p. 15.

2.1 Production and use

(a) Production

In 1970, a patent was issued covering a method of preparing 3,3'-dichloro-4,4'-diaminodiphenyl ether, by treating a suspension of 4,4'-diacetamidodiphenyl ether in acetic acid with chlorine to give *N,N'*-diacetyl-3,3'-dichloro-4,4'-diaminodiphenyl ether, which is hydrolysed with methanolic potassium hydroxide or ethanolic hydrogen chloride (Farbenfabriken Bayer A.-G., 1970).

No evidence was found that 3,3'-dichloro-4,4'-diaminodiphenyl ether has ever been produced commercially in the US or Japan.

No data on its production in Europe were available to the Working Group.

(b) Use

No evidence was found that 3,3'-dichloro-4,4'-diaminodiphenyl ether has ever had any commercial application in the US or Japan.

2.2 Occurrence

3,3'-Dichloro-4,4'-diaminodiphenyl ether is not known to occur as a natural product. No data on its occurrence in the environment were available to the Working Group.

2.3 Analysis

Although no specific studies on this compound were available to the Working Group, the accompanying monographs describe a number of methods generally applicable to the analysis of aromatic amines.

3. Biological Data Relevant to the Evaluation of Carcinogenic Risk to Man

3.1 Carcinogenicity and related studies in animals

Subcutaneous and/or intramuscular administration

Rat: 3,3'Dichloro-4,4'-diaminodiphenyl ether suspended in isotonic saline was injected subcutaneously into 20 male and 20 female Wistar rats. Doses of 250-1000 mg/kg bw per injection were administered once weekly over 190 days (total dose, 10.5 g/kg bw). Within a median induction time of 300 days, 37 rats had a total of 65 malignant and 2 benign tumours; 51 were ear-duct carcinomas and 16 occurred at other unspecified sites. The last treated rat died after 545 days, at which time 4/50 controls injected with saline alone had also died; no tumours were found in controls (Steinhoff & Grundmann, 1970).

3,3'-Dichloro-4,4'-diaminodiphenyl ether suspended in water was injected subcutaneously into 20 male Wistar rats at a dose of 400 mg/kg bw per week. Between the 200th and 340th day, the animals were killed serially, and sections of the ear duct showed multiple circumscribed areas with a high alkaline phosphatase content. After 390 days, the 4 remaining rats had papillomas, papillary carcinomas, squamous-cell carcinomas and carcinomas and adenomas of the Zymbal gland. No such tumours were found in 20 untreated male controls (Herrmann *et al.*, 1973).

3.2 Other relevant biological data

The s.c. LD_{50} in rats is >10 g/kg bw (Steinhoff & Grundmann, 1970).

No data on the metabolism, embryotoxicity, teratogenicity or mutagenicity of 3,3'-dichloro-4,4'-diaminodiphenyl ether were available to the Working Group.

3.3 Case reports and epidemiological studies

No data were available to the Working Group.

4. Comments on Data Reported and Evaluation

4.1 Animal data

3,3'-Dichloro-4,4'-diaminodiphenyl ether is carcinogenic in rats; it produced carcinomas of the ear duct after its subcutaneous injection.

4.2 Human data

No case reports or epidemiological studies were available to the Working Group.

---

[1]See also the section, 'Animal Data in Relation to the Evaluation of Risk to Man' in the introduction to this volume, p. 13.

5. References

Farbenfabriken Bayer A.-G. (1970) 3,3'-Dichloro-4,4'-diaminodiphenyl ethers. French Demande 2,008,760, 23 January

Herrmann, I.F., Schauer, A. & Kamke, W. (1973) Morphologische und histochemische Untersuchungen während der Cancerogenese des äusseren Gehörganges der Ratte, induziert durch 3,3'-Dichlor-4,4'-diamino-diphenyläther. Arch. Oto-Rhino-Laryng., 206, 11-16

Steinhoff, D. & Grundmann, E. (1970) Cancerogene Wirkung von 3,3'-Dichlor-4,4'-diamino-diphenyläther bei Ratten. Naturwissenschaften, 57, 676

2,4'-DIPHENYLDIAMINE

1. Chemical and Physical Data

1.1 Synonyms and trade names

Chem. Abstr. Reg. Serial No.: 492-17-1

Chem. Abstr. Name: (1,1'-Biphenyl)-2,4'-diamine

ortho,para'-Bianiline; 2,4'-biphenyldiamine; 2,4'-diaminobiphenyl; *ortho,para'*-diaminobiphenyl; 2,4'-diaminodiphenyl; *ortho,para'*-dianiline; diphenyline

1.2 Chemical formula and molecular weight

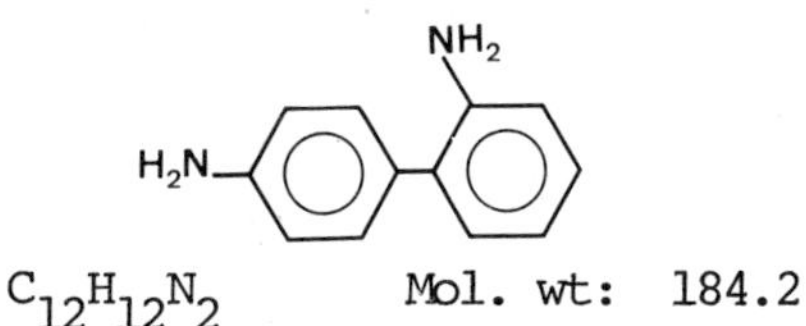

$C_{12}H_{12}N_2$ Mol. wt: 184.2

1.3 Chemical and physical properties of the pure substance

From Weast (1976)

(a) Description: Needles

(b) Boiling-point: 363°C

(c) Melting-point: 54.5°C

(d) Spectroscopy data: λ_{max} 300 nm (E_1^1 = 431), 235 nm (E_1^1 = 861) in ethanol

(e) Solubility: Insoluble in water, soluble in ethanol and ether

(f) Stability: See 'General Remarks on the Substances Considered', p. 27.

1.4 Technical products and impurities

No data were available to the Working Group.

2. Production, Use, Occurrence and Analysis

For background information on this section, see preamble, p. 15.

2.1 Production and use

(a) Production

2,4'-Diphenyldiamine was prepared by Schultz in 1876 by treating 2'-nitro-4-aminodiphenyl with tin and hydrochloric acid (Prager *et al.*, 1930). It is not manufactured commercially, as such, in the US.

(b) Use

2,4'-Diphenyldiamine has reportedly found use in the past as an analytical reagent for the detection of tungsten and tungstate (Lurie, 1964).

2.2 Occurrence

2,4'-Diphenyldiamine is not known to occur as a natural product.

It is formed in quantities of 3 to 25% (depending on temperature and acid concentration used) during the manufacture of benzidine. The reported average concentration in benzidine produced in most factories is about 12%. Most technical grades of benzidine contain about 1% 2,4'-diphenyldiamine as an impurity, while inferior grades contain up to 4% (Lurie, 1964; Marhold *et al.*, 1968). Benzidine is not known to be manufactured commercially, except as an unisolated intermediate, in the US. Its production has been prohibited in Japan since 1973, in Ireland, in the UK and in the USSR (Montesano & Tomatis, 1977) [For further information on benzidine, see IARC, 1972].

No data on the occurrence of 2,4'-diphenyldiamine in the environment were available to the Working Group.

2.3 Analysis

Aromatic substances, including 2,4'-diphenyldiamine, have been identified by paper electrophoresis (Franc & Kovář, 1965). Paper chromatography, using three solvent systems and four spray reagents, has been used to separate and identify 169 amines, including 2,4'-diphenyldiamine (Cee & Gasparič, 1966). Paper and thin-layer chromatography have been used to separate and identify aromatic amines, including 2,4'-diphenyldiamine (Wiesner, 1973).

Gas chromatography has been used to separate and identify 2,4'-diphenyldiamine in the presence of isomeric 2,2'- and 2,3'-diphenyldiamines (Knight, 1971).

3. Biological Data Relevant to the Evaluation of Carcinogenic Risk to Man

3.1 Carcinogenicity and related studies in animals

Oral administration

Rat: A group of 50 male Wistar rats (100 g) received 2 mg 2,4'-diphenyldiamine daily in the diet for life (average total dose, 852 mg); 50 males served as untreated controls. The average survival times were 378 days for the controls and 426 days for the treated groups. No tumours were seen in either group (Marhold *et al.*, 1968) [The Working Group noted the inadequacy of the experiment].

Dog: One female and 1 male mongrel dog were fed 5 mg/kg bw 2,4'-diphenyldiamine daily 6 times per week for 7 years. The female developed a basal-cell carcinoma of the skin after 4½ years, and a bronchiolar carcinoma was found at death at 8½ years. The male had nodular hyperplasia of the spleen at death at 8 years 10 months. Another female dog developed nodular hyperplasia of the spleen at 8 years 10 months after receiving 5 mg/kg bw benzidine and 5 mg/kg bw 2,4'-diphenyldiamine daily for 7 years (the 2,4'-diphenyldiamine was accidentally contaminated with dinitrobiphenyl during 3 weeks in the third year of the experiment) (Marhold *et al.*, 1968).

3.2 Other relevant biological data

(a) Experimental systems

The oral LD_{50} of 2,4'-diphenyldiamine in male rats is 311 mg/kg bw (Marhold *et al.*, 1968).

No data on the metabolism, embryotoxicity, teratogenicity or mutagenicity of this compound were available to the Working Group.

(b) Man

No data were available to the Working Group.

3.3 Case reports and epidemiological studies

No data were available to the Working Group.

4. Comments on Data Reported and Evaluation

4.1 Animal data

2,4'-Diphenyldiamine has been inadequately tested in rats and dogs by oral administration. No evaluation of the carcinogenicity of this compound can be made.

4.2 Human data

No case reports or epidemiological studies were available to the Working Group.

5. References

Cee, A. & Gasparič, J. (1966) Papierchromatographische Trennung und Identifizierung primärer aromatischer Amine in wässrigen Lösungsmittelsystemen. Mikrochim. acta, 1-2, 295-309

Franc, J. & Kovář, V. (1965) Identification of aromatic substances by 'electrophoretic spectra' using paper electrophoresis. J. Chromat., 18, 100-115

IARC (1972) IARC Monographs on the Evaluation of Carcinogenic Risk of Chemicals to Man, Vol. 1, Lyon, pp. 80-86

Knight, J.A. (1971) Gas chromatographic analysis of γ-irradiated aniline for aminoaromatic products. J. Chromat., 56, 201-208

Lurie, A.P. (1964) Benzidine and related diaminobiphenyls. In: Kirk, R.E. & Othmer, D.F., eds, Encyclopedia of Chemical Technology, 2nd ed., Vol. 3, New York, John Wiley and Sons, pp. 408-420

Marhold, J., Matrka, M., Hub, M. & Ruffer, F. (1968) The possible complicity of diphenyline in the origin of tumours in the manufacture of benzidine. Neoplasma, 15, 3-10

Montesano, R. & Tomatis, L. (1977) Legislation concerning chemical carcinogens in several industrialized countries. Cancer Res., 37, 310-316

Prager, B., Jacobson, P., Schmidt, P. & Stern, D., eds (1930) Beilsteins Handbuch der Organischen Chemie, 4th ed., Vol. 13, Syst. No. 1785, Berlin, Springer-Verlag, p. 211

Weast, R.C., ed. (1967) CRC Handbook of Chemistry and Physics, 57th ed., Cleveland, Ohio, Chemical Rubber Co., p. C-210

Wiesner, I. (1973) Chromatographic identification of some intermediates of the reaction of aniline with formaldehyde. Collect. Czech. chem. Commun., 38, 1473-1477

5-NITROACENAPHTHENE

1. Chemical and Physical Data

1.1 Synonyms and trade names

Chem. Abstr. Reg. Services No.: 602-87-9

Chem. Abstr. Name: 1,2-Dihydro-5-nitro-acenaphthylene

5-Nitronaphthalene ethylene

1.2 Chemical formula and molecular weight

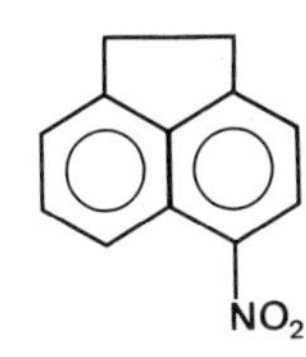

$C_{12}H_9NO_2$ Mol. wt: 199.2

1.3 Chemical and physical properties of the pure substance

(a) Melting-point: 102-103°C (Mitoguchi, 1969)

(b) Spectroscopy data: λ_{max} 262 nm (E^1_1 = 755), 314 nm (E^1_1 = 145), 376 nm (E^1_1 = 361) in chloroform (Webb & Wells, 1975); strong infra-red band at 776 cm^{-1} (Deady *et al.*, 1974). Mass spectral data have been reported (Todd *et al.*, 1973).

1.4 Technical products and impurities

No data were available to the Working Group.

2. Production, Use, Occurrence and Analysis

For background information on this section, see preamble, p. 15.

2.1 Production and use

(a) Production

5-Nitroacenaphthene can be synthesized by the nitration of acenaphthene in acetic anhydride solution using copper or zinc nitrate at 30°C

(Mitoguchi, 1969). It is believed to be produced commercially in Japan by the nitration of acenaphthene with nitric acid in sulphuric acid solution.

No evidence was found that 5-nitroacenaphthene has ever been produced in the US for other than research purposes. Its production in western Europe is estimated to be less than 10 thousand kg annually.

In Japan, 5-nitroacenaphthene has been produced commercially since 1963. Production in 1976 by the two Japanese producers is estimated to have been 220 thousand kg, down from a level of 290 thousand kg in 1973.

(b) Use

5-Nitroacenaphthene has not been used commercially in the US.

In Japan, it is used as a chemical intermediate, without isolation, to produce naphthalimide dyes, which are used as fluorescent whitening agents. A small amount of 5-nitroacenaphthene is also used to produce a paper dye.

2.2 Occurrence

5-Nitroacenaphthene is not known to occur as a natural product. No data on its occurrence in the environment were available to the Working Group.

2.3 Analysis

Polarography of 5-nitroacenaphthene has been investigated (Gupta & Kishore, 1970). No other data on its analysis were available to the Working Group.

3. Biological Data Relevant to the Evaluation of Carcinogenic Risk to Man

3.1 Carcinogenicity and related studies in animals[1]

(a) Oral administration

Rat: Thirty weanling female Wistar rats were fed a 25% protein diet containing 1% 5-nitroacenaphthene for a total of 4 months (average daily intake, 200 mg per animal), with two interruptions, each of 3 weeks, because of toxicity. Twelve rats survived more than 200 days, and the first tumour was found among these 280 days after the start of the experiment. Between the 280th and 500th day, all surviving rats developed malignant tumours, as follows: 1 rhabdomyosarcoma, 2 carcinomas of the ear duct, 5 mammary carcinomas and 10 adenocarcinomas of the small intestine. No neoplasms were observed among 29 controls that survived more than 500 days. In another experiment, in which 20 male Wistar rats were fed the 25% protein diet with 1% 5-nitroacenaphthene continuously for 6 months, animals survived 500 days, and no malignant tumours were observed Takemura *et al.*, 1974).

Hamster: Of 24 weanling female and 10 male Syrian hamsters that received 1% 5-nitroacenaphthene in their diet continuously for 6 months, 13 females survived 270 days after the start of the test, and 7 developed cholangiomas. No tumours were observed in the 20 control females or in 7 treated male hamsters that survived more than 270 days (Takemura *et al.*, 1974).

(b) Intraperitoneal administration

Mouse: In an abstract, it was reported that 20 female dd mice were given twice weekly i.p. injections of 6 mg/kg bw 5-nitroacenaphthene in arachis oil for 18 months. Of 15 surviving mice, 4 developed myeloid leukaemias, 2 developed reticulum-cell sarcomas, and 1 a mammary carcinoma.

---

[1]The Working Group was aware of carcinogenicity studies, in progress or completed, on 5-nitroacenaphthene involving skin application in mice and oral administration in mice and rats (IARC, 1976).

No tumours were observed in 20 female controls (Takemura, 1970) [The Working Group noted that survival times in the controls were not stated].

3.2 Other relevant biological data

(a) Experimental systems

In a carcinogenicity study (see section 3.1), treatment with 1% 5-nitroacenaphthene in the diet shortened the survival time of female rats and significantly impaired weight gain (Takemura *et al.*, 1974).

No data on the metabolism, teratogenicity or embryotoxicity of 5-nitroacenaphthene were available to the Working Group.

Reverse mutation from streptomycin dependence to independence in *Escherichia coli* was not increased by 5-nitroacenaphthene (Szybalski, 1968). The compound induced reverse mutations to histidine independence in *Salmonella typhimurium* TA 98 and TA 100. Addition of rat liver post-mitochondrial supernatant fraction from polychlorinated biphenyl (Aroclor 1254 or Kanechlor 500)-treated animals increased the mutagenic activity (McCann *et al.*, 1975; Yahagi *et al.*, 1975).

(b) Man

No data were available to the Working Group.

3.3 Case reports and epidemiological studies

No data were available to the Working Group.

4. Comments on Data Reported and Evaluation[1]

4.1 Animal data

5-Nitroacenaphthene is carcinogenic in female rats and female hamsters following its oral administration; it produced mainly adenocarcinomas of the small intestine and mammary carcinomas in female but not in male

---

[1]See also the section, 'Animal Data in Relation to the Evaluation of Risk to Man' in the introduction to this volume, p. 13.

rats and cholangiomas in female but not in male hamsters. In addition, it produced leukaemia and reticulum-cell sarcomas in mice following its intraperitoneal injection.

4.2 Human data

No case reports or epidemiological studies were available to the Working Group.

5. References

Deady, L.W., Gray, P.M. & Topsom, R.D. (1974) Infrared spectra of mono- and disubstituted acenaphthenes. Appl. Spectrosc., 28, 552-554

Gupta, S.L. & Kishore, N. (1970) Polarography of 5-nitro-ace-naphthene. Electrochim. acta, 15, 1367-1372

IARC (1976) Information Bulletin on the Survey of Chemicals Being Tested for Carcinogenicity, No. 6, Lyon, pp. 244, 286

McCann, J., Choi, E., Yamasaki, E. & Ames, B.N. (1975) Detection of carcinogens as mutagens in the *Salmonella*/microsome test: assay of 300 chemicals. Proc. nat. Acad. Sci. (Wash.), 72, 5135-5139

Mitoguchi, H. (1969) Synthesis of 5- and 3-nitroacenaphthenes. Nitration of acenaphthene and its derivatives. I. Yuki Gosei Kagaku Kyokai Shi, 27, 642-647

Szybalski, W. (1968) Special microbiological systems. II. Observations on chemical mutagenesis in microorganisms. Ann. N.Y. Acad. Sci., 76, 475-489

Takemura, N. (1970) Carcinogenic action of the intermediates of fluorescent whitening agents, 5-nitroacenaphthene and 5-aminoacenaphthene. (Abstract No. 123). In: Cumley, R.W., ed., Tenth International Cancer Congress, Houston, Texas, Houston, Medical Arts Publ. Co., p. 79

Takemura, N., Hashida, C. & Terasawa, M. (1974) Carcinogenic action of 5-nitroacenaphthene. Brit. J. Cancer, 30, 481-483

Todd, J.F.J., Turner, R.B., Webb, B.C. & Wells, C.H.J. (1973) The positive and negative ion mass spectra of some nitro- and polynitroacenaphthenes. J. chem. Soc. Perkin II, 2, 1167-1171

Webb, B.C. & Wells, C.H.J. (1975) Electronic spectra of nitro- and polynitroacenaphthenes and of nitro- and polynitro-1,8-dimethylnaphthalenes. Spectrochim. acta, 31A, 273-274

Yahagi, T., Shimizu, H., Nagao, M., Takemura, N. & Sugimura, T. (1975) Mutagenicity of 5-nitroacenaphthene in *Salmonella*. Gann, 66, 581-582

1. Chemical and Physical Data

1.1 Synonyms and trade names

Chem. Abstr. Services Reg. No.: 135-88-6

Chem. Abstr. Name: *N*-Phenyl-2-naphthalenamine

Anilinonaphthalene; 2-anilinonaphthalene; *N*-(2-naphthyl)aniline; 2-naphthylphenylamine; β-naphthylphenylamine; 2-phenylamino-naphthalene; phenyl-2-naphthylamine; phenyl(β-naphthyl)amine; phenyl-β-naphthylamine; *N*-phenyl-2-naphthalamine; *N*-phenyl-β-naphthylamine

Aceto PBN; Agerite Powder; Age Rite Powder; Antioxidant 116; Antioxidant PBN; Neosone D; Neozon D; Neozone; Neozone D; Nilox PBNA; Nonox D; PBNA; Stabilizator AR

1.2 Chemical formula and molecular weight

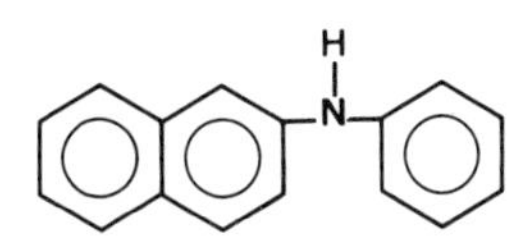

$C_{16}H_{13}N$ Mol. wt: 219.3

1.3 Chemical and physical properties of the pure substance

(a) Description: Grey to tan flakes or powder (B.F. Goodrich Chemical Company, undated)

(b) Boiling-point: 395°C (Kehe & Kouris, 1965)

(c) Melting-point: 108°C (Kehe & Kouris, 1965)

(d) Spectroscopy data: The ultra-violet, fluorescent and phosphorescent spectra have been studied (Seliskar & Brand, 1971).

(e) Solubility: Insoluble in water; soluble in ethanol (50 g/l), benzene (27 g/l) and acetone (640 g/l) (B.F. Goodrich Chemical Company, undated)

(f) Reactivity: May be oxidized to 7-phenyl-7*H*-dibenzo[*c*,*g*]carbazole (Kehe & Kouris, 1965)

(g) Stability: See 'General Remarks on the Substances Considered', p. 27.

1.4 Technical products and impurities

Until recently, *N*-phenyl-2-naphthylamine was available in the US as a commercial grade with the following typical specifications: purity, 97.0% min.; set-point, 106.0°C min.; ash content, 0.5% max.; 2-naphthol content, 0.5% max.

It has been reported that commercial *N*-phenyl-2-naphthylamine is contaminated with 20-30 mg/kg of the carcinogen 2-naphthylamine (NIOSH, 1976) [See also monograph on 2-naphthylamine (IARC, 1974)]. *N*-Phenyl-2-naphthylamine has been available in the US in a variety of formulations, compounded with other antioxidants, e.g., *para*-isopropoxydiphenylamine and diphenyl-*para*-phenylenediamine (Anon., 1975).

In Japan, commercial grades of *N*-phenyl-2-naphthylamine contain aniline, 2-naphthol and 2-naphthylamine as impurities.

In the UK, *N*-phenyl-2-naphthylamine has in the past been contaminated with 15-50 mg/kg 2-naphthylamine; in recent years levels of less than 1 mg/kg have been reported in one commercial product (Veys, 1973).

2. Production, Use, Occurrence and Analysis

For background information on this section, see preamble, p. 15.

2.1 Production and use

(a) Production

N-Phenyl-2-naphthylamine was prepared by Graebe in 1880 by heating 2-naphthol with aniline hydrochloride (Prager *et al.*, 1929). The method of commercial production currently being used in Japan (The Chemical Daily Co. Ltd., 1976) and in the US (Kehe & Kouris, 1965) is condensation of 2-naphthol and aniline in the presence of a catalyst. It is believed that the condensation of 2-naphthylamine and aniline was used for the commercial production of *N*-phenyl-2-naphthylamine by at least one US manufacturer prior to a decision by the company in 1956 to abandon the use of 2-naphthylamine as an intermediate.

The commercial production of *N*-phenyl-2-naphthylamine in the US was first reported in 1928 (US Tariff Commission, 1930). Three US companies reported the production of over 2.24 million kg in 1973 (US International Trade Commission, 1975), 1.37 million kg in 1974 (US International Trade Commission, 1976a) and 708.7 thousand kg in 1975 (US International Trade Commission, 1977a). It is believed that no US companies are currently producing this compound commercially due to the pending regulatory action which would limit employees' exposure to it. US imports of *N*-phenyl-2-naphthylamine through principal US customs districts amounted to 53.6 thousand kg in 1974 (US International Trade Commission, 1976b) and 75.8 thousand kg in 1975 (US International Trade Commission, 1977b).

N-Phenyl-2-naphthylamine is produced commercially in Europe, but production figures were not available to the Working Group.

In Japan, the annual production of *N*-phenyl-2-naphthylamine declined from 5.06 million kg in 1971 to 2.25 million kg in 1974. Four Japanese companies currently produce it; imports are negligible, and there are no exports (The Chemical Daily Company, Ltd, 1976).

(*b*) Use

N-Phenyl-2-naphthylamine is used primarily as an antioxidant (generally at levels of 1-2%) in rubber processing (Kehe & Kouris, 1965) to impart heat, oxidation and flex-cracking resistance in natural rubber, synthetic rubbers and latexes (Anon., 1975). It has also been used as a stabilizer in electrical-insulating silicone enamels (Kehe & Kouris, 1965) and as an antioxidant in other polymers, in greases and in lubricating and transformer oils. It has been described as an effective heat and light stabilizer in a number of applications, as well as a vulcanization accelerator, catalyst and polymerization inhibitor.

It has been used as a component of rocket fuels since the mid-1950's (Mossberg, 1976). Other miscellaneous uses reported in the literature include its use in surgical plasters (Brzezicka-Bak, 1973) and in tin-electroplating baths (Thomas & Harris, 1973).

N-Phenyl-2-naphthylamine has been used as a chemical intermediate in the production of the rubber antioxidant, *N*-phenyl-2-naphthylamine-acetone condensate, and in the production of seven dyes (The Society of Dyers and Colourists, 1971), of which only one, C.I. Acid Blue 98, appears to be of commercial world significance (The Society of Dyers and Colourists, 1975). No evidence was found that C.I. Acid Blue 98 is produced commercially in the US.

All of the *N*-phenyl-2-naphthylamine produced in Japan is used as an antioxidant in rubber processing (The Chemical Daily Co. Ltd, 1976).

In the US, the National Institute for Occupational Safety and Health (NIOSH) has recommended that an emergency temporary standard to regulate occupational exposure to *N*-phenyl-2-naphthylamine be issued, in light of recent findings indicating that it is metabolized in the human body to form the carcinogen, 2-naphthylamine[1]. NIOSH estimates that 15,000 workers in the US are potentially exposed to *N*-phenyl-2-naphthylamine during its manufacture and use (NIOSH, 1976).

2.2 Occurrence

For many years, *N*-phenyl-2-naphthylamine was detected in extracts from a wide variety of animal, plant and microbial lipids. It has been shown, however, that it was a contaminant of the solvents used to extract them, which had come into contact with rubber containing the antioxidant (Brown, 1967).

---

[1]See IARC, 1974.

When samples of 1,3-butadiene rubber were heated to 220°C, the evolved fumes contained 0.012 mg/g (12 mg/kg) of air *N*-phenyl-2-naphthylamine (Stasenkova, 1968). Studies conducted in several synthetic rubber plants in the USSR detected *N*-phenyl-2-naphthylamine in the air (Dvoryaninova & Khorobrykh, 1969; Mazanov & Malakhova, 1973; Mirzoyan & Tsai, 1972; Tsai, 1971).

2.3 Analysis

Ultra-violet spectrophotometry has been used to detect *N*-phenyl-2-naphthylamine in rubber and rubber products (Koldunovich & Blyumina, 1973; Popovici *et al.*, 1968) and in food materials (Hofmann & Ostromow, 1972). Ultra-violet and mass spectrometric methods have been described for the identification and determination of *N*-phenyl-2-naphthylamine in synthetic lubricating oils, without prior separation, in a concentration range of 0.05-2.4 weight percent (Kinder & Neel, 1969).

Paper chromatography, using eight solvent systems and two spray reagents, has been used for the separation and detection of secondary amines, including *N*-phenyl-2-naphthylamine, in solution (Shimizu, 1976).

High-pressure liquid chromatography may be used to separate mixtures of aromatic amine-type antioxidants, including *N*-phenyl-2-naphthylamine; as little as 5 μg could be detected (Majors, 1970).

Gel permeation chromatography in combination with thin-layer chromatography has been used to separate and identify antioxidants, including *N*-phenyl-2-naphthylamine, and other additives, in extracts of vulcanized rubber samples and in mixtures of vulcanization ingredients (Protivová *et al.*, 1974).

Thin-layer chromatography has been used for the separation and identification of antioxidants, including *N*-phenyl-2-naphthylamine, in vulcanized rubber following thermal extraction (McSweeney, 1970), in plastics used as food wraps (Woggon & Jehle, 1968), in liquids which had come into contact with stabilized rubbers, with a sensitivity of 2-5 μg *N*-phenyl-2-naphthylamine (Sluzewska, 1974), in rubber (less than 10% error) (Solodova *et al.*, 1970) and in transformer oils (Dovgopolyi, 1972).

Gas chromatography has been used to determine antioxidants, including *N*-phenyl-2-naphthylamine, in vulcanized rubbers (Styskin, 1973). In a review of detectors suitable for use in gas chromatographic analysis of organonitrogen compounds, including *N*-phenyl-2-naphthylamine, the Coulson conductimetric detector was the preferred method of detection (Adlard & Matthews, 1974).

3. Biological Data Relevant to the Evaluation of Carcinogenic Risk to Man

3.1 Carcinogenicity and related studies in animals[1]

(a) Oral administration

Mouse: Groups of 18 male and 18 female (C57BL/6xC3H/Anf)F_1 mice and 18 male and 18 female (C57BL/6xAKR)F_1 mice received *N*-phenyl-2-naphthylamine according to the following dose schedule: the maximum tolerated dose (464 mg/kg bw) was given in 0.5% aqueous gelatine by stomach tube at 7 days of age and the same amount (not adjusted for increasing body weight) daily up to 28 days of age; subsequently, the mice were given 1206 mg/kg of diet. The experiment was terminated when the mice were about 78 weeks of age, at which time 13, 18, 18 and 17 mice were still alive in the respective groups. The total numbers of mice with tumours among those necropsied were 7/17, 1/18, 7/18 and 3/18, in the four groups. In males of the first strain, the frequency of tumour-bearing animals was significantly greater in comparison with the control group given 0.5% gelatine (7/17 *versus* 0/16, $0.01<P<0.025$), this increase being due mainly to hepatomas (5/17 *versus* 0/16, $0.05<P<0.10$ and 5/17 *versus* 8/79, $0.05<P<0.10$, when the comparison was made with a control group formed by all untreated animals) [see Table 1] (Innes *et al.*, 1969; NTIS, 1968).

Dog: Three dogs were fed *N*-phenyl-2-naphthylamine in daily doses of 540 mg on five days a week for a period of 4½ years. No bladder tumours were observed after this period (Gehrmann *et al.*, 1949).

---

[1]The Working Group was aware of carcinogenicity studies in progress on *N*-phenyl-2-naphthylamine involving oral administration in mice and rats (IARC, 1976).

Table 1[a]

Oral administration of *N*-phenyl-2-naphthylamine

Strain		Total number of mice with tumours		Hepatomas	
(C57BL/6 x C3H/ANF)F_1	male	treated 7/17		treated 5/17	
		total controls	gelatine controls	total controls	gelatine controls
		22/79	0/16	8/79	0/16
		n.s.[b]	$0.01<P<0.025$	$0.05<P<0.10$	$0.05<P<0.10$
	female	treated 1/18		treated 1/18	
		total controls	gelatine controls	total controls	gelatine controls
		8/87	0/16	0/87	0/16
		n.s.	n.s.	n.s.	n.s.
(C57BL/6 x AKR)F_1	male	treated 7/18		treated 3/18	
		total controls	gelatine controls	total controls	gelatine controls
		16/90	3/18	5/90	1/18
		n.s.	n.s.	n.s.	n.s.
	female	treated 3/18		treated 0/18	
		total controls	gelatine controls	total controls	gelatine controls
		7/82	2/17	1/82	0/17
		n.s.	n.s.	n.s.	n.s.

[a] from NTIS (1968)
[b] not significant

Table 2[a]

Subcutaneous administration of *N*-phenyl-2-naphthylamine

Strain		Total number of mice with tumours		Hepatomas	
$(C57BL/6 \times C3H/ANF)F_1$	male	treated 4/17		treated 2/17	
		total controls	DMSO controls	total controls	DMSO controls
		27/141	2/24	9/141	1/24
		n.s.[b]	n.s.	n.s.	n.s.
	female	treated 5/18		treated 1/18	
		total controls	DMSO controls	total controls	DMSO controls
		9/154	1/23	0/154	0/23
		$P<0.01$	n.s.	n.s.	n.s.
$(C57BL/6 \times AKR)F_1$	male	treated 3/18		treated 2/18	
		total controls	DMSO controls	total controls	DMSO controls
		8/161	1/24	1/161	0/24
		n.s.	n.s.	$P<0.05$	n.s.
	female	treated 1/18		treated 0/18	
		total controls	DMSO controls	total controls	DMSO controls
		17/157	3/24	0/157	0/24
		n.s.	n.s.	n.s.	n.s.

[a] from NTIS (1968
[b] not significant

(b) Subcutaneous and/or intramuscular administration

Mouse: Groups of 18 male and 18 female (C57BL/6xC3H/Anf)F_1 mice and 18 male and 18 female (C57BL/6xAKR)F_1 mice were given single s.c. injections of 464 mg/kg bw *N*-phenyl-2-naphthylamine in dimethyl sulphoxide (DMSO) on the 28th day of life and observed up to 80 weeks of age, at which time 16, 17, 16 and 18 mice in the four groups, respectively, were still alive. The total numbers of mice with tumours among those necropsied were 4/17, 5/18, 3/18 and 1/18, in the four groups. In females of the first strain, the frequency of tumour-bearing animals was significantly greater in comparison with control animals (untreated or given DMSO by injection): 5/18 *versus* 9/154, $P<0.01$; there was also a statistically significant increase in the frequency of hepatomas in males of the second strain (2/18 *versus* 1/16, $P<0.05$) [see Table 2] (Innes *et al.*, 1969; NTIS, 1968).

3.2 Other relevant biological data

(a) Experimental systems

The oral LD_{50} in mice is 1450 mg/kg bw, and that in rats 8730 mg/kg bw, when *N*-phenyl-2-naphthylamine is administered in vegetable oil (Kelman, 1966).

Repeated intragastric administrations of *N*-phenyl-2-naphthylamine to rats caused a fall in body weight, depression of the nervous system and disturbance of liver function (Kelman, 1966). Daily intragastric administration of 100 mg/kg bw to rats caused decreases in urinary hippuric acid and adrenal ascorbic acid after 6 months and a drop in urinary function after 18 months; urinary protein was higher after 1 month. Lung and liver weights increased within 1 and 12 months, respectively; changes were observed in the gastrointestinal tract after 6 months, and reproductive function was impaired. Doses of 20 mg/kg bw caused no significant toxic effects (Shimskaya *et al.*, 1971). Daily inhalation by rats of *N*-phenyl-2-naphthylamine aerosol (900 mg/m^3) for 14 days caused weight loss, slight erythrocytopenia and pulmonary emphysema (Kranig, 1964). Mice exposed daily for 4 hours during one month to an atmosphere containing 12 mg/m^3 *N*-phenyl-2-naphthylamine and 339 mg/m^3 1,3-butadiene showed slight leucopenia and irritation of the respiratory tract (Stasenkova, 1968).

In dogs given single oral doses of 5 mg/kg bw [1,4,5,8-^{14}C]-*N*-phenyl-2-naphthylamine alone or following repeated doses of 400 mg/animal unlabelled *N*-phenyl-2-naphthylamine on 5 days a week for 4 weeks, more than 90% of the radioactivity was excreted over 3 days, mainly in the faeces. Only 2.8% of the radioactivity was excreted in the urine. Amounts ranging from 0-10 µg 2-naphthylamine were detected in the urine of dogs given a single dose of 5 mg/kg bw *N*-phenyl-2-naphthylamine, indicating removal of the *N*-phenyl group. No increased excretion of labelled 2-naphthylamine was observed following 4 weeks' pretreatment with 400 mg/animal unlabelled *N*-phenyl-2-naphthylamine before administration of labelled compound. Metabolites of 2-naphthylamine, e.g., 2-naphthylhydroxylamine and 2-amino-1-naphthyl sulphate, were not detected in ether extracts or lyophilized fractions of urine (Batten & Hathway, 1977).

No data on the embryotoxicity, teratogenicity or mutagenicity of *N*-phenyl-2-naphthylamine were available to the Working Group.

(b) Man

Leucoplakia, acne and hypersensitivity to sunlight were observed in 36 workers exposed for prolonged periods to *N*-phenyl-2-naphthylamine (Sielicka-Zuber, 1961).

In 19 volunteers given 10 mg *N*-phenyl-2-naphthylamine containing 8 ng 2-naphthylamine (0.8 mg/kg), from 0.4-3 µg 2-naphthylamine were found in 24-hour urine samples from 7 subjects, 6 of which were non-smokers. In 4 workers exposed to *N*-phenyl-2-naphthylamine dusts (estimated intake, 40 mg), which were estimated to contain 32 ng 2-naphthylamine, 3-8 µg 2-naphthylamine were found in 24-hour urine samples (Kummer & Tordoir, 1975).

2-Naphthylamine was found at a level of 3-4 µg in 24-hour samples of urine from two volunteers who ingested 50 mg *N*-phenyl-2-naphthylamine containing 0.7 µg 2-naphthylamine and from workers (unspecified number) estimated to have inhaled 30 mg *N*-phenyl-2-naphthylamine (Moore *et al.*, 1977).

3.3 Case reports and epidemiological studies[1]

Veys (1973) studied the incidence of bladder tumours among male workers at a large rubber-tire factory in the UK. He demonstrated a risk for workers exposed to materials that contained 2-naphthylamine (a known carcinogen), which were withdrawn from use in that factory in 1949; among 2081 men observed between 1946 and 1970, 23 tumours were observed *versus* 10.3 expected on the basis of local rates. During 1946-1960, 4177 other men worked at the factory but had no known contact with these carcinogenic substances; however, 3301 of these men (79%) had known exposure to *N*-phenyl-2-naphthylamine containing 15-50 mg/kg 2-naphthylamine. No 2-naphthylamine was detected in the atmosphere of the rubber-curing area of the factory (sensitivity, 0.7 $\mu g/m^3$ air) at that time. During 1946-1970, there were 9 cases of bladder cancer among these 4177 men, *versus* 10.0 expected. Of these workers, 1331 (including 1088 exposed to *N*-phenyl-2-naphthylamine) first worked at the factory before 1950 and thus had a follow-up period of over 20 years; among this subgroup, 3 bladder tumours occurred during 1946-1970, *versus* 5.5 expected [The Working Group was unclear about exactly how exposure was defined for certain job categories].

Fox & Collier (1976) studied mortality during 1968-1974 among some 40,000 workers at a number of rubber and cable factories in the UK. They distinguished a large group of workers not exposed to known carcinogenic materials; 33 deaths from bladder cancer occurred in this group *versus* 22.7 expected at national rates, a statistically significant excess [This group of workers would have had mixed exposures to many rubber additives, but it may be assumed that a considerable number of men would have been exposed to *N*-phenyl-2-naphthylamine].

---

[1]The Working Group was aware of studies in progress on the rubber industry in both the US and the UK, in which exposure to specific substances and/or by job categories is being investigated but for which results are not yet available (Muir & Wagner, 1977).

4. Comments on Data Reported and Evaluation[1]

4.1 Animal data

N-Phenyl-2-naphthylamine has been tested in mice by oral and single subcutaneous administration and in a small number of dogs by oral administration. Results of the experiment by oral administration in mice indicate a statistically significant increase in the incidence of all tumours, and in particular of hepatomas, in males of one of the two tested strains. In addition, subcutaneous administration of this compound produced a significant increase in the total incidence of tumours in females of one strain and of hepatomas in males of the other strain. These findings suggest that the compound is carcinogenic but do not allow a conclusive evaluation of the carcinogenicity of this compound to be made.

4.2 Human data

In one study, in one factory, of men with known exposure to *N*-phenyl-2-naphthylamine in rubber processing under controlled conditions, there was no indication of an excess incidence of bladder tumours. However, a broader study of rubber workers who had no exposure to 2-naphthylamine has shown an increased risk of bladder tumours. In the latter study, exposure was mixed but probably included exposure to *N*-phenyl-2-naphthylamine for many workers.

These findings do not permit an assessment of the presence or absence of an elevated risk for cancer of the urinary bladder in persons exposed to *N*-phenyl-2-naphthylamine.

---

[1]Subsequent to the finalization of this evaluation by the Working Group in June 1977, the Secretariat became aware of an abstract of a paper by Checkoway *et al.* (1977) describing a retrospective cohort mortality study of bladder cancers in 4 major US rubber companies between 1964-1973; however, the study cannot be evaluated on the basis of the data presented.

5. References

Adlard, E.R. & Matthews, P.H.D. (1974) Selective nitrogen detectors for the determination of organonitrogen compounds separated by gas chromatography. In: Hodges, D.R., ed., Proceedings of a Symposium, Recent Analytical Developments in the Petroleum Industry, New York, John Wiley and Sons, pp. 59-77

Anon. (1975) Materials and Compounding Ingredients for Rubber, New York, Bill Communications, Inc., pp. 107-108, 125

Batten, P.L. & Hathway, D.E. (1977) Dephenylation of *N*-phenyl-2-naphthylamine in dogs and its possible oncogenic implications. Brit. J. Cancer, 35, 342-346

Brown, B.S. (1967) Booby trap for biochemists: *N*-phenyl-2-naphthylamine. Chem. Brit., 3, 524-526

Brzezicka-Bak, M. (1973) Evaluation of plasters used in medicine. Farm. Pol., 29, 865-872

Checkoway, H., Arp, E., Holbrook, R.H., Jones, F.S., McMichael, A.J., Monson, R., Smith, A.H., Tyroler, H.A. & Van Ert, M. (1977) A case-control study of bladder cancer in the rubber industry (Abstract). Amer. J. Epidemiol., 106, 243

The Chemical Daily Co. Ltd (1976) 6376-Chemicals, Tokyo, p. 536

Dovgopolyi, E.E. (1972) Quantitative determination of antioxidative additives in oils by a thin-layer chromatographic method. Neftepererab. Neftekhim. (Moscow), 2, 23-26

Dvoryaninova, N.K. & Khorobrykh, V.V. (1969) Working conditions in a shop for the formation of SKMS-30 rubber at the Omsk synthetic rubber plant. Nauch. Tr., Omsk. Med. Inst., 88, 62-67

Fox, A.J. & Collier, P.F. (1976) A survey of occupational cancer in the rubber and cablemaking industries: analysis of deaths occurring in 1972-74. Brit. J. industr. Med., 33, 249-264

Gehrmann, G.H., Foulger, J.H. & Fleming, A.J. (1949) Occupational tumours of the bladder. In: Proceedings of the 9th International Congress of Industrial Medicine, London, 1948, Bristol, Wright, pp. 472-475

B.F. Goodrich Chemical Company (undated) Good-Rite A.O. 3102. General Trade Service Bulletin, PCGT-11 (6905), Cleveland, Ohio

Hofmann, W. & Ostromow, H. (1972) Analytical findings of the rubber committee within the scope of the food laws. 11. Kaut. Gummi, Kunstst., 25, 204-206

IARC (1974) IARC Monographs on the Evaluation of Carcinogenic Risk of Chemicals to Man, 4, Some Aromatic Amines, Hydrazine and Related Substances, *N*-Nitroso Compounds and Miscellaneous Alkylating Agents, Lyon, pp. 97-111

IARC (1976) Information Bulletin on the Survey of Chemicals Being Tested for Carcinogenicity, No. 6, Lyon, p. 183

Innes, J.R.M., Ulland, B.M., Valerio, M.G., Petrucelli, L., Fishbein, L., Hart, E.R., Pallotta, A.J., Bates, R.R., Falk, H.L., Gart, J.J., Klein, M., Mitchell, I. & Peters, J. (1969) Bioassay of pesticides and industrial chemicals for tumorigenicity in mice: a preliminary note. J. nat. Cancer Inst., 42, 1101-1114

Kehe, H.J. & Kouris, C.S. (1965) Diarylamines. In: Kirk, R.E. & Othmer, D.F., eds, Encyclopedia of Chemical Technology, 2nd ed., Vol. 7, New York, John Wiley and Sons, pp. 40-49

Kelman, G.Y. (1966) A study of toxicity of certain rubber antioxidants and their hygienic assessment. Gig. i Sanit., 31, 25-28

Kinder, J.F. & Neel, J. (1969) Identification and determination of oxidation inhibitors in synthetic lube oils by ultraviolet and mass spectrometry. Develop. appl. Spectrosc., 7A, 274-285

Koldunovich, E.B. & Blyumina, S.D. (1973) Determination of Thiuram E and Neozone D in Nairit by a spectrophotometric method. Kauch. Rezina, 32, 55-56

Kranig, E.K. (1964) Toxicity of Neosone D. Toksikol. Novykh. Prom. Khim. Veshchestv., 6, 37-45

Kummer, R. & Tordoir, W.F. (1975) Phenyl-betanaphthylamine (PBNA), another carcinogenic agent? T. soc. Geneesk., 53, 415-419

Majors, R.E. (1970) High speed liquid chromatography of antioxidants and plasticizers using solid core supports. J. chromat. Sci., 8, 338-345

Mazanov, G.N. & Malakhova, T.V. (1973) Atmospheric conditions in the weighing department of a technical rubber product plant. Sb. Nauchn. Tr. Sanit. Tekh., 5, 166-169

McSweeney, G.P. (1970) Micromethod for the rapid identification of compounding ingredients in rubbers by thermal extraction and thin-layer chromatography. J. Inst. Rubber Industr., 4, 245-246

Mirzoyan, I.M. & Tsai, L.M. (1972) Working conditions and hygienic study of the polymerization works of the Karaganda synthetic rubber plant. In: Filin, A.P., ed., Proceedings of a Scientific Conference on the Problems of Industrial Hygiene and Occupational Diseases, 1971, Karaganda, Research Institute on the Problems of Industrial Hygiene and Occupational Diseases, pp. 250-252

Moore, R.M., Jr, Woolf, B.S., Stein, H.P., Thomas, A.W. & Finklea, J.F. (1977) Metabolic precursors of a known human carcinogen. Science, 195, 344

Mossberg, W. (1976) US muls curb on rocket-fuel chemical due to possible cancer link for workers, New York Times, December 21

Muir, C.S. & Wagner, G., eds (1977) Directory of On-going Research in Cancer Epidemiology 1977, Lyon (IARC Scientific Publications No. 17), pp. 217 (Abstract 444), 230 (Abstract 477), 265 (Abstract 569)

NIOSH (National Institute for Occupational Safety and Health) (1976) Current Intelligence Bulletin: Metabolic Precursors of a Known Human Carcinogen, *beta*-Naphthylamine, Rockville, Maryland, pp. 1-3

NTIS (National Technical Information Service) (1968) Evaluation of Carcinogenic, Teratogenic and Mutagenic Activities of Selected Pesticides and Industrial Chemicals, Vol. 1, Carcinogenic Study, Washington DC, US Department of Commerce

Popovici, M., Petrescu, A.D., Isopescu, A., Redes, A. & Stanciu, D. (1968) Fluorescence of rubber components and its use for investigating and testing rubber and rubber products. Ind. Usoara, 15, 690-693

Prager, B., Jacobson, P., Schmidt, P. & Stern, D., eds (1929) Beilsteins Handbuch der Organischen Chemie, 4th ed., Vol. 12, Syst. No. 1725, Berlin, Springer-Verlag, p. 1275

Protivová, J., Pospíšil, I. & Holčík, J. (1974) Antioxidants and stabilizers. XLVIII. Analysis of the components of stabilization and vulcanization mixtures for rubbers by gel permeation and thin-layer chromatographic methods. J. Chromat., 92, 361-370

Seliskar, C.J. & Brand, L. (1971) Electronic spectra of 2-aminonaphthalene-6-sulfonate and related molecules. I. General properties and excited-state reactions. J. Amer. chem. Soc., 93, 5405-5414

Shimizu, I. (1976) The paper chromatography of a group of secondary amines. J. Chromat., 118, 96-98

Shumskaya, N.I., Ivanov, V.N. & Tolgskaya, M.S. (1971) Remote after-effects of the chronic poisoning of white rats with Neozone D. Toksikol. Novykh. Prom. Khim. Veshchestv., 12, 86-93

Sielicka-Zuber, L. (1961) Skin changes due to *N*-phenyl-β-naphthylamine. Polski Tygod. Lekar., 16, 1483-1486

Sluzewska, L. (1974) Detection of antioxidants used in production of rubber and investigations on the degree of their migration from rubber products. Rocz. Panstw. Zakl. Hig., 25, 495-504

The Society of Dyers and Colourists (1971) Colour Index, 3rd ed., Vol. 4, Yorkshire, UK, pp. 4223, 4453, 4454, 4808

The Society of Dyers and Colourists (1975) Colour Index, revised 3rd ed., Vol. 5, Yorkshire, UK, pp. 5033, 5034, 5084

Solodova, G.M., Malyshev, A.I. & Rostovtseva, E.E. (1970) Use of thin-layer chromatography for determining antioxidants in rubber. Probl. Analit. Khim., 1, 91-94

Stasenkova, K.P. (1968) Experimental materials for evaluating the toxicity of 1,3-butadiene rubber. Toksikol. Novykh. Prom. Khim. Veshchestv., 10, 90-99

Styskin, E.L. (1973) Rapid micromethod for determining the type of anti-oxidant in rubbers and vulcanizates. Kauch. Rezina, 32, 49-51

Thomas, B.M. & Harris, A. (1973) Tin-plating additives and baths. British Patent 1,339,133, November 28 [to Ciba-Geigy (UK) Ltd]

Tsai, L.M. (1971) Hygienic characteristics of the rubber industry. Zdravokkr. Kas., 30, 57-58

US International Trade Commission (1975) Synthetic Organic Chemicals, US Production and Sales, 1973, ITC Publication 728, Washington DC, US Government Printing Office, pp. 136, 139

US International Trade Commission (1976a) Synthetic Organic Chemicals, US Production and Sales, 1974, USITC Publication 776, Washington DC, US Government Printing Office, pp. 42, 136, 139

US International Trade Commission (1976b) Imports of Benzoid Chemicals and Products, 1974, USITC Publication 762, Washington DC, US Government Printing Office, p. 25

US International Trade Commission (1977a) Synthetic Organic Chemicals, US Production and Sales, 1975, USITC Publication 804, Washington DC, US Government Printing Office, pp. 131, 134

US International Trade Commission (1977b) Imports of Benzoid Chemicals and Products, 1975, USITC Publication 806, Washington DC, US Government Printing Office, p. 25

US Tariff Commission (1930) Census of Dyes and of Other Synthetic Organic Chemicals, 1928, Tariff Information Series No. 38, Washington DC, US Government Printing Office, p. 35

Veys, C.A. (1973) A Study on the Incidence of Bladder Tumours in Rubber Workers (Thesis for doctorate of medicine, Faculty of Medicine, University of Liverpool, Liverpool, UK)

Woggon, H. & Jehle, D. (1968) Specifications of plastics. Detection of antioxidants and ultraviolet-absorbing materials in plastics. Z. Lebensmitt. Untersuch.-Forsch., 136, 77-82

4,4'-THIODIANILINE

1. Chemical and Physical Data

1.1 Synonyms and trade names

Chem. Abstr. Services Reg. No.: 139-65-1

Chem. Abstr. Name: 4,4'-Thiobisbenzenamine

Bis(4-aminophenyl)sulphide; bis(*para*-aminophenyl)sulphide; 4,4-diaminodiphenyl sulphide; 4,4'-diaminodiphenyl sulphide; 4,4-diaminodiphenylsulphide; 4,4'-diaminodiphenylsulphide; *para,para*'-diaminodiphenyl sulphide; di(*para*-aminophenyl)-sulphide; thioaniline; 4,4'-thioaniline; *para,para*'-thiodi-aniline; thiodi-*para*-phenylenediamine

1.2 Chemical formula and molecular weight

H_2N–C_6H_4–S–C_6H_4–NH_2

$C_{12}H_{12}N_2S$ Mol. wt: 216.3

1.3 Chemical and physical properties of the pure substance

From Prager *et al.* (1930)

(a) Description: Needles

(b) Melting-point: 108°C

(c) Solubility: Sparingly soluble in cold water; slightly soluble in hot water; soluble in ethanol, ether and hot benzene

(d) Stability: See 'General Remarks on the Substances Considered', p. 27.

1.4 Technical products and impurities

No data were available to the Working Group.

2. Production, Use, Occurrence and Analysis

For background information on this section, see preamble, p. 15.

2.1 Production and use

(a) Production

4,4'-Thiodianiline was prepared by Merz and Weith in 1871 by boiling sulphur with aniline for several days (Prager *et al.*, 1930). Information on the method used for commercial production of 4,4'-thiodianiline is not available.

The commercial production of 4,4'-thiodianiline was first reported in the US in 1941-1943 (US Tariff Commission, 1945). In 1974, the latest year in which production was reported, only one US company reported an undisclosed amount (see preamble, p. 15) (US International Trade Commission, 1976).

No evidence was found that 4,4'-thiodianiline has ever been produced commercially in Japan.

(b) Use

4,4'-Thiodianiline can be used as an intermediate in the production of three dyes: C.I. Mordant Yellow 16, Milling Red G and Milling Red FR (The Society of Dyers and Colourists, 1971); however, only the first is believed to be of commercial significance in the US (US International Trade Commission, 1977).

4,4'-Thiodianiline has been investigated experimentally for a variety of biological activities (Goldsworthy & Gertler, 1949; Gupta *et al.*, 1959; Kamboj & Kar, 1966; Mukerjee *et al.*, 1961).

2.2 Occurrence

4,4'-Thiodianiline is not known to occur as a natural product. No data on its occurrence in the environment were available to the Working Group.

2.3 Analysis

Thin-layer chromatography using four different solvent systems has been used to separate and identify a group of aromatic diamines, including 4,4'-thiodianiline, and *N*-benzamides (Krasnov *et al.*, 1970); paper chromatography has been used to separate and identify a group of aromatic *para,para*'-diamines, including 4,4'-thiodianiline (Gasparic & Snobl, 1971).

Gas chromatography has been used to separate and identify one group of high-boiling aromatic amines, including 4,4'-thiodianiline, with a relative error of not more than 10-12% (Kazinik *et al.*, 1971a), and another group of aromatic amines, with a relative error of not more than 5.8% (Kazinik *et al.*, 1971b).

3. Biological Data Relevant to the Evaluation of Carcinogenic Risk to Man

3.1 Carcinogenicity and related studies in animals[1]

Oral administration

Rat: Twenty female Sprague-Dawley rats, 40 days of age, were given 10 doses of 40 mg/animal 4,4'-thiodianiline in 1 ml sesame oil every 3 days by gastric intubation; 140 control rats received the vehicle alone. Animals were killed after 9 months; of 12 surviving rats autopsied, 3 had developed mammary carcinomas. When the total dose was reduced to 300 mg/animal, 1 mammary carcinoma was found in 10 rats (7 survived 9 months, 8 were autopsied). Among 132 controls autopsied, of which 127 survived 9 months, 3 had mammary carcinomas and 1 a mammary fibroadenoma. The increased incidence of tumours of the mammary gland in treated animals that received the higher dose is statistically significant (Griswold *et al.*, 1968) [$P<0.01$].

---

[1]The Working Group was aware of carcinogenicity tests in progress on 4,4'-thiodianiline involving oral administration in mice and rats (IARC, 1976).

3.2 Other relevant biological data

4,4'-Thiodianiline given orally to mice on days 1-5 of pregnancy at doses of 100 and 150 mg/kg bw completely prevented foetal implantation; a dose of 50 mg/kg bw slightly reduced the number of implantation sites (Kamboj & Kar, 1966).

No data on the toxicity, metabolism or mutagenicity of 4,4'-thiodianiline were available to the Working Group.

3.3 Case reports and epidemiological studies

No data were available to the Working Group.

4. Comments on Data Reported and Evaluation[1]

4.1 Animal data

4,4'-Thiodianiline is carcinogenic in rats after its oral administration; it produced an increased incidence of mammary carcinomas.

4.2 Human data

No case reports or epidemiological studies were available to the Working Group.

---

[1]See also the section, 'Animal Data in Relation to the Evaluation of Risk to Man' in the introduction to this volume, p. 13.

5. References

Gasparic, J. & Snobl, D. (1971) Identification of organic compounds. LXXIII. Paper chromatography of aromatic *p,p*'-diamines. Sb. Ved. Pr., Vys. Sk. Chemickotechnol., Pardubice, 25, 33-40

Goldsworthy, M.C. & Gertler, S.I. (1949) Fungicidal and phytotoxic properties of 506 synthetic organic compounds. US Dept. Agr., Plant Disease Reptr., Suppl. 182, 89-109

Griswold, D.P., Jr, Casey, A.E., Weisburger, E.K. & Weisburger, J.H. (1968) The carcinogenicity of multiple intragastric doses of aromatic and heterocyclic nitro or amino derivatives in young female Sprague-Dawley rats. Cancer Res., 28, 924-933

Gupta, S.K., Mathur, I.S. & Mukerji, B. (1959) The therapeutic activity of some sulfides, quinazolones, and pyrimidines in experimental tuberculosis of guinea pigs. J. Sci. Ind. Res. (India), 18C, 1-5

IARC (1976) Information Bulletin on the Survey of Chemicals Being Tested for Carcinogenicity, No. 6, Lyon, p. 166

Kamboj, V.P. & Kar, A.B. (1966) Antiimplantation effect of some aromatic sulfur derivatives. Indian J. exp. Biol., 4, 120-121

Kazinik, E.M., Gudkova, G.A. & Shcheglova, T.A. (1971a) Gas-liquid chromatographic analysis of some high-boiling aromatic amines. Zh. analit. Khim., 26, 154-157

Kazinik, E.M., Gudkova, G.A., Mesh, L.Y., Shcheglova, T.A. & Ivanov, A.V. (1971b) Gas-chromatographic analysis of high-boiling diamines - diphenyl, diphenyl sulfide, and diphenyl sulfone derivatives. Zh. analit. Khim., 26, 1920-1923

Krasnov, E.P., Logunova, V.I. & Shumakova, V.D. (1970) Thin-layer chromatography of aromatic diamines and *N*-benzamides. In: Pakshver, A.B., ed., Synthetic Polymer Fibres, Moscow, Khimiya, pp. 292-294

Mukerjee, N., Kundu, S. & Bose, R. (1961) The effect of diaminodiphenyl sulfide on experimental rat leprosy. Bull. Calcutta School Trop. Med., 9, 14-15

Prager, B., Jacobson, P., Schmidt, P. & Stern, D., eds (1930) Beilsteins Handbuch der Organischen Chemie, 4th ed., Vol. 12, Syst. No. 1853, Berlin, Springer-Verlag, pp. 535-536

The Society of Dyers and Colourists (1971) Colour Index, 3rd ed., Vol. 4, Yorkshire, UK, p. 4704

US International Trade Commission (1976) Synthetic Organic Chemicals, US Production and Sales, 1974, USITC Publication 776, Washington DC, US Government Printing Office, p. 44

US International Trade Commission (1977) Synthetic Organic Chemicals, US Production and Sales, 1975, USITC Publication 804, Washington DC, US Government Printing Office, p. 71

US Tariff Commission (1945) Synthetic Organic Chemicals, US Production and Sales, 1941-43, Report No. 153, Second Series, Washington DC, US Government Printing Office, p. 20

ortho-TOLUIDINE (HYDROCHLORIDE)

1. Chemical and Physical Data

ortho-Toluidine

1.1 Synonyms and trade names

Chem. Abstr. Services Reg. No.: 95-53-4

Chem. Abstr. Name: 2-Methylbenzenamine

1-Amino-2-methylbenzene; 2-amino-1-methylbenzene; 2-aminotoluene; *ortho*-aminotoluene; 1-methyl-2-aminobenzene; 2-methyl-1-aminobenzene; 2-methylaniline; *ortho*-methylaniline; *ortho*-methylbenzenamine; 2-toluidine; *ortho*-tolylamine

1.2 Chemical formula and molecular weight

NH2 CH3

C_7H_9N Mol. wt: 107.2

1.3 Chemical and physical properties of the pure substance

From Weast (1976), unless otherwise specified

(a) Description: Light-yellow liquid (Windholz, 1976)

(b) Boiling-point: 200.2°C

(c) Melting-point: -14.7°C (β-form)

(d) Density: d_4^{20} 0.9984

(e) Refractive index: n_D^{20} 1.5725

(f) Spectroscopy data: λ_{max} 232.5 nm (E_1^1 = 758.6), 281.5 nm (E_1^1 = 144.5); infra-red, nuclear magnetic resonance and mass spectral data have been tabulated (Grasselli, 1973).

(g) Solubility: Slightly soluble in water; miscible with ethanol, ether and carbon tetrachloride

(h) Volatility: Vapour pressure is 1 mm at 44°C and 10 mm at 81.4°C.

(i) Stability: See 'General Remarks on the Substances Considered', p. 27.

1.4 Technical products and impurities

ortho-Toluidine is available in the US as a technical grade, with a minimum purity of 99.5% (as determined by gas chromatography). It contains *meta* (0.4% max.) and/or *para* (0.1% max.) isomers, 1-methyl-2-nitrobenzene and moisture (0.1% max.) as impurities (E.I. du Pont de Nemours & Co. Inc., 1974). Stabilized technical grade *ortho*-toluidine may also contain less than 0.5% of an unidentified stabilizing ingredient which prevents darkening upon storage.

In western Europe, *ortho*-toluidine has a minimum purity of 99.5% and contains a maximum of 0.5% *para*-toluidine and aniline and 0.05% methylnitrobenzene isomers.

ortho-Toluidine is available in Japan with a minimum purity of 99%, a maximum moisture content of 0.2%, and may contain *para*-toluidine and 1-methyl-2-nitrobenzene as impurities.

ortho-Toluidine hydrochloride

1.1 Synonyms and trade names

Chem. Abstr. Services Reg. No.: 636-21-5

Chem. Abstr. Name: 2-Methylbenzenamine hydrochloride

1-Amino-2-methylbenzene hydrochloride; 2-amino-1-methylbenzene hydrochloride; 2-aminotoluene hydrochloride; *ortho*-aminotoluene hydrochloride; 1-methyl-2-aminobenzene hydrochloride; 2-methyl-1-aminobenzene hydrochloride; 2-methylaniline hydrochloride; *ortho*-methylaniline hydrochloride; *ortho*-methylbenzenamine hydrochloride; 2-toluidine hydrochloride; *ortho*-tolylamine hydrochloride

1.2 Empirical formula and molecular weight

$C_7H_9N.HCl$ Mol. wt: 143.6

1.3 Chemical and physical properties

From Weast (1976), unless otherwise specified

(a) Description: Monoclinic prisms

(b) Boiling-point: 242.2°C

(c) Melting-point: 215°C

(d) Spectroscopy data: λ_{max} 284 nm, 266 nm, 258 nm in methanol; infra-red and nuclear magnetic resonance spectral data have been tabulated (Grasselli, 1973).

(e) Solubility: Soluble in water and ethanol; insoluble in ether and benzene

1.4 Technical products and impurities

No data were available to the Working Group.

2. Production, Use, Occurrence and Analysis

For background information on this section, see preamble, p. 15.

2.1 Production and use

(a) Production

ortho-Toluidine was prepared by Muspratt & Hoffmann in 1844 by the reduction of 1-methyl-2-nitrobenzene with alcoholic ammonium sulphide (Prager *et al.*, 1929). It is believed to be manufactured currently by the catalytic hydrogenation of 1-methyl-2-nitrobenzene. In the UK, the standard method of production has been the reduction of 1-methyl-2-nitrobenzene with iron filings.

ortho-Toluidine has been produced commercially in the US for over 50 years (US Tariff Commission, 1922), and commercial production of *ortho*-toluidine hydrochloride was first reported in 1956 (US Tariff Commission, 1957). In 1975, two US companies reported an undisclosed amount (see preamble, p. 15) of *ortho*-toluidine, and one reported an undisclosed amount of *ortho*-toluidine hydrochloride (US International Trade Commission, 1977a).

US imports through principal US customs districts were reported as 11.8 thousand kg *ortho*-toluidine and 21.9 thousand kg *ortho*-toluidine hydrochloride in 1974 (US International Trade Commission, 1976a) and one thousand kg of the hydrochloride in 1975 (US International Trade Commission, 1977b).

ortho-Toluidine was first produced commercially in the UK in 1880. It is believed that at least nine companies currently produce it in western Europe, including four companies in the UK. Total production in western Europe amounts to 1-5 million kg per year. Less than 10 thousand kg are produced annually in Benelux and in Austria; 10-100 thousand kg in Spain; 100 thousand-1 million kg in France, in Italy and in Switzerland; and 1-5 million kg in the Federal Republic of Germany and in the UK. Less than 10,000 kg are imported annually into the Federal Republic of Germany and into the UK; 10-100 thousand kg into Austria, into Benelux, into France, into Italy and into Scandinavia; and 100 thousand-1 million kg into Spain and into Switzerland. Less than 10 thousand kg are exported annually from Benelux; 10-100 thousand kg from France, from Italy, from Spain and from Switzerland; and 100 thousand-1 million kg from the Federal Republic of Germany and from the UK.

ortho-Toluidine was first produced commercially in Japan in 1926. Production during the period of 1971-1975 amounted to about 1 million kg per year, and there are currently three commercial producers (Japan Dyestuff Industry Association, 1976). Japanese imports of *ortho*-toluidine in 1971 and 1972 were negligible but amounted to 80 thousand kg in 1973 and 202 thousand kg in 1974; no imports were reported in 1975. Japanese exports of *ortho*-toluidine amounted to 38 thousand kg in 1973 and 10 thousand kg in

1974; none were reported in 1975 (Japan Tariff Association, 1976). Small amounts of *ortho*-toluidine hydrochloride are sold by minor Japanese dye manufacturing companies.

It is believed that in some countries *ortho*-toluidine has frequently been produced in the same plant area as that used for production of other aromatic amines, especially aniline (BIOS, 1946).

(b) Use

A major use for *ortho*-toluidine and *ortho*-toluidine hydrochloride appears to be as intermediates in the manufacture of dyes. They can be used to produce more than 50 dyes (The Society of Dyers and Colourists, 1971), of which 17 are now believed to be produced commercially in the US (US International Trade Commission, 1976b). Of these, six are produced from *ortho*-toluidine (The Society of Dyers and Colourists, 1971); the most significant commercially in the US is C.I. Vat Red 1, of which over 38.6 thousand kg were reportedly sold in 1975 (US International Trade Commission, 1977a).

ortho-Toluidine and *ortho*-toluidine hydrochloride are believed to be used as intermediates in the production of 4-*ortho*-tolylazo-*ortho*-toluidine (*ortho*-aminoazotoluene[1]) and 4-(4-amino-*meta*-tolylazo)-*meta*-toluenesulphonic acid, which are believed to be used as raw materials for the production of 11 dyes in the US. The most important of these are: C.I. Acid Red 115 (1974 US sales, 12.7 thousand kg), C.I. Direct Red 72 (1975 US production, 62.6 thousand kg) and C.I. Solvent Red 26 (1974 US sales, 58.1 thousand kg) (US International Trade Commission, 1976b, 1977a).

It is believed that *ortho*-toluidine is also used to produce special dyes for use in colour photography, as an antioxidant in the manufacture of rubber and as an ingredient in a clinical laboratory reagent used to test blood samples for glucose. Other reported uses include the printing of textiles blue-black and making colours fast to acids (Windholz, 1976).

---

[1]See IARC (1975).

It is estimated that 1-5 million kg *ortho*-toluidine are used annually in western Europe in the manufacture of antioxidants, dyes and pigments. In Switzerland, *ortho*-toluidine is used as an agricultural and industrial chemical intermediate.

In Japan, *ortho*-toluidine is used in the manufacture of dyes (50%), pigments (20%) and antioxidants (20%) and in other applications (10%).

According to the US Occupational Safety and Health Administration's health standards for air contaminants, an employee's exposure to *ortho*-toluidine should not exceed an eight-hour time-weighted average of 5 ppm (22 mg/m^3) in the workplace air in any eight-hour workshift of a forty-hour work week (US Occupational Safety and Health Administration, 1976). The work environment hygiene standards for *ortho*-toluidine in terms of an eight-hour time-weighted average are 10 mg/m^3 (1973) in the Federal Republic of Germany, 22 mg/m^3 (1974) in the German Democratic Republic and 5 mg/m^3 (1969) in Czechoslovakia (Winell, 1975). The acceptable ceiling concentration of *ortho*-toluidine in the USSR (1972) (Winell, 1975) and in Rumania (as reported by the Institute of Hygiene and Public Health in Bucharest) is 3 mg/m^3 (Goldstein, 1973). In Japan, the workplace tolerance concentration of *ortho*-toluidine is 5 ppm (22 mg/m^3).

ortho-Toluidine and its hydrochloride are not permitted for use in cosmetics in the European Economic Communities (EEC, 1976).

2.2 Occurrence

ortho-Toluidine has been reported to be a component of the tar produced by low-temperature carbonization of coal (Pichler & Hennenberger, 1969). It has also been detected in the gasoline fraction of hydro-cracked Arlan petroleum (Ben'kovskii *et al.*, 1972), in tobacco smoke (Pailer *et al.*, 1966), in the steam volatiles from the distillation of leaves of Latakia tobacco, which is believed to be widely used in pipe tobaccos (Irvine & Saxby, 1969), and in the volatile aroma components of black tea (Vitzthum *et al.*, 1975).

The concentration of *ortho*-toluidine measured in the air of the working environment in a toluidine manufacturing plant was in the range of 0.5-28.6 mg/m^3 (Khlebnikova *et al.*, 1970).

2.3 Analysis

Primary amines, including *ortho*-toluidine, have been determined in amounts greater than 1 mg by atomic absorption spectrophotometry of the copper-amine complex (Mitsui & Fujimura, 1974).

Nuclear magnetic resonance spectroscopy has been used to determine *ortho*-toluidine and its isomers, with an error of less than 1.3% (Yasuda & Kakiyama, 1974).

Spectrophotometry has been used to analyse aromatic amines, including *ortho*-toluidine, in prepared mixtures by measuring the maximum absorbance of the 1,3,5-trinitrobenzene-amine complexes; concentrations of 2-35 g/l *ortho*-toluidine have been determined (Sharma & Tiwari, 1972). It has been measured colorimetrically by oxidation with peroxydisulphate (Gupta & Srivastava, 1971); in natural waters as the 4-aminoantipyrine-amine complex, with a detection limit of 0.1 mg/l (El-Dib *et al.*, 1975); and by diazotization coupling with a phenolic compound, with a limit of detection of 0.1 mg/l (El-Dib, 1971).

Thin-layer chromatography has been used to separate aromatic amines, including *ortho*-toluidine, especially from their isomers: on films impregnated with 1,1'-diazido-4,4'-disulphostilbene and visualization by ultra-violet light (Thielemann, 1972); on silica gel-calcium oxalate with two solvent systems and visualization by exposure to nitrogen oxides (Srivastava & Dua, 1975); on silica gel using seven different solvents and visualization with potassium peroxydisulphate (Srivastava *et al.*, 1972); on silica gel containing silver oxide using two solvent systems and visualization by ultra-violet irradiation or with ninhydrin reagent (Tabak *et al.*, 1970); on silica gel impregnated with cadmium sulphate with five solvent systems using three detection methods (Yasuda, 1971); and on silica gel impregnated with cadmium acetate with four solvent systems (Yasuda, 1972). Visualization with ultra-violet light after reaction with glutaconic aldehyde in pyridine has also been used. The limit of detection is 0.1-0.3 μg (Ohkuma & Sakai, 1975).

Column chromatography on anion-exchange resin (Funasaka *et al.*, 1972) and gel permeation chromatography (Protivová & Pospíšil, 1974) have been used to separate aromatic amines, including *ortho*-toluidine.

Gas chromatography using porous-layer open-tube columns has been used to separate a mixture of *ortho*-toluidine, pyridine, picolines and lutidines (Franken *et al.*, 1971); and gas chromatography on barium chloride has been used for separation of nitrogen-substituted benzene compounds, including *ortho*-toluidine (Grob *et al.*, 1970).

A method for sampling and analysing air for aniline compounds, including *ortho*-toluidine, has been described. The compound is adsorbed on silica gel tubes, desorbed with ethanol and determined by gas chromatography. The limit of detection for *ortho*-toluidine is 1.03 mg/m^3 (Wood & Anderson, 1975).

The liquid crystal material, 4,4'-biphenylylene bis[*para*-(heptyloxy)-benzoate], has been used as a gas chromatographic stationary phase for the separation of isomeric toluidines (Richmond, 1971).

3. Biological Data Relevant to the Evaluation of Carcinogenic Risk to Man

3.1 Carcinogenicity and related studies in animals[1]

(a) Oral administration

Mouse: In an abstract, it was reported that male and female Charles River (HA/ICR) CG-1 mice given *ortho*-toluidine (probably as the hydrochloride) at two dose levels in the diet developed vascular tumours (Russfield *et al.*, 1973a) [No further details were available to the Working Group].

---

[1]The Working Group was aware of carcinogenicity studies in progress on *ortho*-toluidine hydrochloride involving oral administration in mice and rats (IARC, 1976).

Rat: In this and another abstract, it was reported that feeding of *ortho*-toluidine (probably as the hydrochloride) to 50 male Charles River (Sprague-Dawley derived) CD rats at two dose levels in the diet for two years increased the incidence of subcutaneous fibromas and fibrosarcomas to 83% from 16% that occurred spontaneously in 111 control rats. Urinary bladder cancers and hepatomas were also present in treated animals (Russfield *et al.*, 1973a,b) [No further details were available to the Working Group].

(b) Subcutaneous and/or intramuscular administration

Various species: Fifteen rabbits and 10 guinea-pigs were injected subcutaneously with 1.0 and 0.5 ml, respectively, of a 2% solution of *ortho*-toluidine in olive oil 6 times a week. Animals that survived more than 100 days developed papillomas in the bladder (number of animals not specified) (Morigami & Nisimura, 1940). Papillomas in the bladder were found in 2/2 rats, 4/5 rabbits and 5/8 guinea-pigs injected subcutaneously with *ortho*-toluidine (dose and duration unspecified) in olive oil or alcohol (Satani *et al.*, 1941) [The Working Group noted the inadequate reporting of these experiments].

3.2 Other relevant biological data

(a) Experimental systems

The oral LD_{50} by intubation in young male Sprague-Dawley rats of undiluted *ortho*-toluidine is 900 mg/kg bw (Jacobson, 1972). The oral LD_{50} of *ortho*-toluidine hydrochloride administered to young male Osborne-Mendel rats as a single dose in water by stomach tube is 2951 mg/kg bw (Lindstrom *et al.*, 1969). Following oral administration of *ortho*-toluidine in oil solutions, the LD_{50} values in mice, rats and rabbits were 515, 670 and 843 mg/kg bw, respectively (Lunkin, 1967). The LD_{50} for male rabbits by skin application is 3250 mg/kg bw (Smyth *et al.*, 1962).

All mice given one fifth of the intragastric LD_{50} daily for 20 days survived. When rats were given one-twentieth of the intragastric LD_{50} for 2½ months, the animals showed reticulocytosis, methaemoglobinemia and erythropenia (Lunkin, 1967).

ortho-Toluidine was fed to 6 male and 6 female albino rats in initial dosages of 2 g of a 7.5% solution in peanut oil (reduced after 64 days to half the dose) added to a synthetic diet for up to 91 days; it produced epithelial changes in the bladder, which included keratosis, metaplasia and, in 3 cases, a tendency to incipient papillomatosis (Ekman & Strömbeck, 1947, 1949).

No data on the embryotoxicity or teratogenicity of *ortho*-toluidine were available to the Working Group.

ortho-Toluidine did not induce reverse mutations in *Salmonella typhimurium* TA 1535, TA 1537, TA 1538, TA 98 or TA 100 in the presence or absence of a rat liver preparation (Garner & Nutman, 1977; McCann *et al.*, 1975).

(b) Man

ortho-Toluidine is absorbed *via* the respiratory tract and skin (ILO, 1971). Clinical signs of intoxication include physiological and psychical disturbances and marked irritation of the kidneys and bladder (Hamblin, 1963; Sax, 1968). Acute poisoning with *ortho*-toluidine may be accompanied by methaemoglobinemia and haematuria (ILO, 1971).

3.3 Case reports and epidemiological studies

Case & Pearson (1954) and Case *et al.* (1954) reported a historical cohort study of bladder tumours among workers in the British chemical industry engaged in the manufacture of dyestuff intermediates. Groups were defined in terms of exposure to aniline, benzidine, 1- and 2-naphthylamine, auramine and magenta. In fact, many aniline workers were also exposed to toluidines produced in the same plant. The 21 firms in the survey included the major British producers of aniline, and the groups at risk were observed from 1921-1952. There were 812 men classified as having exposure to aniline but not to the other suspected carcinogens. The death certificate for one of these mentioned a bladder tumour, although 0.52 was expected at national rates [The Working Group noted that exposure of these workers to *ortho*-toluidine is not specified in this paper but can be inferred from knowledge of aniline production processes, see 2.1(b), p. 353].

Uebelin & Pletscher (1954) studied the occurrence of bladder tumours in workers in a factory in Switzerland producing dyestuff intermediates. Among 300 workers exposed to 2-naphthylamine and/or benzidine alone or with other aromatic amines for unspecified periods during 1924-1953, 97 cases were found. A further 650 workers had been exposed only to aniline or other aromatic amines but not to 2-naphthylamine or benzidine; 3 of these men developed bladder tumours, but their exact exposures were uncertain. The authors distinguished a subgroup of 35 men who prepared *para*-chloro-*ortho*-toluidine from *ortho*-toluidine; no bladder tumours were found among these men [The Working Group noted that insufficient details were provided concerning person-years at risk or follow-up to evaluate the significance of this observation].

Gropp (1958) described 98 cases of bladder tumours that occurred from 1903-1955 among workers at a German factory, and Oettel (1959, 1967) and Oettel *et al.* (1968) have reported both these and some further cases. Most of Gropp's cases are reported as having had exposure to 2-naphthylamine alone (23 men) or with other amines, including benzidine, aniline and toluidines (45 men). However, 11 cases were reported as having had exposure only to aniline/toluidine; the population at risk is not known. These 11 cases were all diagnosed between 1905 and 1935 and had their first exposure between 1874 and 1905; they had longer latent periods and later ages at onset than the cases associated with exposure to 2-naphthylamine and benzidine [The Working Group noted that this was a period when the aniline produced was a crude substance, and the process residues were shown to contain 4-aminobiphenyl and naphthylamine derivatives (Walpole *et al.*, 1952)].

Khlebnikova *et al.* (1970) reported bladder tumours in workers engaged on the production of *ortho*-toluidine and/or *para*-toluidine in the USSR during the 1960's. The concentration range of *ortho*-toluidine measured in the air of the working environment was 0.5-28.6 mg/m^3. Two cases of bladder tumours (papillomas) were found when 75 out of 81 current toluidine workers were examined cystoscopically. The 81 workers comprised 19 women and 62 men, aged 22-55; 35 were operators, 18 were fitters, and 10 were cleaners. Nine had worked on *ortho*-toluidine production for less than one year, and 31 for 1-5 years; the remaining 41 workers had longer exposure and had all been in contact with *para*-toluidine as well as *ortho*-toluidine, as had the fitters and cleaners. One worker with a bladder tumour, aged 26, had been exposed only to *para*-toluidine and only for 20 months; the other worker with a bladder tumour, aged 49, had worked in contact with both isomers for 23 years. Six other cases of bladder tumours, of which 4 were carcinomas (1 was found to be a papilloma and 1 a multiple papilloma), had been found earlier, upon cystoscopic examination of 16 former workers who had worked with the toluidines for periods ranging from 12-17 years [The Working Group noted that the information contained in the paper with regard to person-years of exposure and to the other substances to which these workers may have had prior or concomitant exposure is insufficient to evaluate the significance of these observations].

4. Comments on Data Reported and Evaluation

4.1 Animal data

The data available to the Working Group concerning *ortho*-toluidine was from experiments that have been insufficiently described. Although tumours have been reported at various sites after administration of the compound by various routes, the inadequacy of reporting does not allow an evaluation of the carcinogenicity of this compound to be made.

4.2 Human data

The epidemiological studies available to the Working Group dealt with workers who had mixed exposures. The results of one study in Germany were equivocal. A second, in Switzerland, had negative results and was inadequate in terms of persons-years exposure. A third, probably representing a large propertion of all workers in the UK who were exposed to *ortho*-toluidine during the production process, demonstrated one observed case *versus* 0.52 expected. A fourth, in the USSR, appeared to involve occupational exposure specifically to toluidines; it was suggestive of a carcinogenic effect, but information was incomplete.

These findings, taken together, did not allow the Working Group to draw any firm conclusion about the carcinogenicity of *ortho*-toluidine to humans.

5. References

Ben'kovskii, V.G., Baikova, A.Y., Bulatova, B.T., Lyubopytova, N.S. & Popov, Y.N. (1972) Nitrogenous bases in gasoline from the hydrocracking of Arlan petroleum. Neftekhimiya, 12, 454-459

BIOS (British Intelligence Objectives Sub-Committee) (1946) Final Report No. 1141, German Chemical Industry with Special Reference to the Design of Plant for Dyestuffs and Intermediates, London, HMSO

Case, R.A.M. & Pearson, J.T. (1954) Tumours of the urinary bladder in workmen engaged in the manufacture and use of certain dyestuff intermediates in the British chemical industry. II. Further consideration of the role of aniline and of the manufacture of auramine and magenta (fuchsine) as possible causative agents. Brit. J. industr. Med., 11, 213-216

Case, R.A.M., Hosker, M.E., McDonald, D.B. & Pearson, J.T. (1954) Tumours of the urinary bladder in workmen engaged in the manufacture and use of certain dyestuff intermediates in the British chemical industry. I. The role of aniline, benzidine, *alpha*-naphthylamine, and *beta*-naphthylamine. Brit. J. industr. Med., 11, 75-104

EEC (1976) Législation. Journal Officiel des Communautés Européennes, No. L 262, p. 175

Ekman, B. & Strömbeck, J.P. (1947) Demonstration of tumorigenic decomposition products of 2,3-azotoluene. Acta physiol. scand., 14, 43-50

Ekman, B. & Strömbeck, J.P. (1949) The effect of some splitproducts of 2,3'-azotoluene on the urinary bladder in the rat and their excretion on various diets. Acta path. microbiol. scand., 26, 447-467

El-Dib, M.A. (1971) Colorimetric determination of aniline derivatives in natural waters. J. Ass. off. analyt. Chem., 54, 1383-1387

El-Dib, M.A., Abdel-Rahman, M.O. & Aly, O.A. (1975) 4-Aminoantipyrine as a chromogenic agent for aromatic amine determination in natural water. Water Res., 9, 513-516

Franken, J.J., Vidal-Madjar, C. & Guiochon, G. (1971) Gas-solid chromatographic analysis of aromatic amines, pyridine, picolines, and lutidines on cobalt phthalocyanine with porous-layer open-tube columns. Analyt. Chem., 43, 2034-2037

Funasaka, W., Hanai, T., Fujimura, K. & Ando, T. (1972) Non-aqueous solvent chromatography. II. Separation of benzene derivatives in the anion-exchange and *n*-butyl alcohol system. J. Chromat., 72, 187-191

Garner, R.C. & Nutman, C.A. (1977) Testing of some azo dyes and their reduction products for mutagenicity using *Salmonella typhimurium* TA 1538. Mutation Res. (in press)

Goldstein, I. (1973) Studies on MAC values of nitro- and amino-derivatives of aromatic hydrocarbons. In: Horvath, M., ed., Adverse Effects of Environmental Chemicals and Psychotropic Drugs, Vol.1, New York, Elsevier, pp. 153-154

Grasselli, J.G., ed. (1973) CRC Atlas of Spectral Data and Physical Constants for Organic Compounds, Cleveland, Ohio, Chemical Rubber Co., pp. B-942-B-943

Grob, R.L., Gondek, R.J. & Scales, T.A. (1970) A study of the strontium and barium halides as column packings in gas-solid chromatography. J. Chromat., 53, 477-486

Gropp, D. (1958) Zur Ätiologie des sogenannten Anilin-Blasenkrebses. (Inaugural-Dissertation Thesis, Johannes-Gutenberg Universität, Mainz, FRG)

Gupta, R.C. & Srivastava, S.P. (1971) Oxidation of aromatic amines by peroxodisulphate ion. II. Identification of aromatic amines on the basis of absorption maxima of coloured oxidation products. Z. analyt. Chem., 257, 275-277

Hamblin, D.O. (1963) Aromatic nitro and amino compounds. In: Patty, F.A., ed., Industrial Hygiene and Toxicology, 2nd ed., Vol. II, New York, Interscience, p. 2155

IARC (1975) IARC Monographs on the Evaluation of Carcinogenic Risk of Chemicals to Man, 8, Some Aromatic Azo Compounds, Lyon, pp. 61-74

IARC (1976) Information Bulletin on the Survey of Chemicals Being Tested for Carcinogenicity, No. 6, Lyon, p. 184

ILO (International Labour Office) (1971) Encyclopedia of Occupational Health and Safety, Vol. 1, New York, McGraw-Hill, p. 96

Irvine, W.J. & Saxby, M.J. (1969) Steam volatile amines of Latakia tobacco leaf. Phytochemistry, 8, 473-476

Jacobson, K.H. (1972) Acute oral toxicity of mono- and di-alkyl ring-substituted derivatives of aniline. Toxicol. appl. Pharmacol., 22, 153-154

Japan Dyestuff Industry Association (1976) Statistics of Dyestuffs (1971-1975), Tokyo

Japan Tariff Association (1976) Japan Exports and Imports, Commodity by Country (1971-1975), Tokyo

Khlebnikova, M.I., Gladkova, E.V., Kurenko, L.T., Pshenitsyn, A.V. & Shalin, B.M. (1970) Problems of industrial hygiene and health status of workers engaged in the production of *o*-toluidine. Gig. Tr. Prof. Zabol., 14, 7-10

Lindstrom, H.V., Bowie, W.C., Wallace, W.C., Nelson, A.A. & Fitzhugh, O.G. (1969) The toxicity and metabolism of mesidine and pseudocumidine in rats. J. Pharmacol. exp. Ther., 167, 223-234

Lunkin, V.N. (1967) Information for the hygienic establishment of the maximum allowable concentration of *para*- and *ortho*-toluidines in inland waters. Ref. Zh. Otd. Vyp. Farmakol. Khimioter. Sredstva. Toksikol., No. 12.54.1096

McCann, J., Choi, E., Yamasaki, E. & Ames, B.N. (1975) Detection of carcinogens as mutagens in the *Salmonella*/microsome test: assay of 300 chemicals. Proc. nat. Acad. Sci. (Wash.), 72, 5135-5139

Mitsui, T. & Fujimura, Y. (1974) Indirect determination of primary amines by atomic absorption spectrophotometry. Bunseki Kagaku, 23, 1309-1314

Morigami, S. & Nisimura, I. (1940) Experimental studies on aniline bladder tumors. Gann, 34, 146-147

Oettel, H. (1959) Zur Frage des Berufskrebses durch Chemikalien. In: Verhandlungen der Deutschen Gesellschaft für Pathologie, 43 Tagung, Stuttgart, Gustav Fischer-Verlag, pp. 313-320

Oettel, H. (1967) Bladder cancer in Germany. In: Lampe, K.F., ed., Bladder Cancer, A Symposium, Birmingham, Alabama, Aesculapius Publishing Company, pp. 196-199

Oettel, H., Thiess, A.M. & Uhl, C. (1968) Beitrag zur Problematik berufsbedingter Lungenkrebse. Zbl. Arbeitsmed. Arbeitsschutz, 18, 291-303

Ohkuma, S. & Sakai, I. (1975) Detection of aromatic primary amines by a photochemical reaction with pyridine. Bunseki Kagaku, 24, 385-387

Pailer, M., Huebsch, W.J. & Kuhn, H. (1966) Aromatic amines occurring in tobacco smoke (preliminary communication). Monatsh. Chem., 97, 1448-1451

Pichler, H. & Hennenberger, P. (1969) Untersuchung flüssiger und gasförmiger Produkte der Steinkohlenschwelung. III. Zusammensetzung eines Innenabsaugöls, eines Vakuumschwelteers und eines Rohbenzols der Hochtemperaturverkokung. Brennst.-Chem., 50, 341-346

E.I. du Pont de Nemours & Co. Inc. (1974) Sales Specification: *o*-Toluidine Technical, *o*-Toluidine-S2 Technical, Wilmington, Delaware, Dyes and Chemicals Division

Prager, B., Jacobson, P., Schmidt, P. & Stern, D., eds (1929) Beilsteins Handbuch der Organischen Chemie, 4th ed., Vol. 12, Syst. No. 1671-1672, Berlin, Springer-Verlag, p. 772

Protivova, J. & Pospisil, J. (1974) Antioxidants and stabilizers. XLVII. Behaviour of amine antioxidants and antiozonants and model compounds in gel permeation chromatography. J. Chromat., 88, 99-107

Richmond, A.B. (1971) Use of liquid crystals for the separation of position isomers of disubstituted benzenes. J. chromat. Sci., 9, 571-574

Russfield, A.B., Homburger, F., Weisburger, E.K. & Weisburger, J.H. (1973a) Further studies on carcinogenicity of environmental chemicals including simple aromatic amines (Abstract No. 20). Toxicol. appl. Pharmacol., 25, 446-447

Russfield, A.B., Boger, E., Homburger, F., Weisburger, E.K. & Weisburger, J.H. (1973b) Effects of structure of seven methyl anilines on toxicity and on incidence of subcutaneous and liver tumors in Charles River rats (Abstract No. 3467). Fed. Proc., 32, 833

Satani, Y., Tanimura, T., Nishimura, I. & Isikawa, Y. (1941) Klinische und experimentelle Untersuchung des Blasenpapilloma. Gann, 35, 275-276

Sax, N.I. (1968) Dangerous Properties of Industrial Materials, 4th ed., New York, Van Nostrand-Reinhold, pp. 1175-1176

Sharma, J.P. & Tiwari, R.D. (1972) Charge-transfer complexes and their applications; distinction and determination of some aromatic amines. Microchem. J., 17, 151-159

Smyth, H.F., Jr, Carpenter, C.P., Weil, C.S., Pozzani, U.C. & Striegel, J.A. (1962) Range-finding toxicity data: List VI. Amer. industr. Hyg. Ass. J., 23, 95-107

The Society of Dyers and Colourists (1971) Colour Index, 3rd ed., Vol. 4, Yorkshire, UK, pp. 4853, 4857, 4859

Srivastava, S.P. & Dua, V.K. (1975) TLC [thin-layer chromatography] separation of closely related amines. Fresenius' Z. analyt. Chem., 276, 382

Srivastava, S.P., Gupta, R.C. & Gupta, A. (1972) Peroxydisulfate oxidations. VII. Identification of aromatic amines and phenols by thin-layer chromatography. Fresenius' Z. analyt. Chem., 262, 31-32

Tabak, S., Mauro, A.E. & Del'Acqua, A. (1970) Argentation thin-layer chromatography with silver oxide. II. Amines, unsaturated and aromatic carboxylic acids. J. Chromat., 52, 500-504

Thielemann, H. (1972) Separation and identification of the three isomeric toluidines with 1,1'-diazido-4,4'-disulfostilbene-impregnated film in thin-layer chromatography. Z. Chem., 12, 343

Uebelin, F. & Pletscher, A. (1954) Ätiologie und Prophylaxe gewerblicher Tumoren in der Farbstoffindustrie. Schweiz. med. Wschr., 84, 917-928

US International Trade Commission (1976a) Imports of Benzenoid Chemicals and Products, 1974, ITC Publication 762, Washington DC, US Government Printing Office, p. 28

US International Trade Commission (1976b) Synthetic Organic Chemicals, US Production and Sales, 1974, USITC Publication 776, Washington DC, US Government Printing Office, pp. 45, 51, 54, 57, 58, 61, 65, 66, 69, 70, 78, 81

US International Trade Commission (1977a) Synthetic Organic Chemicals, US Production and Sales, 1975, USITC Publication 804, Washington DC, US Government Printing Office, pp. 42, 50, 53, 63, 75

US International Trade Commission (1977b) Imports of Benzenoid Chemicals and Products, 1975, USITC Publication 806, Washington DC, US Government Printing Office, p. 27

US Occupational Safety and Health Administration (1976) Air contaminants. US Code of Federal Regulations, Title 29, part 1910.1000, p. 585

US Tariff Commission (1922) Census of Dyes and Other Synthetic Organic Chemicals, 1921, Tariff Information Series No. 26, Washington DC, US Government Printing Office, p. 26

US Tariff Commission (1957) Synthetic Organic Chemicals, US Production and Sales, 1956, Report No. 200, Second Series, Washington DC, US Government Printing Office, p. 79

Vitzthum, O.G., Werkhoff, P. & Hubert, P. (1975) New volatile constituents of black tea aroma. J. agric. Fd Chem., 23, 999-1003

Walpole, A.L., Williams, M.H.C. & Roberts, D.C. (1952) The carcinogenic action of 4-aminodiphenyl and 3:2'-dimethyl-4-aminodiphenyl. Brit. J. industr. Med., 9, 255-263

Weast, R.C., ed. (1976) CRC Handbook of Chemistry and Physics, 57th ed., Cleveland, Ohio, Chemical Rubber Co., pp. C-519, D-198

Windholz, M., ed. (1976) The Merck Index, 9th ed., Rahway, NJ, Merck & Co., p. 1226

Winell, M. (1975) An international comparison of hygienic standards for chemicals in the work environment. Ambio, 4, 34-36

Wood, G.O. & Anderson, R.G. (1975) Personal air sampling for vapors of aniline compounds. Amer. industr. Hyg. Ass. J., 36, 538-548

Yasuda, K. (1971) Thin-layer chromatography of aromatic amines on cadmium sulphate-impregnated silica gel thin layers. J. Chromat., 60, 144-149

Yasuda, K. (1972) Thin-layer chromatography of aromatic amines on cadmium acetate impregnated silica gel thin layers. J. Chromat., 72, 413-420

Yasuda, S. & Kakiyama, H. (1974) Determination of the three isomers of cresol, toluic acid, and toluidine by nuclear magnetic resonance. Bunseki Kagaku, 23, 615-620

2,4-XYLIDINE (HYDROCHLORIDE)

1. Chemical and Physical Data

2,4-Xylidine

1.1 Synonyms and trade names

Chem. Abstr. Services Reg. No.: 95-68-1

Chem. Abstr. Name: 2,4-Dimethylbenzenamine

1-Amino-2,4-dimethylbenzene; 4-amino-1,3-dimethylbenzene; 4-amino-3-methyltoluene; 4-amino-1,3-xylene; 2,4-dimethyl aniline; 2,4-dimethylaniline; 4-methyl-*ortho*-toluidine; 2-methyl-*para*-toluidine; *meta*-xylidine; *meta*-4-xylidine

1.2 Chemical formula and molecular weight

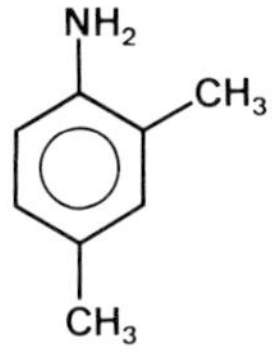

NH2, CH3, CH3

$C_8H_{11}N$ Mol. wt: 121.2

1.3 Chemical and physical properties of the pure substance

From Weast (1976), unless otherwise specified

(a) Description: Colourless liquid at room temperature

(b) Boiling-point: 214°C at 760 mm; 91°C at 10 mm

(c) Melting-point: 16°C

(d) Density: d_4^{20} 0.9723

(e) Refractive index: n_D^{20} 1.5569

(f) Spectroscopy data: λ_{max} 290 (E_1^1 = 132) in ethanol; infra-red, ultra-violet, nuclear magnetic resonance and mass spectral data have been tabulated (Grasselli, 1973).

(g) Solubility: Slightly soluble in water; soluble in ethanol, ether and benzene

(h) Volatility: Vapour pressure is 1 mm at 52.6°C and 10 mm at 93°C.

(i) Stability: See 'General Remarks on the Substances Considered', p. 27.

1.4 Technical products and impurities

In the UK, 2,4-xylidine is available in a mixture of xylidine (essentially 2,4- and 2,6-isomers) that contains a minimum of 99.5% diazotizable amine, with a distillation range of 90% within 3°C.

2,4-Xylidine is available in Japan in a grade with a minimum purity of 99% and 0.2% max. moisture content. Other xylidine isomers are present as impurities.

2,4-Xylidine hydrochloride

1.1 Synonyms and trade names

Chem. Abstr. Services Reg. No.: 21436-96-4

Chem. Abstr. Name: 2,4-Dimethylbenzenamine hydrochloride

1-Amino-2,4-dimethylbenzene hydrochloride; 4-amino-1,3-dimethylbenzene hydrochloride; 4-amino-3-methyltoluene hydrochloride; 4-amino-1,3-xylene hydrochloride; 2,4-dimethyl aniline hydrochloride; 2,4-dimethylaniline hydrochloride; 4-methyl-*ortho*-toluidine hydrochloride; 2-methyl-*para*-toluidine hydrochloride; *meta*-xylidine hydrochloride; *meta*-4-xylidine hydrochloride

1.2 Empirical formula and molecular weight

$C_8H_{11}N.HCl$ Mol. wt: 157.7

1.3 Chemical and physical properties of the pure substance

From Prager *et al.* (1929)

(a) Description: Prisms

(b) Boiling-point: 253.1°C at 728 mm, 255.1°C at 760 mm

(c) Melting-point: 235°C

(d) Solubility: Practically insoluble in cold water

1.4 Technical products and impurities

No data were available to the Working Group.

2. Production, Use, Occurrence and Analysis

For background information on this section, see preamble, p. 15.

2.1 Production and use

(a) Production

2,4-Xylidine was prepared by Beilstein & Deumelandt in 1867 by the reduction of asymmetrical 2,4-dimethylnitrobenzene with iron and acetic acid (Prager *et al.*, 1929). It is produced commercially by nitrating and reducing commercial xylene and is separated from the resulting isomeric mixture of xylidines by formation of the acetate salt (Kouris & Northcott, 1963).

The commercial production of an unspecified xylidine isomer and its salt in the US was first reported in 1917 (US Tariff Commission, 1919), and that of 2,4-xylidine, specifically, in 1920 (US Tariff Commission, 1921). In 1975, only one US company reported an undisclosed amount (see preamble, p. 15) of 2,4-xylidine (US International Trade Commission, 1977). US imports through principal US customs districts were reported as 64.3 thousand kg in 1974 (US International Trade Commission, 1976).

2,4-Xylidine was first produced commercially in the UK in 1910. It is believed that at least 6 companies currently produce this compound in western Europe. Annual production of a mixture essentially of 2,4- and 2,6-xylidine amounts to 10-100 thousand kg, respectively, in France, Italy, Switzerland and the UK and 100 thousand to 1 million kg in the Federal Republic of Germany. Less than 10 thousand kg are imported annually, respectively, into Austria, the Federal Republic of Germany, Italy and Scandinavia and 10-100 thousand kg, respectively, into Benelux, France and Switzerland. The UK exports 10-100 thousand kg annually.

2,4-Xylidine has been produced commercially in Japan since 1969. In 1975, one Japanese company produced 20-30 thousand kg, all of which was used as an intermediate without isolation.

No evidence was found that 2,4-xylidine hydrochloride has ever been produced commercially in the US or Japan.

(b) Use

2,4-Xylidine can be used as an intermediate in the production of 8 dyes which are believed to be of commercial world significance (The Society of Dyers and Colourists, 1971). Three of these are produced commercially in the US (US International Trade Commission, 1977): C.I. Acid Red 26, C.I. Direct Violet 14 and C.I. Solvent Orange 7.

Less than 10 thousand kg of a mixture essentially of 2,4- and 2,6-xylidine are used annually in Austria, in Spain and in Scandinavia, 10-100 thousand kg in Benelux, in Switzerland and in the UK and 100 thousand to 1 million kg in France and in the Federal Republic of Germany. In Sweden, it is used for the production of leather dyes and in the UK for the production of antioxidants (85%) and pigments (15%).

In Japan, 2,4-xylidine is used in the manufacture of a diazo yellow pigment.

2,4-Xylidine and its salts are believed to be used as dyestuff intermediates in Switzerland. They are not permitted for use in cosmetics in the European Economic Communities (EEC, 1976).

The current work environment hygiene standards for xylidine (unspecified isomer) are 5 ppm (25 mg/m^3) (1974) in the USA, 25 mg/m^3 (1974) in the Federal Republic of Germany, 10 mg/m^3 (1973) in the German Democratic Republic and 5 mg/m^3 (1969) in Czechoslovakia (all in terms of 8-hour time-weighted average values). The acceptable ceiling concentration of xylidine in the USSR is 3 mg/m^3 (1972) (Winell, 1975). In Japan, the workplace tolerance concentration of 2,4-xylidine is 5 ppm (25 mg/m^3).

2.2 Occurrence

2,4-Xylidine and 2,4-xylidine hydrochloride are not known to occur as natural products. No data on their occurrence in the environment were available to the Working Group.

2.3 Analysis

Aromatic amines, including 2,4-xylidine, have been determined colorimetrically by reaction with potassium peroxydisulphate (Gupta & Srivastava, 1971), in natural waters at concentrations of 0.1 mg/l by forming a coloured 4-amino-antipyrine complex (El-Dib *et al.*, 1975) or by diazotization and coupling with resorcinol or 1-naphthol (El-Dib, 1971).

Thin-layer chromatography has been used to separate and identify aromatic amines, including 2,4-xylidine, on various adsorbents such as silica gel buffered with sodium acetate (Bassl *et al.*, 1967), silica gel and alumina containing zinc, cadmium and nickel nitrates (Shimomura & Walton, 1968) and silica gel layers impregnated with cadmium acetate (Yasuda, 1972).

A method for the sampling and analysis of aromatic amines, including 2,4-xylidine, in air has been described. Pollutants are adsorbed on silica gel, desorbed with ethanol and determined by gas chromatography. Concentrations as low as 1.05 mg/m^3 were detected (Wood & Anderson, 1975). Gas chromatography has also been used to separate and identify primary and secondary amines, including 2,4-xylidine, as their trifluoroacetyl derivatives (Dove, 1967; Irvine & Saxby, 1969). Gas chromatography has also been used to separate and determine amines, including 2,4-xylidine, as their *N*-substituted 2,5-dimethylpyrroles, with a sensitivity of 0.1 g/l (Walle, 1968).

3. Biological Data Relevant to the Evaluation of Carcinogenic Risk to Man

3.1 Carcinogenicity and related studies in animals[1]

Oral administration

Rat: In an abstract, it was reported that in a 2-year feeding study with 50 male Charles River (Sprague-Dawley) rats, subcutaneous fibromas or fibrosarcomas occurred in 39% of animals fed 2,4-xylidine and in 16% of controls. An excess of hepatomas also occurred in treated rats (Russfield *et al.*, 1973) [No further details were available to the Working Group].

3.2 Other relevant biological data

(a) Experimental systems

The oral LD_{50} of 2,4-xylidine hydrochloride in male Osborne-Mendel rats is 1259 mg/kg bw (Lindstrom *et al.*, 1969).

A single i.v. injection of 20 mg/rat 2,4-xylidine increased the blood methaemoglobin level from 1.5% to 3.5% after one hour (Lindstrom *et al.*, 1969). Dietary concentrations of 2500 mg/kg 2,4-xylidine reduced the weight gain of rats, and 10,000 mg/kg caused cholangiofibrosis, bile-duct proliferation, hepatic-cell necrosis and hyperplasia and kidney damage, including tubular atrophy, interstitial fibrosis, inflammation and papillary oedema (Lindstrom *et al.*, 1963).

Dogs administered daily oral doses of 50 mg/kg bw 2,4-xylidine orally for 4 weeks had slightly enlarged livers with fatty degeneration (Magnusson *et al.*, 1971).

It was reported in an abstract that mice given a single oral dose of 158 mg/kg bw 2,4-xylidine hydrochloride had acidophilic granules in hepatocytes; electron microscopic examination of these cells showed increased

---

[1]The Working Group was aware of carcinogenicity studies in progress on 2,4-xylidine hydrochloride involving oral administration in rats (IARC, 1976).

numbers of lysosomes and focal degeneration. 2,4-Xylidine and *N*-hydroxy-2,4-xylidine released lysosomal enzymes from the large-granule fraction of liver (Takahashi *et al.*, 1974).

Male Osborne-Mendel rats given 200 mg/kg bw/day 2,4-xylidine hydrochloride by gastric intubation excreted small amounts of unchanged xylidine and the corresponding sulphamate and acetanilide. The predominant metabolite was 3-methyl-4-aminobenzoic acid, which was excreted to a small extent in the unconjugated form and as its glycine conjugate, but mainly as 3-methyl-4-acetamidobenzoic acid; no aminophenols were detected (Lindstrom, 1961).

No data on the embryotoxicity, teratogenicity or mutagenicity of 2,4-xylidine or 2,4-xylidine hydrochloride were available to the Working Group.

(b) Man

No data were available to the Working Group.

3.3 Case reports and epidemiological studies

No data were available to the Working Group.

4. Comments on Data Reported and Evaluation

4.1 Animal data

Data reported in an abstract indicate that 2,4-xylidine produced an increased incidence of subcutaneous and liver tumours in rats following its oral administration. The inadequacy of reporting does not allow an evaluation of the carcinogenicity of this compound to be made.

4.2 Human data

No case reports or epidemiological studies were available to the Working Group.

5. References

Bassl, A., Heckemann, H.J. & Baumann, E. (1967) Thin-layer chromatography of primary aromatic amines. I. J. Prakt. Chem., 36, 265-273

Dove, R.A. (1967) Separation and determination of aniline and the toluidine, xylidine, ethylaniline, and *N*-methyltoluidine isomers by gas chromatography of their *N*-trifluoroacetyl derivatives. Analyt. Chem., 39, 1188-1190

EEC (1976) Législation. Journal Officiel des Communautés Européennes, No. L 262, p. 175

El-Dib, M.A. (1971) Colorimetric determination of aniline derivatives in natural waters. J. Ass. off. analyt. Chem., 54, 1383-1387

El-Dib, M.A., Abdel-Rahman, M.O. & Aly, O.A. (1975) 4-Aminoantipyrine as a chromogenic agent for aromatic amine determination in natural water. Water Res., 9, 513-516

Grasselli, J.G., ed. (1973) CRC Atlas of Spectral Data and Physical Constants for Organic Compounds, Cleveland, Ohio, Chemical Rubber Co., p. B-206

Gupta, R.C. & Srivastava, S.P. (1971) Oxidation of aromatic amines by peroxodisulphate ion. II. Identification of aromatic amines on the basis of absorption maxima of coloured oxidation products. Z. analyt. Chem., 257, 275-277

IARC (1976) Information Bulletin on the Survey of Chemicals Being Tested for Carcinogenicity, No. 6, Lyon, p. 115

Irvine, W.J. & Saxby, M.J. (1969) Gas chromatography of primary and secondary amines as their trifluoroacetyl derivatives. J. Chromat., 43, 129-131

Kouris, C.S. & Northcott, J. (1963) Aniline. In: Kirk, R.E. & Othmer, D.F., eds, Encyclopedia of Chemical Technology, 2nd ed., Vol. 2, New York, John Wiley and Sons, p. 421

Lindstrom, H.V. (1961) The metabolism of FD & C Red No. 1. I. The fate of 2,4-*meta*-xylidine in rats. J. Pharmacol. exp. Ther., 132, 306-310

Lindstrom, H.V., Hansen, W.H., Nelson, A.A. & Fitzhugh, O.G. (1963) The metabolism of FD & C Red No. 1. II. The fate of 2,5-*para*-xylidine and 2,6-*meta*-xylidine in rats and observations on the toxicity of xylidine isomers. J. Pharmacol. exp. Ther., 142, 257-264

Lindstrom, H.V., Bowie, W.C., Wallace, W.C., Nelson, A.A. & Fitzhugh, O.G. (1969) The toxicity and metabolism of mesidine and pseudocumidine in rats. J. Pharmacol. exp. Ther., 167, 223-234

Magnusson, G., Bodin, N.-O. & Hansson, E. (1971) Hepatic changes in dogs and rats induced by xylidine isomers. Acta path. microbiol. scand., Sect. A, 79, 639-648

Prager, B., Jacobson, P., Schmidt, P. & Stern, D., eds (1929) Beilsteins Handbuch der Organischen Chemie, 4th ed., Vol. 12, Syst. No. 1704, Berlin, Springer-Verlag, pp. 1111-1114

Russfield, A.B., Boger, E., Homburger, F., Weisburger, E.K. & Weisburger, J.H. (1973) Effect of structure of seven methyl anilines on toxicity and on incidence of subcutaneous and liver tumors in Charles River rats (Abstract No. 3467). Fed. Proc., 32, 833

Shimomura, K. & Walton, H.F. (1968) Thin-layer chromatography of amines by ligand exchange. Separation Sci., 3, 493-499

The Society of Dyers and Colourists (1971) Colour Index, 3rd ed., Vol. 4, Yorkshire, UK, p. 4863

Takahashi, A., Omori, Y. & Takeuchi, M. (1974) Early biochemical and morphological changes in the liver of mice after a single oral administration of 2,4-xylidine (Abstract No. 38). Jap. J. Pharmacol., 24, 41

US International Trade Commission (1976) Imports of Benzenoid Chemicals and Products, 1974, USITC Publication 762, Washington DC, US Government Printing Office, p. 29

US International Trade Commission (1977) Synthetic Organic Chemicals, US Production and Sales, 1975, USITC Publication 804, Washington DC, US Government Printing Office, pp. 43, 56, 64, 73

US Tariff Commission (1919) Report on Dyes and Related Coal-Tar Chemicals, 1918, revised ed., Washington DC, US Government Printing Office, p. 31

US Tariff Commission (1921) Census of Dyes and Coal-Tar Chemicals, 1920, Tariff Information Series No. 23, Washington DC, US Government Printing Office, p. 28

Walle, T. (1968) Quantitative gas-chromatographic determination of primary amines in submicrogram quantities after condensation with 2,5-hexanedione. Acta pharm. suecica, 5, 353-366

Weast, R.C., ed. (1976) CRC Handbook of Chemistry and Physics, 57th ed., Cleveland, Ohio, Chemical Rubber Co., pp. C-146, D-200

Winell, M. (1975) An international comparison of hygienic standards for chemicals in the work environment. Ambio, 4, 34-36

Wood, G.O. & Anderson, R.G. (1975) Personal air sampling for vapors of aniline compounds. Amer. industr. Hyg. Ass. J., 36, 538-548

Yasuda, K. (1972) Thin-layer chromatography of aromatic amines on cadmium acetate impregnated silica gel thin layers. J. Chromat., 72, 413-420

1. Chemical and Physical Data

2,5-Xylidine

1.1 Synonyms and trade names

Chem. Abstr. Services Reg. No.: 95-78-3

Chem. Abstr. Name: 2,5-Dimethylbenzenamine

1-Amino-2,5-dimethylbenzene; 3-amino-1,4-dimethylbenzene; 5-amino-1,4-dimethylbenzene; 2-amino-4-methyltoluene; 2-amino-1,4-xylene; 2,5-dimethyl aniline; 2,5-dimethylaniline; 3,5-dimethylbenzenamine; 5-methyl-*ortho*-toluidine; 6-methyl-*meta*-toluidine; *para*-xylidine

1.2 Chemical formula and molecular weight

NH_2 CH_3 CH_3

$C_8H_{11}N$ Mol. wt: 121.2

1.3 Chemical and physical properties of the pure substance

From Weast (1976), unless otherwise specified

(a) Description: Colourless to yellow oil (Kouris & Northcott, 1963)

(b) Boiling-point: 214°C at 760 mm; 97-101°C at 10 mm

(c) Melting-point: 15.5°C

(d) Density: d_4^{21} 0.9790

(e) Refractive index: n_D^{21} 1.5591

(f) Spectroscopy data: λ_{max} 236 nm (E_1^1 = 632), 288 nm (E_1^1 = 204); infra-red and nuclear magnetic resonance spectra have been tabulated (Grasselli, 1973).

(g) Solubility: Slightly soluble in water and ethanol, soluble in ether (Kouris & Northcott, 1963)

(h) Stability: See 'General Remarks on the Substances Considered', p. 27.

1.4 Technical products and impurities

2,5-Xylidine is available in the US as a laboratory grade of at least 98% purity (Aldrich Chemical Company, Inc., 1976).

In the UK, it is available with a purity of 98% and contains 2,6-dimethylaniline and other isomers as impurities.

2,5-Xylidine hydrochloride

1.1 Synonyms and trade names

Chem. Abstr. Services Reg. No.: 51786-53-9

Chem. Abstr. Name: 2,5-Dimethylbenzenamine hydrochloride

1-Amino-2,5-dimethylbenzene hydrochloride; 3-amino-1,4-dimethylbenzene hydrochloride; 5-amino-1,4-dimethylbenzene hydrochloride; 2-amino-4-methyltoluene hydrochloride; 2-amino-1,4-xylene hydrochloride; 2,5-dimethyl aniline hydrochloride; 2,5-dimethylaniline hydrochloride; 3,5-dimethylbenzenamine hydrochloride; 5-methyl-*ortho*-toluidine hydrochloride; 6-methyl-*meta*-toluidine hydrochloride; *para*-xylidine hydrochloride

1.2 Empirical formula and molecular weight

$C_8H_{11}N.HCl$ Mol. wt: 157.7

1.3 Chemical and physical properties of the pure substance

From Prager *et al.* (1929)

(a) Boiling-point: 247.4°C

(b) Melting-point: 228°C

1.4 Technical products and impurities

2,5-Xylidine hydrochloride is available in the US as a laboratory grade (Pfaltz & Bauer, Inc., 1976).

2. Production, Use, Occurrence and Analysis

For background information on this section, see preamble, p. 15.

2.1 Production and use

(a) Production

2,5-Xylidine was prepared by Jannasch in 1875 by the reduction of 1,4-dimethylnitrobenzene with tin and hydrochloric acid (Prager *et al.*, 1929). 2,5-Xylidine is produced commercially by the following method: (1) nitrating and reducing commercial xylene to an isomeric mixture of xylidines; (2) after 2,4-xylidine is removed by formation of the acetate salt, 2,5-xylidine hydrochloride is precipitated by the addition of hydrochloric acid (Kouris & Northcott, 1963), and this can be converted to 2,5-xylidine by the addition of base.

The commercial production of an unspecified isomer of xylidine and its salt in the US was first reported in 1917 (US Tariff Commission, 1919). Commercial production of 2,5-xylidine, specifically, was first reported in the US in 1927 (US Tariff Commission, 1928) and that of 2,5-xylidine hydrochloride in 1930 (US Tariff Commission, 1931). Commercial production of 2,5-xylidine hydrochloride has not been reported in the US since 1958 (US Tariff Commission, 1959). In 1971, the last year in which commercial production was reported, only one US company reported the manufacture of an undisclosed amount (see preamble, p. 15) (US Tariff Commission, 1973). US imports of 2,5-xylidine through principal US customs districts were reported as 3700 kg in 1974 (US International Trade Commission, 1976).

In the UK, 10-100 thousand kg 2,5-xylidine are imported annually. It has been produced in negligible quantities in Japan.

(b) Use

2,5-Xylidine can be used as an intermediate for the production of 6 dyes which are believed to be of commercial world significance (The Society of Dyers and Colourists, 1971). Two of these, C.I. Direct Violet 7 and Solvent Red 26, are produced commercially in the US (US International Trade Commission, 1977).

In the UK, 10-100 thousand kg 2,5-xylidine are used annually for dyestuff manufacture. In Switzerland, 10-100 thousand kg 2,5-xylidine and its salts are used annually as dyestuff intermediates. 2,5-Xylidine and its salts are not permitted for use in cosmetics in the European Economic Communities (EEC, 1976).

The current work environment hygiene standards for xylidine (unspecified isomer) are 5 ppm (25 mg/m^3) (1974) in the USA, 25 mg/m^3 (1974) in the Federal Republic of Germany, 10 mg/m^3 (1973) in the German Democratic Republic and 5 mg/m^3 in Czechoslovakia (all in terms of 8-hour time-weighted average values). The acceptable ceiling concentration of xylidine in the USSR is 3 mg/m^3 (1972) (Winell, 1975).

2.2 Occurrence

2,5-Xylidine has been detected in the steam distillate of cured Latakia tobacco, which is used in pipe tobaccos (Irvine & Saxby, 1969a). No data on the occurrence of 2,5-xylidine and 2,5-xylidine hydrochloride in the environment were available to the Working Group.

2.3 Analysis

Aromatic amines, including 2,5-xylidine, have been determined colorimetrically by reaction with potassium peroxydisulphate (Gupta & Srivastava, 1971), in natural waters at concentrations of 0.1 mg/l by forming a coloured 2-aminoantipyrine complex (El-Dib *et al.*, 1975) or by diazotization and coupling with resorcinol or 1-naphthol (El-Dib, 1971).

Aniline and 16 of its alkyl and alkoxy derivatives, including 2,5-xylidine, have been separated using paper chromatography and 10 solvent systems (Przborowska, 1971).

Thin-layer chromatography has been used to separate and identify aromatic amines, including 2,5-xylidine, on various adsorbents such as silica gel buffered with sodium acetate (Bassl *et al.*, 1967), silica gel layers impregnated with cadmium acetate (Yasuda, 1972) or alumina to separate organic products of the coking industry (Medvedev *et al.*, 1973).

Gas chromatography has been used to separate and identify primary and secondary amines, including 2,5-xylidine, as their trifluoroacetyl derivatives in laboratory mixtures (Irvine & Saxby, 1969b), in volatile amines from cured Latakia tobacco leaf (Irvine & Saxby, 1969a) and in commercial, mixed xylidine (Dove, 1967).

3. Biological Data Relevant to the Evaluation of Carcinogenic Risk to Man

3.1 Carcinogenicity and related studies in animals

Oral administration

Rat: In an abstract, it was reported that in a 2-year feeding study, 50 male Charles River (Sprague-Dawley) rats treated with 2,5-xylidine developed hepatomas. Subcutaneous fibromas or fibrosarcomas were also seen in 24% of treated animals and in 16% of controls (Russfield *et al.*, 1973) [No further details were available to the Working Group].

3.2 Other relevant biological data

(a) Experimental systems

A single i.v. injection of 20 mg/rat 2,5-xylidine increased the blood methaemoglobin level from 1.5% to 3.5% after 3 hours (Lindstrom *et al.*, 1969).

When doses of 200 mg/kg bw/day 2,5-xylidine hydrochloride were given by stomach tube to male Osborne-Mendel rats, the urines contained 4-hydroxy-2,5-dimethylaniline and its conjugates as the main metabolites; some unchanged amine, 4-methyl-2-aminobenzoic acid and 4-methyl-3-aminobenzoic acid were also found (Lindstrom *et al.*, 1963).

No data on the embryotoxicity, teratogenicity or mutagenicity of 2,5-xylidine or its hydrochloride were available to the Working Group.

(b) Man

No data were available to the Working Group.

3.3 Case reports and epidemiological studies

No data were available to the Working Group.

4. Comments on Data Reported and Evaluation

4.1 Animal data

Data reported in an abstract indicate that 2,5-xylidine produced an increased incidence of subcutaneous and liver tumours in rats following its oral administration. The inadequacy of the reporting does not allow an evaluation of the carcinogenicity of this compound to be made.

4.2 Human data

No case reports or epidemiological studies were available to the Working Group.

5. References

Aldrich Chemical Company, Inc. (1976) The 1977-1978 Aldrich Catalog/ Handbook of Organic and Biochemicals, Catalog No. 18, Milwaukee, Wisconsin, p. 310

Bassl, A., Heckemann, H.J. & Baumann, E. (1967) Thin-layer chromatography of primary aromatic amines. I. J. Prakt. Chem., 36, 265-273

Dove, R.A. (1967) Separation and determination of aniline and the toluidine, xylidine, ethylaniline, and *N*-methyltoluidine isomers by gas chromatography of their *N*-trifluoroacetyl derivatives. Analyt. Chem., 39, 1188-1190

EEC (1976) Législation. Journal Officel des Communautés Européennes, No. L 262, p. 175

El-Dib, M.A. (1971) Colorimetric determination of aniline derivatives in natural waters. J. Ass. off. analyt. Chem., 54, 1383-1387

El-Dib, M.A., Abdel-Rahman, M.O. & Aly, O.A. (1975) 4-Aminoantipyrine as a chromogenic agent for aromatic amine determination in natural water. Water Res., 9, 513-516

Grasselli, J.G., ed. (1973) CRC Atlas of Spectral Data and Physical Constants for Organic Compounds, Cleveland, Ohio, Chemical Rubber Co., p. B-206

Gupta, R.C. & Srivastava, S.P. (1971) Oxidation of aromatic amines by peroxodisulphate ion. II. Identification of aromatic amines on the basis of absorption maxima of coloured oxidation products. Z. analyt. Chem., 257, 275-277

Irvine, W.J. & Saxby, M.J. (1969a) Steam volatile amines of Latakia tobacco leaf. Phytochemistry, 8, 473-476

Irvine, W.J. & Saxby, M.J. (1969b) Gas chromatography of primary and secondary amines as their trifluoroacetyl derivatives. J. Chromat., 43, 129-131

Kouris, C.S. & Northcott, J. (1963) Aniline. In: Kirk, R.E. & Othmer, D.F., eds, Encyclopedia of Chemical Technology, 2nd ed., Vol. 2, New York, John Wiley and Sons, pp. 421-422

Lindstrom, H.V., Hansen, W.H., Nelson, A.A. & Fitzhugh, O.G. (1963) The metabolism of FD & C Red No. 1. II. The fate of 2,5-*para*-xylidine and 2,6-*meta*-xylidine in rats and observations on the toxicity of xylidine isomers. J. Pharmacol. exp. Ther., 142, 257-264

Lindstrom, H.V., Bowie, W.C., Wallace, W.C., Nelson, A.A. & Fitzhugh, O.G. (1969) The toxicity and metabolism of mesidine and pseudocumidine in rats. J. Pharmacol. exp. Ther., 167, 223-234

Medvedev, V.A., Davydov, V.D., Stupa, L.R. & Vasyutin, L.F. (1973) Thin-layer chromatography of organic products of the by-product coking industry. Koks Khim., 11, 37-39

Pfaltz & Bauer, Inc. (1976) Research Chemicals Catalog, Stamford, Conn., p. 409

Prager, B., Jacobson, P., Schmidt, P. & Stern, D., eds (1929) Beilsteins Handbuch der Organischen Chemie, 4th ed., Vol. 12, Syst. No. 1704, Berlin, Springer-Verlag, pp. 1135-1141

Przborowska, M. (1971) Effect of molecular structure on partition of aniline derivatives in the liquid-liquid chromatography systems. I. Alkyl and alkoxy derivatives of aniline in formamide systems. Chem. Anal. (Warsaw), 16, 1217-1223

Russfield, A.B., Boger, E., Homburger, F., Weisburger, E.K. & Weisburger, J.H. (1973) Effect of structure of seven methyl anilines on toxicity and on incidence of subcutaneous and liver tumors in Charles River rats (Abstract No. 3467). Fed. Proc., 32, 833

The Society of Dyers and Colourists (1971) Colour Index, 3rd ed., Vol. 4, Yorkshire, UK, p. 4863

US International Trade Commission (1976) Imports of Benzenoid Chemicals and Products, 1974, USITC Publication 762, Washington DC, US Government Printing Office, p. 29

US International Trade Commission (1977) Synthetic Organic Chemicals, US Production and Sales, 1975, USITC Publication 804, Washington DC, US Government Printing Office, pp. 63, 73

US Tariff Commission (1919) Report on Dyes and Related Coal-Tar Chemicals, 1918, revised ed., Washington DC, US Government Printing Office, p. 31

US Tariff Commission (1928) Census of Dyes and of Other Synthetic Organic Chemicals, 1927, Tariff Information Series No. 37, Washington DC, US Government Printing Office, p. 33

US Tariff Commission (1931) Census of Dyes and of Other Synthetic Organic Chemicals, 1930, Report No. 19, Second Series, Washington DC, US Government Printing Office, p. 30

US Tariff Commission (1959) Synthetic Organic Chemicals, US Production and Sales, 1958, Report No. 205, Second Series, Washington DC, US Government Printing Office, p. 77

US Tariff Commission (1973) Synthetic Organic Chemicals, US Production and Sales, 1971, TC Publication 614, Washington DC, US Government Printing Office, p. 52

Weast, R.C., ed. (1976) CRC Handbook of Chemistry and Physics, 57th ed., Cleveland, Ohio, Chemical Rubber Co., p. C-146

Winell, M. (1975) An international comparison of hygienic standards for chemicals in the work environment. Ambio, 4, 34-36

Yasuda, K. (1972) Thin-layer chromatography of aromatic amines on cadmium acetate impregnated silica gel thin layers. J. Chromat., 72, 413-420

SUPPLEMENTARY CORRIGENDA TO VOLUMES 1 - 15

Corrigenda covering Volumes 1 - 6 appeared in Volume 7, others appeared in Volumes 8, 10, 11, 12, 13 and 15.

Volume 10

p. 256	9th line	*replace* "In addition to its natural occurrence in coffee and tea, tannic acid ..." *by* "In addition to the natural occurrence of tannins in coffee and tea, tannic acid ...".

Volume 12

p. 209	Title of monograph	*replace* SEMICARBAZIDE (HYDROCHLORIDE) *by* SEMICARBAZIDE HYDROCHLORIDE

Volume 13

p. 229	3.1 Rat 1st line	*replace* adequately *by* inadequately

CUMULATIVE INDEX TO IARC MONOGRAPHS ON THE EVALUATION OF CARCINOGENIC RISK OF CHEMICALS TO MAN

Numbers underlined indicate volume, and numbers in italics indicate page. References to corrigenda are given in parentheses.